T0324757

Cognitive Science and Technology

Series editor

David M.W. Powers, Adelaide, Australia

More information about this series at http://www.springer.com/series/11554

Qi Zhao
Editor

Computational and Cognitive Neuroscience of Vision

Editor
Qi Zhao
Electrical and Computer Engineering
National University of Singapore
Singapore
Singapore

ISSN 2195-3988 ISSN 2195-3996 (electronic)
Cognitive Science and Technology
ISBN 978-981-10-0211-3 ISBN 978-981-10-0213-7 (eBook)
DOI 10.1007/978-981-10-0213-7

Library of Congress Control Number: 2016948094

Printed on acid-free paper

This Springer imprint is published by Springer Nature
The registered company is Springer Science+Business Media Singapore Pte Ltd.

Contents

Neural Mechanisms of Saliency, Attention, and Orienting 1
Brian J. White and Douglas P. Munoz

Insights on Vision Derived from Studying Human Single Neurons 25
Jan Kamiński and Ueli Rutishauser

Recognition of Occluded Objects 41
Hanlin Tang and Gabriel Kreiman

Towards a Theory of Computation in the Visual Cortex 59
David A. Mély and Thomas Serre

Invariant Recognition Predicts Tuning of Neurons in Sensory Cortex 85
Jim Mutch, Fabio Anselmi, Andrea Tacchetti, Lorenzo Rosasco, Joel Z. Leibo and Tomaso Poggio

Speed Versus Accuracy in Visual Search: Optimal Performance and Neural Implementations 105
Bo Chen and Pietro Perona

The Pupil as Marker of Cognitive Processes 141
Wolfgang Einhäuser

Social Saliency 171
Shuo Wang and Ralph Adolphs

Vision and Memory: Looking Beyond Immediate Visual Perception . . . 195
Cheston Tan, Stephane Lallee and Bappaditya Mandal

Approaches to Understanding Visual Illusions 221
Chun Siong Soon, Rachit Dubey, Egor Ananyev and Po-Jang Hsieh

Impact of Neuroscience in Robotic Vision Localization and Navigation 235
Christian Siagian and Laurent Itti

Attention and Cognition: Principles to Guide Modeling 277
John K. Tsotsos

Computational Neuroscience of Vision: Visual Disorders 297
Clement Tan

Summary

One of the great mysteries of the human brain is its remarkable processing power to parse and summarize data from the sensory input. With nearly 30 % of our cortical surface representing information that is predominantly visual, vision is the most critical and highly developed sense. In fact, its seemingly effortless nature beguiles the profound complexity of the processes underlying vision, and the study of vision has attracted attention from researchers in a wide range of science and engineering disciplines. Neurobiologists and psychologists have spent centuries to study the neural and cognitive mechanisms of the visual system, while computer scientists, statisticians, and engineers aim to develop artificial visual systems that show the same degree of autonomy and adaptability as biological systems.

With many points of interaction, it is believed that a concerted effort from all these research areas to characterize human perceptual and cognitive abilities promises new discoveries and will lead to substantial near-term technological benefits. The information derived from understanding of the visual processes and the development of engineering tools can also help prevent and treat visual disorders that were never possible before. Furthermore, a recent trend of interdisciplinary research encompassing brain science, computer science, engineering, and medicine has successfully demonstrated its impact to the society and has influenced public policy, research funding, and scientific publications to further encourage interdisciplinary research.

Despite a plethora of professional readings devoted to vision research and the trend of integrative research, a practical difficulty is the barrier between disciplines. To bridge such a gap, this book attempts to provide a systematic and comprehensive overview of vision from various perspectives, ranging from neuroscience to cognition, and from computational principles to engineering developments. With an emphasis on the link across multiple disciplines and the impact such synergy can lead to in terms of both scientific breakthroughs and technology innovations, the book targets both professional readers and nonspecialists.

Neural Mechanisms of Saliency, Attention, and Orienting

Brian J. White and Douglas P. Munoz

Abstract Active vision involves a continual re-orienting of the line of sight with stimuli pertinent to current goals. In humans and other primates, such orienting behavior relies on a distributed network of cortical and subcortical brain areas. The neural basis of orienting is theorized to be under the control of two general mechanisms: One mechanism transforms complex visual input into a spatial map that highlights the most visually conspicuous locations (saliency map). A second mechanism integrates bottom-up saliency with internal goals for flexible orienting towards behaviorally relevant stimuli (priority map). We review evidence for such mechanisms in the primate brain and raise novel issues and insights that challenge current views about the neural basis of saliency, attention, and orienting.

1 Overview

The act of looking involves a continual re-orienting of the line of sight with stimuli pertinent to current goals via movements of the eyes and/or head (consider the pattern of eye movements as you read this text). This orienting response can be triggered by salient stimuli or elicited voluntarily. This is indicative of the fact that orienting behavior is under the control of a distributed network of brain areas involved in sensory, motor, and executive functions, including the occipital, parietal, and frontal cortices, basal ganglia, thalamus, cerebellum, superior colliculus and brainstem reticular formation (Fig. 1; for comprehensive reviews of the eye movement system see Liversedge et al. (2011), Van Gompel et al. (2007). Within each region are multiple populations of neurons and subnuclei that perform specific operations for the sensory-to-motor transformations that guide orienting. It is the-

B.J. White (✉) · D.P. Munoz
Centre for Neuroscience Studies, Queen's University, Kingston, ON K7L 3N6, Canada
e-mail: brian.white@queensu.ca

D.P. Munoz
e-mail: doug.munoz@queensu.ca

Q. Zhao (ed.), *Computational and Cognitive Neuroscience of Vision*, Cognitive Science and Technology, DOI 10.1007/978-981-10-0213-7_1

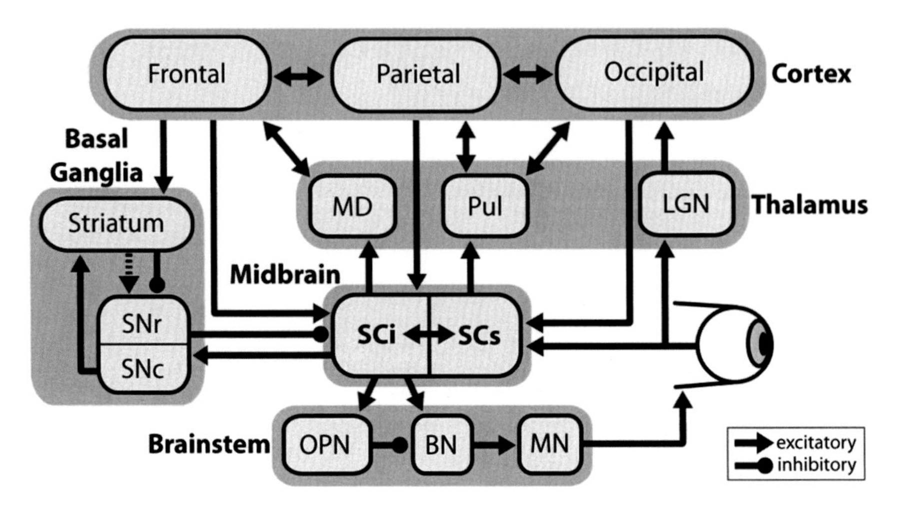

Fig. 1 Visual Orienting Network. Abbreviations: BN: burst neurons, LGN: lateral geniculate nucleus, MD: mediodorsal nucleus of the thalamus, MN: abducens motor neurons, OPN: omnipause neurons, Pul: Pulvinar, SCs: superior colliculus superficial layers, SCi: superior colliculus intermediate layers, SNr: substantia nigra pars reticulata, SNc: substantia nigra pars compacta. *Solid arrows* excitatory connections, ovals: inhibitory connections. *Dashed arrow* indirect pathway. Some brain areas omitted for brevity

orized that orienting behavior is governed by (i) a bottom-up saliency mechanism that transforms complex visual input into a spatial map of visual conspicuity (Itti and Koch 2001; Itti et al. 1998; Koch and Ullman 1985), and/or (ii) a mechanism that integrates visual and goal-related processes to prioritize which stimuli will gain access to downstream orienting circuitry to control the moment-by-moment locus of gaze (Fecteau and Munoz 2006; Serences and Yantis 2006). Here, we present an overview of the core brain areas that comprise the primate visual orienting network, and review recent research that elucidates their functional roles with respect to saliency coding, attention and orienting. We highlight recent research in the midbrain superior colliculus (SC), which contains the structure and function ideally suited for the role of a saliency and priority mechanism compartmentalized within its visual and visuomotor layers, respectively.

2 The Visual Orienting Network

2.1 *Superior Colliculus*

Most brain areas in the visual orienting network converge in the midbrain SC (Fig. 1; optic tectum in non-mammals), an evolutionarily old hub of the orienting system that plays a central role in various components of orienting behavior

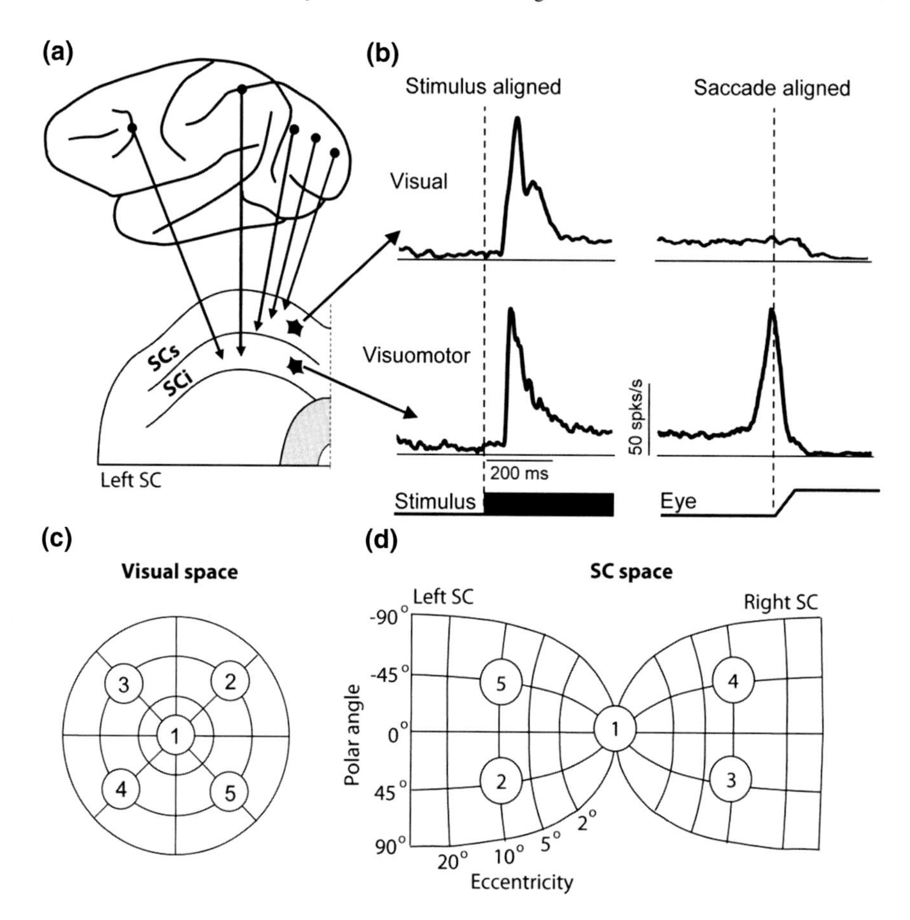

Fig. 2 The Superior Colliculus. a Coronal view of the SC illustrating the functional layers and their dominant cortical inputs. **b** Responses of two common types of SC neurons. **c-d** Illustration of the log-polar mapping between visual space and SC space

including eye and head movements (for reviews see Corneil and Munoz (2014), Gandhi and Katnani (2011), Hall and Moschovakis (2004), Krauzlis et al. (2013), White and Munoz (2011a), and non-spatial components such as pupil dilation (Wang et al. 2012; Wang and Munoz 2015). The SC is multilayered but can be divided into two dominant functional layers (Fig. 2a, b): a superficial visual-only layer (SCs), and an intermediate/deeper multisensory, cognitive, and motor-related layer (SCi).

The SCs receives dominant inputs from the retina and early visual cortex (V1, V2, V3, V4, middle temporal area, MT; (Cerkevich et al. 2014); (Lock et al. 2003) (Fig. 2a), with the projections terminating systematically deeper as one moves from striate to extrastriate areas (Tigges and Tigges 1981). Some SCs neurons project to the SCi directly (Saito and Isa 2005). Visually-responsive neurons in the SC discharge a burst of action potentials approximately 50 ms after the appearance of a visual stimulus in the neuron's response field (RF) (Fig. 2b, upper panels). Some SCs neurons show sustained activation in the presence of continued visual stimulation, but generally cease discharging during the execution of a saccade (Marino et al. 2012a; White et al. 2009).

The SCi is the major point of convergence of descending saccade-related signals from fronto-parietal cortices (Leichnetz et al. 1981; Lynch et al. 1985; Shook et al. 1990; Stanton et al. 1988) (Fig. 2a), and the basal ganglia (Hikosaka et al. 2014; Shires et al. 2010; Watanabe and Munoz 2011), as well as converging visual inputs from the SCs and extrastriate visual cortical areas (Cerkevich et al. 2014; Lock et al. 2003). One subtype of SCi neurons discharge a high-frequency burst of action potentials just before and during a saccade into the neuron's preferred RF (Munoz and Wurtz 1995; Sparks and Mays 1980; Wurtz and Goldberg 1972). A second subtype of SCi neurons discharge a burst of action potentials following the appearance of a visual stimulus in the RF, and a separate burst of action potentials associated with a saccade into the neuron's RF (Mohler and Wurtz 1976) (Fig. 2b, lower panels). Many SCi neurons also exhibit "prelude" or "build-up" activity that precedes the saccade (Glimcher and Sparks 1992; Munoz and Wurtz 1995), which is associated with motor preparation (Corneil et al. 2007; Dorris and Munoz 1998), covert attention (Ignashchenkova et al. 2004; Kustov and Robinson 1996), and reward (Basso and Wurtz 1998; Thevarajah et al. 2009).

SCs neurons are organized into a retinotopically coded map of contralateral visual space (Ottes et al. 1986) (Fig. 2c, d). SCi neurons are organized into a retinotopically coded map specifying saccade vectors into the contralateral visual field (Fig. 2c, d) (Gandhi and Katnani 2011; Marino et al. 2008; Sparks and Mays 1980). As such, local stimulation of the SCi causes the eyes to move with a fixed vector of a specific direction and amplitude (Robinson 1972). Importantly, SC visual and motor maps are spatially aligned (Marino et al. 2008). This allows for a fast sensory-to-motor transformation because the visual response is mapped directly onto the appropriate SCi output neurons, which in turn project to the brainstem premotor circuitry that triggers the saccade (Rodgers et al. 2006) and/or head movement (Corneil et al. 2007).

In addition, the vertical connectivity between the SCs and the premotor layers of the SCi (Behan and Appell 1992; Helms et al. 2004) has been proposed to mediate visually-guided orienting (Isa and Saito 2001). There are also horizontal long-range connections across the SCs and SCi (Munoz and Istvan 1998; Meredith and Ramoa 1998; Phongphanphanee et al. 2014), which play an important role in shaping the spatiotemporal dynamics across the maps. This has important functional consequences for how spatially competing visual and goal-related signals interact to determine orienting behavior. The convergence of signals from cortical and subcortical brain regions shape the activation across the maps of the SCs and SCi, and the manner in which these competitive signals are integrated in the SCi is thought to determine various aspects of orienting behavior (Dorris et al. 2007; Li and Basso 2005; Marino et al. 2012a; McPeek et al. 2003; White et al. (2012, 2013)).

2.2 Occipital Cortex

The occipital cortex is the brain's central visual processing hub (Werner and Chalupa 2014), and is the major source of visual input to the orienting system. Briefly, from the retina, visual signals traverse three parallel pathways devoted to specific chromatic and spatio-temporal aspects of the visual input (so-called magno-, parvo-, and konio-cellular paths). These pathways terminate in layers 2/3 and 4 of the primary visual cortex (V1). A hallmark of neurons in V1, and the visual cortex in general, is their selectivity for visual features such as color, orientation, or direction of motion. As one proceeds up the visual processing hierarchy the complexity of visual processing and tuning properties increases. In addition, visual signals split into two broadly construed processing streams (Ungerleider and Mishkin 1982; Goodale and Westwood 2004)—a ventral stream associated with the processing and perception of objects, and a dorsal stream primarily concerned with the control of actions, including eye movements. Saccadic selection of stimuli in complex visual environments would not be possible without the ability to distinguish features that define behaviorally relevant stimuli. Because oculomotor brain areas such as superior colliculus (SC) and frontal eye fields (FEF) are not selective for visual features (Davidson and Bender 1991; Marrocco and Li 1977; Schall and Thompson 1999; White et al. 2009), saccade target selection must involve important communication between these areas and the visual cortex. Interestingly, SC and FEF receive direct inputs from visual cortical areas belonging to both the dorsal and ventral streams (Cerkevich et al. 2014; Lock et al. 2003; Schall et al. 1995).

2.3 Fronto-Parietal Cortices

Several nuclei in the primate frontal and parietal cortices play an important role in the control of visual attention and voluntary gaze behavior (Fig. 1). Like the SCi, the FEF projects to the brainstem saccade circuitry (Segraves 1992; Shook et al. 1990; Stanton et al. 1988), and contributes to the transformation of visual signals into saccadic commands. Microstimulation of the FEF elicits contralateral saccades (Bruce et al. 1985). The FEF also projects directly to the SCi (Stanton et al. 1988), and is important for voluntary saccade control (Everling and Munoz 2000; Hanes and Wurtz 2001; Munoz and Everling 2004; Segraves and Goldberg 1987). Like the SCi, the FEF also has visual and saccade-related responses, and has retinotopically organized RFs of contralateral space, but the precise mapping is not as well understood as in the SC.

The supplementary eye fields (SEF) also project directly to the brainstem and SCi (Shook et al. 1990). The SEF exhibits visual and saccade-related activation, and low current microstimulation of SEF can evoke saccades (Schlag and Schlag-Rey 1987). However, the precise function of SEF is not well understood.

SEF is thought to play an important role in the executive control of saccades (Coe et al. 2002; Husain et al. 2003; Stuphorn and Schall 2006), but its activation patterns are sometimes very different from FEF and SCi. For example, SEF is involved in oculomotor performance monitoring; e.g., evaluating the consequences of oculomotor errors, successes, and reinforcement (Purcell et al. 2012a; Stuphorn et al. 2000).

Some of the descending command signals from FEF and SEF may bypass the SCi and influence the brainstem saccade generator directly (Segraves 1992; Shook et al. 1990; Stanton et al. 1988). In support of a bypass hypothesis, Schiller and colleagues (Schiller and Sandell 1983; Schiller et al. 1980) found that saccades were impaired but not abolished with the physical removal of either SCi or FEF alone, while removal of both structures abolished saccades. However, these experiments took place weeks to months after the surgery, and it has been argued that this might have led to significant neuroplastic reorganization and recovery of function (Hanes and Wurtz 2001). To test this hypothesis, Hanes and Wurtz pharmacologically deactivated the SCi and found that saccades could no longer be elicited with FEF microstimulation (Hanes and Wurtz 2001), supporting the serial hypothesis. Thus, most of the effect of FEF on saccades appears to be mediated through the SC.

Lastly, a region of the posterior parietal cortex known as the lateral intraparietal area (LIP) plays an important role in higher order visual processes associated with orienting and attention (Bisley and Goldberg 2003; Bisley and Goldberg 2010; Goldberg et al. 2002). Although LIP has direct projections to both FEF and SCi, it has been argued that LIP does not participate directly in the production of saccades (Paré and Dorris 2011), because (i) the magnitude of pre-saccadic activity is significantly reduced when saccades are made in the absence of a visual stimulus (Ferraina et al. 2002; Paré and Wurtz 1997), (ii) a large amount of microstimulation current is required to elicit a saccade (Kurylo and Skavenski 1991; Keating et al. 1983), and (iii) ablation of LIP does not impair saccade production (Lynch and McLaren 1989).

2.4 Basal Ganglia

The basal ganglia represent an important group of subcortical nuclei involved in voluntary motor control, learning, and reward (Hikosaka et al. 2014; Shires et al. 2010; Watanabe and Munoz 2011) (Fig. 1). With respect to visual orienting, the basal ganglia mediate pathways between cortex and the SCi, and influence the SCi through direct inhibitory projections from the substantia nigra pars reticulata (SNr), a major output node of the basal ganglia (Jayaraman et al. 1977). The nigrotectal projection regulates saccades burst initiation by imposing a blanket of tonic GABAergic inhibition over the SCi (Hikosaka et al. 2000), the release of which allows the SCi to trigger downstream premotor circuitry to drive the appropriate orienting response.

2.5 *Brainstem*

Finally, the brainstem reticular formation and upper cervical spinal cord contain a group of nuclei that form the central saccade generating mechanism (Fig. 1; for reviews see Munoz et al. 2000; Sparks 2002; Scudder et al. 2002). The paramedian pontine reticular formation (PPRF) and rostral interstitial nucleus of the medial longitudinal fasciculus (riMLF) determine the horizontal and vertical components of the saccade, respectively. Tonic discharge from omnipause neurons (OPNs) act as a gate that inhibits saccade burst neurons (BNs, both excitatory and inhibitory) in the PPRF and riMLF to control when a saccade may be initiated. A saccade is initiated by a trigger signal, from the SC (Moschovakis 1996), which disinhibits the OPNs for a brief moment, allowing BNs to generate the pulse of activation required to produce a saccade of a specific amplitude. Tonic neurons in the nucleus prepositus hypoglossi provide the lower-frequency step signal that holds the eyes in place (fixation) at the end of each saccade.

3 Neural Representations of Visual Saliency

Several theories and models of visual attention postulate the existence of a saliency map (Borji and Itti 2013) (Fig. 3a), which acts to transform visual input into a feature- and behavior-agnostic representation of visual conspicuity. Since its introduction almost 30 years ago (Koch and Ullman 1985), saliency map theory has attracted wide spread attention, with an explosion of applications not only in neuroscience and psychology, but also in machine vision, surveillance, defense, transportation, medical diagnosis, design and advertising (Borji and Itti 2013).

There have been reports of the neural equivalent of a saliency map in various, predominantly cortical, brain regions (e.g., V1 (Li et al. 2006; Li 2002; Zhaoping 2008; Zhang et al. 2012); V4 (Burrows and Moore 2009; Mazer and Gallant 2003); LIP (Arcizet et al. 2011; Bisley and Goldberg 2010; Gottlieb et al. 1998; Kusunoki et al. 2000; Leathers and Olson 2012); FEF (Purcell et al. 2012b; Thompson and Bichot 2005), and the SC (Asadollahi et al. 2010; Knudsen 2011; Mysore et al. 2011; White et al. 2014)). To be clear, several of these studies use the term saliency to refer to a combination of sensory and goal-related processes (Treue 2003)—what others have termed a priority map (Fecteau and Munoz 2006; Serences and Yantis 2006). Although sensory and goal-related processes certainly overlap in many areas of the visual orienting network, it is hypothesized that the visual system first generates a feature- and behavior-agnostic representation of the visual input based purely on low-level visual attributes (Itti and Koch 2001; Itti et al. 1998; Koch and Ullman 1985; Fig. 3a).

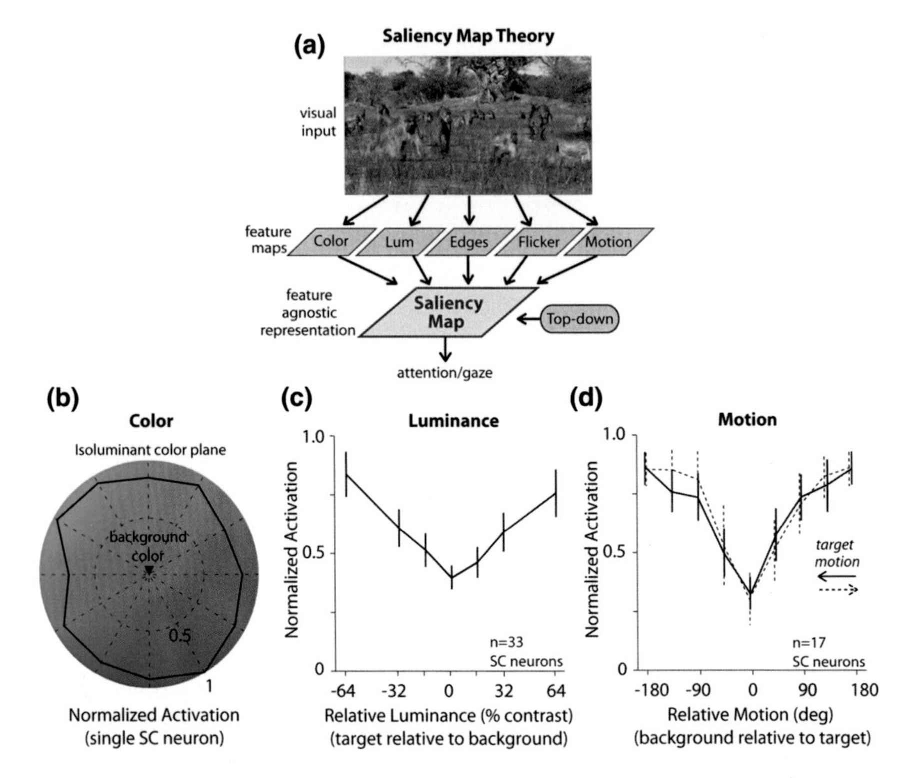

Fig. 3 Feature-agnostic visual representation in the Superior Colliculus. **a** Illustration of saliency map theory (Itti and Koch 2001; Itti et al. 1998; Koch and Ullman 1985). Briefly, visual input is decomposed into several high-level feature maps (color, luminance, motion etc.) with center-surround characteristics that allow certain stimuli the stand out from their neighbors. The feature maps are combined to create a feature-agnostic representation of visual saliency that when combined with the top-down input determines the locus of attention and gaze. **b-d** Feature-agnostic visual representation of SC neurons to stimuli defined by (**b**) isoluminant color (isoluminant with a grey background whose color is defined by the center of the isoluminant plane; adapted from White et al. 2009), **c** relative luminance (positive contrast stimulus *brighter* than background, negative contrast stimulus *darker* than background; zero contrast stimulus differs only in chromaticity; adapted from White et al. (2009), supplementary material), **d** relative motion (adapted from Davidson and Bender 1991)

We hypothesized that the primate SCs represents an ideal candidate for the role of a visual saliency map (Boehnke and Munoz 2008; Fecteau and Munoz 2006; White and Munoz 2011a; White et al. 2014) because: (1) It is heavily interconnected with early visual areas (Cerkevich et al. 2014; Lock et al. 2003); (2) It encodes stimuli in a featureless manner (Davidson and Bender 1991; Marrocco and Li 1977; White et al. 2009); (3) It has the long-range center-surround organization well-suited for a saliency mechanism (Munoz and Istvan 1998; Meredith and Ramoa 1998; Phongphanphanee et al. 2014); (4) There is strong evidence of such a mechanism in the pre-mammalian optic tectum (Asadollahi et al. 2010; Knudsen 2011; Mysore et al. 2011), which pre-dates the evolution of neocortex.

Indeed, there is evidence that the SC encodes relative saliency irrespective of the visual feature that defines the stimulus. For example, SC neurons are sensitive to isoluminant color stimuli (White et al. 2009) (i.e., isoluminant relative to the background), which indicates that these neurons receive input from chromatic pathways, but they do not show strong preference for any one color (Fig. 3b). SC neurons are also sensitive to luminance contrast (Bell et al. 2006; Marino et al. 2012b), but they have no preference for whether the stimulus is defined by positive contrast (brighter then background) or negative contrast (darker than the background; White et al. 2009; Fig. 3c). Similarly, SC neurons are sensitive to relative motion (i.e., the motion of a local stimulus relative to background motion), but they are not tuned to any specific motion direction (Davidson and Bender 1991; Fig. 3d). In other words, SC neurons are sensitive to the difference between a stimulus relative to its surround irrespective of the feature that defines that difference, which is a principle tenant of the salience map theory (Fig. 3a).

In addition, these low-level characteristics of visual stimuli have very predictable effects on SC activity and subsequent oculomotor behavior (Bell et al. 2004; Dorris et al. 2002; Fecteau et al. 2004; Fecteau and Munoz 2005; Marino et al. 2012b; White et al. 2009; White and Munoz 2011b). For example, reducing stimulus contrast decreases the magnitude of a visual response and increases response onset latency (Fig. 4a) leading to a systematic increase in saccade reaction time (SRT; Fig. 4b) (Marino et al. 2012b). SC visual responses are also delayed for isoluminant color stimuli relative to luminance-defined stimuli (Fig. 4c), and this delay results in a subsequent delay in SRT (Fig. 4d; White et al. 2009), which is similarly observed in human SRTs (White et al. 2006). Differences in processing latency for different visual features has important functional consequences for the target selection process (White and Munoz 2011b), which we discuss below.

Recently, we performed a test of the saliency hypothesis by comparing directly the output of a computational saliency model (Itti and Koch 2001; Itti et al. 1998; Koch and Ullman 1985), with the activation of neurons in the SC during free viewing of natural scenes (White et al. 2014). The activation of SC neurons was well-predicted by the model. Notably, the SCs, whose dominant inputs arise from visual cortex (Fig. 1, 2a), encoded model-predicted saliency irrespective of saccade goals (consistent with a bottom-up saliency map). In contrast, the SCi, which has strong inputs from fronto-parietal cortices and the basal ganglia (Fig. 1, 2a), only encoded model-predicted saliency of stimuli that were also the goal of a subsequent saccade. These results provide compelling evidence that the SCs embodies the role of a saliency map, whereas the SCi acts to prioritize which stimuli gain access to downstream orienting circuitry to control gaze. This is in agreement with a recent study that compared the output of a similar computational saliency model with the activation of FEF neurons during free search in natural scenes (Fernandes et al. 2014). Like the SCi, the correlation between model saliency and the discharge of FEF neurons was weak and mostly goal-dependent (inconsistent with a bottom-up saliency map). Thus, saliency may be computed in the SCs, and relayed to the SCi where it is under greater top-down control. This signal might also be relayed to

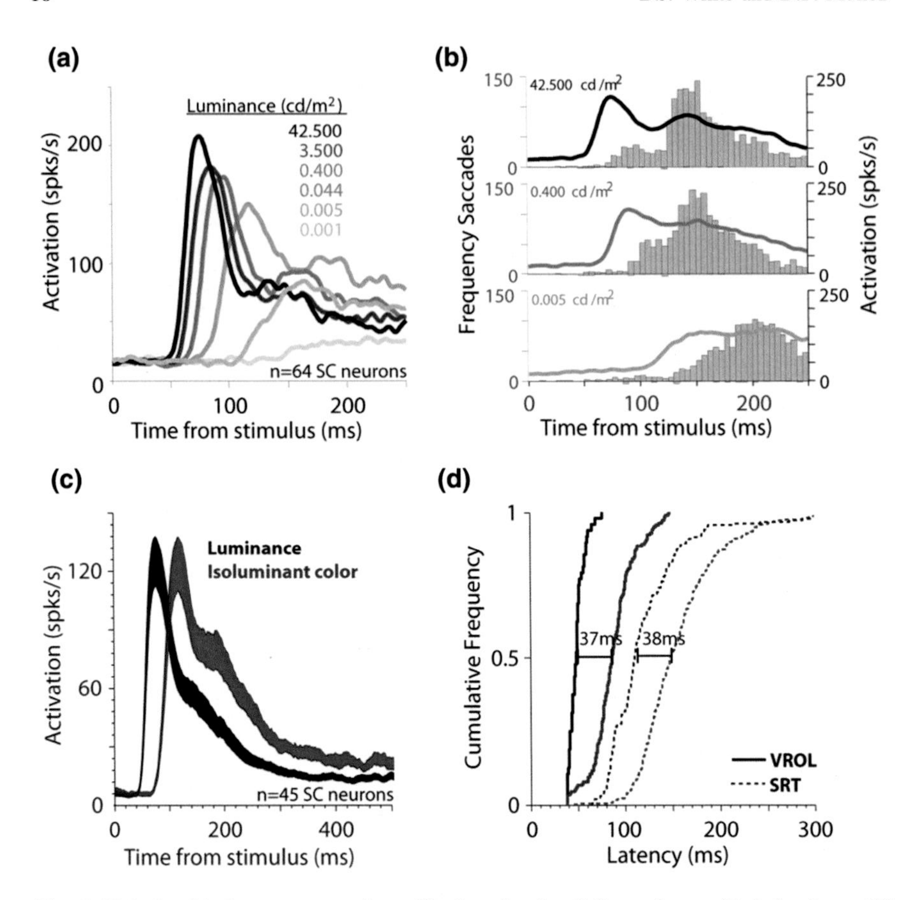

Fig. 4 Relationship between superior colliculus visual activity and saccadic behavior. **a** SC visual responses to stimuli at various luminance levels (adapted from Marino et al. (2012b), with permission). **b** Distributions of saccadic reaction times (SRT) and superimposed SC visual responses associated with high, medium and low luminance. Note the systematic delay in the visual response onset latency (VROL) and SRT with decreasing luminance. **c** Comparison of SC visual responses to stimuli defined by luminance versus isoluminant color (adapted from White et al. 2009). **d** Comparison of VROL (*solid lines*) and SRT (*dashed lines*). Note the equalivalent delay in VROL and SRT for color relative to luminance

other brain areas via tecto-thalamic pathways (Berman and Wurtz 2011; Berman and Wurtz 2010; Casanova 2004).

4 Neural Representations of Behavioral Priority

Several brain areas in the visual orienting network are hypothesized to embody the role of a priority map (Fecteau and Munoz 2006; Serences and Yantis 2006), which is defined by the combined representation of visual salience and behavioral relevance associated with a given stimulus/location in the visual field, and whose output determines orienting behavior. Two popular areas of research illustrate the neural representation of priority: (i) spatial attention, and (ii) target selection.

4.1 Spatial Attention

Covert spatial attention has been described by such metaphors as a 'filter' or 'spotlight' (Broadbent 1958; Posner et al. 1980; Treisman 1969) that enhances the processing of certain stimuli while suppressing others. As such, spatial attention is often described as a regulator of sensory input (Carrasco 2011; Driver 2001). The most common neural signature of attention is the enhancement of activation (usually sensory) of neurons representing the retinotopic location where the observer is required to perform some sort of visual detection or discrimination (McAlonan et al. 2008; McAdams and Reid 2005; Moran and Desimone 1985; Motter 1993; Reynolds et al. 2000; Treue and Maunsell 1996; for reviews see Desimone and Duncan 1995; Reynolds and Chelazzi 2004). The source of this enhancement is believed to arise from feedback from frontal and parietal areas (Bisley 2011; Kastner and Ungerleider 2000; Noudoost and Moore 2011; Squire et al. 2013), which act to increase the excitability of neurons in the visual cortex representing the attended stimulus, while suppressing unattended stimuli. This enhancement is also thought to rely on the rhythmic synchronization of neurons, within and between brain areas of the network, which are tuned to the spatial and/or feature attributes of the attended stimulus (Bosman et al. 2012; Womelsdorf and Fries 2007).

A number of studies are beginning to call into question the view that attention is under the exclusive control of a fronto-parietal network (Asadollahi et al. 2010; Goddard et al. 2012; Knudsen 2011; Krauzlis et al. 2013; Mysore and Knudsen 2011; Mysore et al. 2011). This idea has grown especially out of work in the avian brain (Knudsen 2011), which centers the orienting network around a set of fundamental subcortical components: (1) a saliency map in the optic tectum visual layers (OTv/SCs) that combines information across visual features; (2) a multisensory/motor map in the optic tectum deeper layers (OTd/SCi) that issues orienting commands based on integrated saliency and behavioral relevance; (3) Specialized cholinergic and GABAergic circuits, formed by the isthmic nuclei (Wang et al. 2006), or parabigeminal nucleus in primates (Graybiel 1978), which play a central role in mediating local and global spatial interactions across the OT/SC maps; (4) A basal ganglia circuit that gates commands for orienting movements (substantia nigra pars reticulata; Hikosaka et al. 2000), and signals the occurrence of behaviorally relevant events/stimuli to the forebrain (substantia nigra pars compacta; Comoli et al. 2003; Redgrave and Gurney 2006; Redgrave et al. 2010) crucial for reinforcement learning.

There is extensive research showing that the primate SC has discharge patterns associated with the control of visuospatial attention (Bell et al. 2004; Dorris et al.

2002; Dorris et al. 2007; Fecteau et al. 2004; Gattass and Desimone 1996; Ignashchenkova et al. 2004; Krauzlis et al. 2013; Kustov and Robinson 1996; Lovejoy and Krauzlis 2010; Muller et al. 2005; Zenon and Krauzlis 2012). For example, it has been known for some time that SC neurons show an enhancement of visually-evoked activation at an attended location (Goldberg and Wurtz 1972; Ignashchenkova et al. 2004; Li and Basso 2008; Wurtz and Mohler 1976). Subthreshold microstimulation of the SCi (below levels that produce a saccade) can facilitate visual discrimination at the spatially selective location represented by the stimulated site, which is indicative of a covert shift of visual attention (Muller et al. 2005). In addition, pharmacological inactivation of a local region of the SC disrupts monkeys' ability to perceptually judge the direction of a motion cue in the affected region, which is indicative of a deficit in covert attention (Lovejoy and Krauzlis 2010).

The fact that SC inactivation disrupts motion perception has suggested that a circuit through the SC might mediate the visual cortical signatures of attention (Zenon and Krauzlis 2012), for example, in motion selective area MT/MST (Treue and Maunsell 1996). This hypothesis was tested by examining attentional modulation of motion selective neurons in MT and MST before and after SC inactivation during a motion change detection task (Zenon and Krauzlis 2012). Despite the deficit in motion change detection (i.e., the behavioral signature of attention) after SC deactivation, the neural signature of attention in area MT and MST remained unaffected. These results imply that the SC plays a critical role in the behavioral consequences of spatial attention, which operates independently of the cortical signature of attention. This has led to a controversial hypothesis that the filter-like properties described as attention, rather than acting as a cause, may be the byproduct of circuits centered on the basal ganglia involved in value-based decision making (Krauzlis et al. 2014; Zenon and Krauzlis 2014).

4.2 Target Selection

In order to understand the neural mechanisms of orienting, we must understand how the brain selects one stimulus from an array of alternatives. In the laboratory, saccade target selection is often studied using visual search or selection tasks in which an observer must select a target stimulus from several distractor stimuli. In such tasks, a target might differ from distractors along one or more visual dimensions (e.g., color, orientation, direction of motion). Here, we focus on recent advances in the neural basis of target selection in non-human primates.

The temporal progression of the target selection process is straightforward (for reviews see Schall and Cohen 2011; Schall and Thompson 1999): Immediately following the appearance of an array of visual stimuli, neurons that exhibit target selection initially respond equivalently whether a target- or distractor-stimulus falls in the RF of a given neuron (Fig. 5a). Shortly thereafter, but before an overt orienting response (e.g., a saccade), activation associated with the target increases

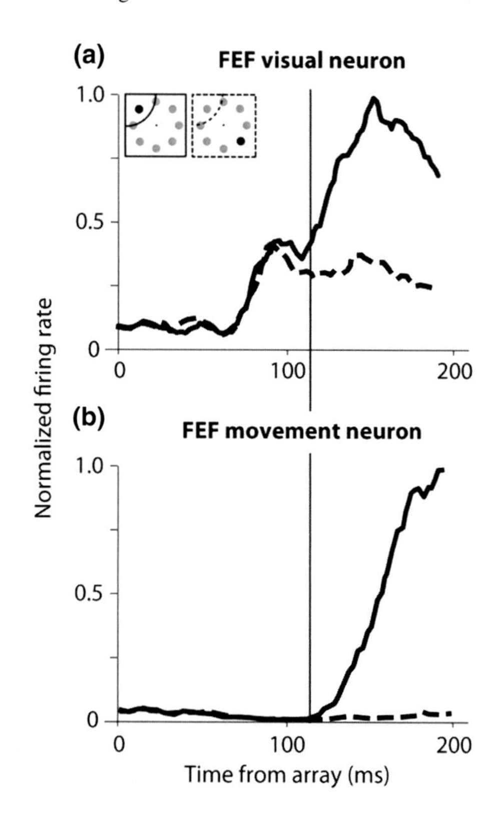

Fig. 5 Visual- and movement-related FEF neurons during target selection. Characteristic response pattern of **a** an FEF visual neuron, and **b** an FEF movement neuron during a visual search/selection task (adapted from Schall et al. 2011, with permission). *Solid* versus *dashed* traces represent FEF activation when the target versus the opposite distractor appeared in the center of the neuron's response field, respectively. The *vertical solid line* represents the approximate time in which the visual neuron discriminated the target from the distractor. Note how the movement neuron only begins to accumulate after the visual neuron has discriminated the target

while activation associated with the distractors decreases. A similar target discrimination process is exhibited by neurons throughout the visual orienting network, including visual area V4 (Ogawa and Komatsu 2004), LIP (Ipata et al. 2006; Thomas and Paré 2007), FEF (Bichot and Schall 1999; Bichot et al. 2001b; Cohen et al. 2009; Sato et al. 2001; Schall and Hanes 1993; Thompson et al. 1996), and the SC (Horwitz and Newsome 1999; Kim and Basso 2008; McPeek and Keller 2002; Shen and Paré 2007; Shen et al., 2011; White and Munoz, 2011b; White et al. 2013).

Exactly how these brain areas interact and uniquely contribute to the target selection process is not completely understood, but we know from deactivation studies that FEF (Wardak et al. 2006) and SC (McPeek and Keller 2004) play particularly important roles. Both FEF and SC are not selective for visual features (Davidson and Bender 1991; Marrocco and Li 1977; Schall and Thompson 1999; White et al. 2009), so the selection process observed there must depend on afferent input from feature selective areas in visual cortex. Also, FEF and SC have projections back to visual cortex (e.g., SC-> pulvinar-> cortex: Berman and Wurtz 2010; Casanova 2004; Stepniewska et al. 2000; FEF- > V4: Noudoost and Moore 2011; Schall et al. 1995; Squire et al. 2013), which places them in an ideal position to influence visual cortical representations of target selection. Although we know little about the functional role of SC-> pulvinar-> cortex pathways (Berman and Wurtz 2011), the FEF is known to play an important role in modulating activity in visual cortex (Noudoost and Moore 2011; Squire et al. 2013), and this has been shown to facilitate the selection of a target whose features are known in advance (Zhou and Desimone 2011).

It is also important to note that the target selection process observed in FEF (and SC) does not operate in isolation or in any obvious sequence, but most likely involves feedback between areas of the network (Schall and Cohen 2011), as well as competitive spatial interactions within areas of the network (e.g., SC: Dorris et al. 2007; Marino et al. 2012a; White et al. 2013). Nonetheless, target selection involves at least two sequential processes: (i) a visual selection process, and (ii) a motor process associated with the shift in gaze towards the stimulus (Thompson et al. 1996). This is evidenced by the fact that target selection is observed even when the animal is required to withhold a saccade (Thompson et al. 1997) or report the target location with a manual response (Thompson and Bichot 2005). These two processes are observed in the response patterns of FEF visual and movement neurons, respectively (Fig. 5a, b), which is detailed in a recent model of target selection (Purcell et al. 2010; Purcell et al. 2012b; Schall et al. 2011). According to this gated accumulator model, visually responsive neurons in FEF (and possibly SC) provide the major drive to the target selection process. That is, visual evidence supporting a saccade to the target or one of the distractors is derived directly from visually responsive neurons. These visual signals are integrated by movement neurons, which accumulate the evidence for each of the alternatives. This is consistent with the fact that FEF movement neurons begin accumulating only after the target is successfully discriminated by FEF visual neurons (vertical line passing through Fig. 5a, b). A saccade is initiated when accumulated support for a particular location/stimulus reaches a fixed threshold (Hanes and Schall 1996; see however Jantz et al. 2013). This model was inspired by the response pattern of FEF visual- and movement-related neurons, but it remains to be seen whether it accurately describes target selection in other brain areas like the SC, where target selection is often represented by visuo-movement neurons that have both visual and motor responses (McPeek and Keller 2002; White and Munoz 2011b). However, a recent study (Shen et al. 2011) reported that a subset of SCs visual neurons also show target discrimination before saccades, so these neurons may act in a manner similar to FEF visual neurons presumed to drive the target selection process.

Also, variability in the visual or motor components of the target selection process in the SC and FEF is highly correlated with gaze behavior. For example, when target conspicuity is low, target discrimination is weaker and delayed in the SC and FEF (Bichot et al. 2001a; Cohen et al. 2009; Marino et al. 2012b; Sato et al. 2001; White and Munoz 2011b), which is in turn associated with longer SRTs. Similarly, greater motor preparatory activity is associated with faster SRTs (Dorris and Munoz 1995; Dorris et al. 1997; Hanes and Schall 1996). Also, internal factors such as trading accuracy over speed can affect multiple aspects of the selection process in FEF, including reducing pre-stimulus baseline activation, attenuating the magnitude of the visually evoked response, extending the duration of the visual selection process, and slowing the accumulation of the motor preparatory process (Heitz and Schall 2012). Thus, the target selection process is highly malleable and subject to the top-down demands of a task, resulting in diverse effects on the neural dynamics of the priority map and subsequent behavior.

Finally, because the SC and FEF rely on inputs from feature selective regions of visual cortex, processing latencies of different features have important functional consequences for the target selection process (White and Munoz 2011b). For example, the circuitry underlying primate color vision is relatively independent of luminance and motion (Solomon and Lennie 2007), and has slower processing latency (Maunsell et al. 1999; Schmolesky et al. 1998). This delay for color trickles all the way down to motor systems, resulting in a delay in SRTs (White et al. 2006), and manual RTs (Kane et al. 2011). As discussed above, this delay in SRT is due to a delay in the arrival of isoluminant color signals in the SC (White et al. 2009; Fig. 4b). Similarly, during target selection with isoluminant color stimuli (White and Munoz 2011b), the initial visual response of SCi neurons is delayed relative to the same stimuli with an added luminance pedestal (Fig. 6a, b). Yet these neurons discriminate the target from distractors at approximately the same time in both luminant and isoluminant conditions (Fig. 6a, c). Interestingly, SRTs are most closely tied to the arrival time (Fig. 6d), not discrimination time, of the visual signals in the SC, provided target discriminability is easy (Bichot et al. 2001a;

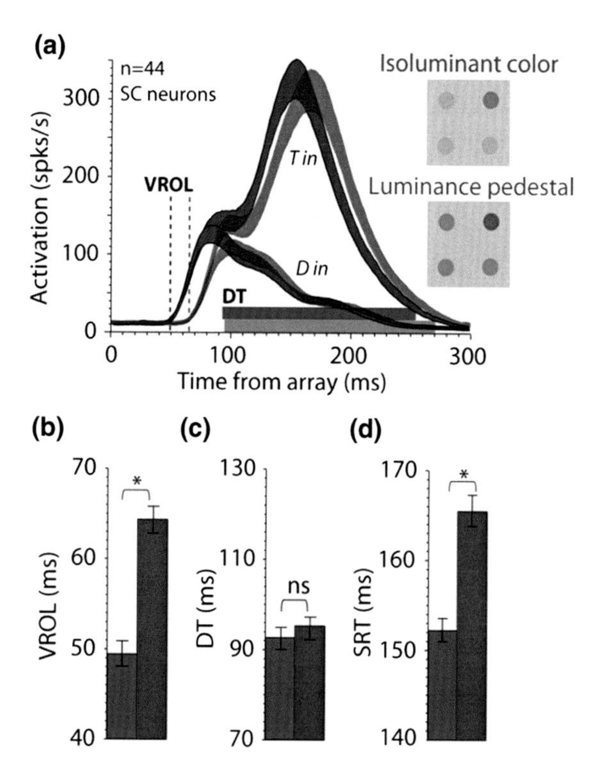

Fig. 6 Visual feature timing and target selection. a Response pattern of SC visuomotor neurons during a visual search/selection task using pure chromatic stimuli (*red curves*, isoluminant color condition), versus stimuli with the same difference in chromaticity between target and distractors but with an increment in luminance contrast relative to the background (*grey curves*, luminance pedestal condition) (adapted from White and Munoz 2011b). **b** VROL: visual response onset latency. **c** DT: discrimination time. Note how the delay in SRT for isoluminant color **d** is associated with the delay in VROL (**b**), not DT (**c**)

Cohen et al. 2009; Sato et al. 2001; White and Munoz 2011b). Thus, the relative latency of visual features is important for the dynamics of saliency and priority maps and should be implemented into saliency models that describe processes operating across multiple visual features (Itti and Koch 2001; Itti et al. 1998; Koch and Ullman 1985), and models of target selection in which the relative timing of visual features should be critical (Purcell et al. 2012b; Purcell et al. 2010; Schall et al. 2011).

5 Conclusion

We presented a selective review of the neural mechanisms of saliency, attention, and orienting, highlighting recent advances in key oculomotor areas. We briefly surveyed the brain areas that comprise the visual orienting network, illustrating the complexity of the wider system underlying the sensory, cognitive, and motor processes associated with orienting behavior. We highlighted evidence that the evolutionarily old superior colliculus contains the structure and function suited for the role of a visual saliency and behavioral priority mechanism compartmentalized within its visual and visuomotor layers, respectively. Many questions remain unanswered, especially with respect to the unique contributions of, and interactions between, areas of the network. Advancing our knowledge along this front will require multifaceted approaches that include the exploitation of multi-neuron recording techniques in multiple brain areas simultaneously, the use of natural stimuli in tasks that mimic real world conditions, and computational approaches to interpret such complex neural and behavioral data.

References

Arcizet F, Mirpour K, Bisley JW (2011) A pure salience response in posterior parietal cortex. Cereb Cortex 21:2498–2506

Asadollahi A, Mysore SP, Knudsen EI (2010) Stimulus-driven competition in a cholinergic midbrain nucleus. Nat Neurosci 13:889–895

Basso MA, Wurtz RH (1998) Modulation of neuronal activity in superior colliculus by changes in target probability. J Neurosci 18:7519–7534

Behan M, Appell PP (1992) Intrinsic circuitry in the cat superior colliculus: Projections from the superficial layers. J Comp Neurol 315:230–243

Bell AH, Fecteau JH, Munoz DP (2004) Using auditory and visual stimuli to investigate the behavioral and neuronal consequences of reflexive covert orienting. J Neurophysiol 91:2172–2184

Bell AH, Meredith MA, Van Opstal AJ, Munoz DP (2006) Stimulus intensity modifies saccadic reaction time and visual response latency in the superior colliculus. Exp Brain Res 174:53–59

Berman RA, Wurtz RH (2011) Signals conveyed in the pulvinar pathway from superior colliculus to cortical area MT. J Neurosci 31:373–384

Berman RA, Wurtz RH (2010) Functional identification of a pulvinar path from superior colliculus to cortical area MT. J Neurosci 30:6342–6354
Bichot NP, Schall JD (1999) Effects of similarity and history on neural mechanisms of visual selection. Nat Neurosci 2:549–554
Bichot NP, Chenchal Rao S, Schall JD (2001a) Continuous processing in macaque frontal cortex during visual search. Neuropsychologia 39:972–982
Bichot NP, Thompson KG, Chenchal Rao S, Schall JD (2001b) Reliability of macaque frontal eye field neurons signaling saccade targets during visual search. J Neurosci 21:713–725
Bisley JW (2011) The neural basis of visual attention. J Physiol 589:49–57
Bisley JW, Goldberg ME (2010) Attention, intention, and priority in the parietal lobe. Annu Rev Neurosci 33:1–21
Bisley JW, Goldberg ME (2003) Neuronal activity in the lateral intraparietal area and spatial attention. Science 299:81–86
Boehnke SE, Munoz DP (2008) On the importance of the transient visual response in the superior colliculus. Curr Opin Neurobiol 18(6):544–551
Borji A, Itti L (2013) State-of-the-art in visual attention modeling. IEEE Trans Pattern Anal Mach Intell 35:185–207
Bosman CA, Schoffelen JM, Brunet N, Oostenveld R, Bastos AM, Womelsdorf T, Rubehn B, Stieglitz T, De Weerd P, Fries P (2012) Attentional stimulus selection through selective synchronization between monkey visual areas. Neuron 75:875–888
Broadbent D (1958) Perception and communication. Pergamon Press, London
Bruce CJ, Goldberg ME, Bushnell MC, Stanton GB (1985) Primate frontal eye fields. II. physiological and anatomical correlates of electrically evoked eye movements. J Neurophysiol 54:714–734
Burrows BE, Moore T (2009) Influence and limitations of popout in the selection of salient visual stimuli by area V4 neurons. J Neurosci 29:15169–15177
Carrasco M (2011) Visual attention: the past 25 years. Vision Res 51:1484–1525
Casanova C (2004) The visual functions of the pulvinar. In: Werner JS (ed) The visual neurosciences. MIT Press, Cambridge, MA
Cerkevich CM, Lyon DC, Balaram P, Kaas JH (2014) Distribution of cortical neurons projecting to the superior colliculus in macaque monkeys. Eye Brain 2014:121–137
Coe B, Tomihara K, Matsuzawa M, Hikosaka O (2002) Visual and anticipatory bias in three cortical eye fields of the monkey during an adaptive decision-making task. J Neurosci 22:5081–5090
Cohen JY, Heitz RP, Woodman GF, Schall JD (2009) Neural basis of the set-size effect in frontal eye field: Timing of attention during visual search. J Neurophysiol 101:1699–1704
Comoli E, Coizet V, Boyes J, Bolam JP, Canteras NS, Quirk RH, Overton PG, Redgrave P (2003) A direct projection from superior colliculus to substantia nigra for detecting salient visual events. Nat Neurosci 6:974–980
Corneil BD, Munoz DP (2014) Overt responses during covert orienting. Neuron 82:1230–1243
Corneil BD, Munoz DP, Olivier E (2007) Priming of head premotor circuits during oculomotor preparation. J Neurophysiol 97:701–714
Davidson RM, Bender DB (1991) Selectivity for relative motion in the monkey superior colliculus. J Neurophysiol 65:1115–1133
Desimone R, Duncan J (1995) Neural mechanisms of selective visual attention. Annu Rev Neurosci 18:193–222
Dorris MC, Munoz DP (1998) Saccadic probability influences motor preparation signals and time to saccadic initiation. J Neurosci 18:7015–7026
Dorris MC, Munoz DP (1995) A neural correlate for the gap effect on saccadic reaction times in monkey. J Neurophysiol 73:2558–2562
Dorris MC, Olivier E, Munoz DP (2007) Competitive integration of visual and preparatory signals in the superior colliculus during saccadic programming. J Neurosci 27:5053–5062
Dorris MC, Paré M, Munoz DP (1997) Neuronal activity in monkey superior colliculus related to the initiation of saccadic eye movements. J Neurosci 17:8566–8579

Dorris MC, Klein RM, Everling S, Munoz DP (2002) Contribution of the primate superior colliculus to inhibition of return. J Cogn Neurosci 14:1256–1263
Driver J (2001) A selective review of selective attention research from the past century. Br J Psychol 92:53–78
Everling S, Munoz DP (2000) Neuronal correlates for preparatory set associated with pro-saccades and anti-saccades in the primate frontal eye field. J Neurosci 20:387–400
Fecteau JH, Munoz DP (2006) Salience, relevance, and firing: A priority map for target selection. Trends Cogn Sci 10:382–390
Fecteau JH, Munoz DP (2005) Correlates of capture of attention and inhibition of return across stages of visual processing. J Cogn Neurosci 17:1714–1727
Fecteau JH, Bell AH, Munoz DP (2004) Neural correlates of the automatic and goal-driven biases in orienting spatial attention. J Neurophysiol 92:1728–1737
Fernandes HL, Stevenson IH, Phillips AN, Segraves MA, Kording KP (2014) Saliency and saccade encoding in the frontal eye field during natural scene search. Cereb Cortex 24:3232–3245
Ferraina S, Paré M, Wurtz RH (2002) Comparison of cortico-cortical and cortico-collicular signals for the generation of saccadic eye movements. J Neurophysiol 87:845–858
Gandhi NJ, Katnani HA (2011) Motor functions of the superior colliculus. Annu Rev Neurosci 34:205–231
Gattass R, Desimone R (1996) Responses of cells in the superior colliculus during performance of a spatial attention task in the macaque. Rev Bras Biol 56(2):257–279
Glimcher PW, Sparks DL (1992) Movement selection in advance of action in the superior colliculus. Nature 355:542–545
Goddard CA, Sridharan D, Huguenard JR, Knudsen EI (2012) Gamma oscillations are generated locally in an attention-related midbrain network. Neuron 73:567–580
Goldberg ME, Wurtz RH (1972) Activity of superior colliculus in behaving monkey. II. effect of attention on neuronal responses. J Neurophysiol 35:560–574
Goldberg ME, Bisley J, Powell KD, Gottlieb J, Kusunoki M (2002) The role of the lateral intraparietal area of the monkey in the generation of saccades and visuospatial attention. Ann N Y Acad Sci 956:205–215
Goodale MA, Westwood DA (2004) An evolving view of duplex vision: separate but interacting cortical pathways for perception and action. Curr Opin Neurobiol 14(2):203–211
Gottlieb JP, Kusunoki M, Goldberg ME (1998) The representation of visual salience in monkey parietal cortex. Nature 391:481–484
Graybiel AM (1978) A satellite system of the superior colliculus: the parabigeminal nucleus and its projections to the superficial collicular layers. Brain Res 145:365–374
Hall WC, Moschovakis A (2004) The superior colliculus: New approaches for studying sensorimotor integration. CRC Press, Boca Raton
Hanes DP, Wurtz RH (2001) Interaction of the frontal eye field and superior colliculus for saccade generation. J Neurophysiol 85:804–815
Hanes DP, Schall JD (1996) Neural control of voluntary movement initiation. Science 274:427–430
Heitz RP, Schall JD (2012) Neural mechanisms of speed-accuracy tradeoff. Neuron 76:616–628
Helms MC, Ozen G, Hall WC (2004) Organization of the intermediate gray layer of the superior colliculus. I. intrinsic vertical connections. J Neurophysiol 91:1706–1715
Hikosaka O, Takikawa Y, Kawagoe R (2000) Role of the basal ganglia in the control of purposive saccadic eye movements. Physiol Rev 80:953–978
Hikosaka O, Kim HF, Yasuda M, Yamamoto S (2014) Basal ganglia circuits for reward value-guided behavior. Annu Rev Neurosci 37:289–306
Horwitz GD, Newsome WT (1999) Separate signals for target selection and movement specification in the superior colliculus. Science 284:1158–1161
Husain M, Parton A, Hodgson TL, Mort D, Rees G (2003) Self-control during response conflict by human supplementary eye field. Nat Neurosci 6:117–118

Ignashchenkova A, Dicke PW, Haarmeier T, Thier P (2004) Neuron-specific contribution of the superior colliculus to overt and covert shifts of attention. Nat Neurosci 7:56–64
Ipata AE, Gee AL, Goldberg ME, Bisley JW (2006) Activity in the lateral intraparietal area predicts the goal and latency of saccades in a free-viewing visual search task. J Neurosci 26:3656–3661
Isa T, Saito Y (2001) The direct visuo-motor pathway in mammalian superior colliculus; novel perspective on the interlaminar connection. Neurosci Res 41:107–113
Itti L, Koch C (2001) Computational modelling of visual attention. Nat Rev Neurosci 2:194–203
Itti L, Koch C, Niebur E (1998) A model of saliency-based visual attention for rapid scene analysis. IEEE Trans Pattern Anal Mach Intell 20:1254–1259
Jantz JJ, Watanabe M, Everling S, Munoz DP (2013) Threshold mechanism for saccade initiation in frontal eye field and superior colliculus. J Neurophysiol 109:2767–2780
Jayaraman A, Batton RR, Carpenter MB (1977) Nigrotectal projections in the monkey: An autoradiographic study. Brain Res 135:147–152
Kane A, Wade A, Ma-Wyatt A (2011) Delays in using chromatic and luminance information to correct rapid reaches. J Vis 11(10):1–18
Kastner S, Ungerleider LG (2000) Mechanisms of visual attention in the human cortex. Annu Rev Neurosci 23:315–341
Keating EG, Gooley SG, Pratt SE, Kelsey JE (1983) Removing the superior colliculus silences eye movements normally evoked from stimulation of the parietal and occipital eye fields. Brain Res 269:145–148
Kim B, Basso MA (2008) Saccade target selection in the superior colliculus: A signal detection theory approach. J Neurosci 28:2991–3007
Knudsen EI (2011) Control from below: The role of a midbrain network in spatial attention. Eur J Neurosci 33(11):1961–1972
Koch C, Ullman S (1985) Shifts in selective visual attention: Towards the underlying neural circuitry. Hum Neurobiol 4:219–227
Krauzlis RJ, Lovejoy LP, Zenon A (2013) Superior colliculus and visual spatial attention. Annu Rev Neurosci 36:165–182
Krauzlis RJ, Bollimunta A, Arcizet F, Wang L (2014) Attention as an effect not a cause. Trends Cogn Sci 18:457–464
Kurylo DD, Skavenski AA (1991) Eye movements elicited by electrical stimulation of area PG in the monkey. J Neurophysiol 65:1243–1253
Kustov AA, Robinson DL (1996) Shared neural control of attentional shifts and eye movements. Nature 384:74–77
Kusunoki M, Gottlieb J, Goldberg ME (2000) The lateral intraparietal area as a salience map: the representation of abrupt onset, stimulus motion, and task relevance. Vis Res 40:1459–1468
Leathers ML, Olson CR (2012) In monkeys making value-based decisions, LIP neurons encode cue salience and not action value. Science 338:132–135
Leichnetz GR, Spencer RF, Hardy SG, Astruc J (1981) The prefrontal corticotectal projection in the monkey; an anterograde and retrograde horseradish peroxidase study. Neuroscience 6:1023–1041
Li W, Piech V, Gilbert CD (2006) Contour saliency in primary visual cortex. Neuron 50:951–962
Li X, Basso MA (2008) Preparing to move increases the sensitivity of superior colliculus neurons. J Neurosci 28:4561–4577
Li X, Basso MA (2005) Competitive stimulus interactions within single response fields of superior colliculus neurons. J Neurosci 25:11357–11373
Li Z (2002) A saliency map in primary visual cortex. Trends Cogn Sci 6:9–16
Liversedge S, Gilchrist I, Everling S (2011) The Oxford handbook of eye movements. Oxford University Press, New York
Lock TM, Baizer JS, Bender DB (2003) Distribution of corticotectal cells in macaque. Exp Brain Res 151:455–470
Lovejoy LP, Krauzlis RJ (2010) Inactivation of primate superior colliculus impairs covert selection of signals for perceptual judgments. Nat Neurosci 13:261–266

Lynch JC, McLaren JW (1989) Deficits of visual attention and saccadic eye movements after lesions of parietooccipital cortex in monkeys. J Neurophysiol 61:74–90
Lynch JC, Graybiel AM, Lobeck LJ (1985) The differential projection of two cytoarchitectonic subregions of the inferior parietal lobule of macaque upon the deep layers of the superior colliculus. J Comp Neurol 235:241–254
Marino RA, Rodgers CK, Levy R, Munoz DP (2008) Spatial relationships of visuomotor transformations in the superior colliculus map. J Neurophysiol 100:2564–2576
Marino RA, Trappenberg TP, Dorris M, Munoz DP (2012a) Spatial interactions in the superior colliculus predict saccade behavior in a neural field model. J Cogn Neurosci 24:315–336
Marino RA, Levy R, Boehnke S, White BJ, Itti L, Munoz DP (2012b) Linking visual response properties in the superior colliculus to saccade behavior. Eur J Neurosci 35:1738–1752
Marrocco RT, Li RH (1977) Monkey superior colliculus: Properties of single cells and their afferent inputs. J Neurophysiol 40:844–860
Maunsell JH, Ghose GM, Assad JA, McAdams CJ, Boudreau CE, Noerager BD (1999) Visual response latencies of magnocellular and parvocellular LGN neurons in macaque monkeys. Vis Neurosci 16:1–14
Mazer JA, Gallant JL (2003) Goal-related activity in V4 during free viewing visual search. evidence for a ventral stream visual salience map. Neuron 40:1241–1250
McAdams CJ, Reid RC (2005) Attention modulates the responses of simple cells in monkey primary visual cortex. J Neurosci 25:11023–11033
McAlonan K, Cavanaugh J, Wurtz RH (2008) Guarding the gateway to cortex with attention in visual thalamus. Nature 456:391–394
McPeek RM, Keller EL (2004) Deficits in saccade target selection after inactivation of superior colliculus. Nat Neurosci 7:757–763
McPeek RM, Keller EL (2002) Saccade target selection in the superior colliculus during a visual search task. J Neurophysiol 88:2019–2034
McPeek RM, Han JH, Keller EL (2003) Competition between saccade goals in the superior colliculus produces saccade curvature. J Neurophysiol 89:2577–2590
Meredith MA, Ramoa AS (1998) Intrinsic circuitry of the superior colliculus: Pharmacophysiological identification of horizontally oriented inhibitory interneurons. J Neurophysiol 79:1597–1602
Mohler CW, Wurtz RH (1976) Organization of monkey superior colliculus: Intermediate layer cells discharging before eye movements. J Neurophysiol 39:722–744
Moran J, Desimone R (1985) Selective attention gates visual processing in the extrastriate cortex. Science 229:782–784
Moschovakis AK (1996) The superior colliculus and eye movement control. Curr Opin Neurobiol 6:811–816
Motter BC (1993) Focal attention produces spatially selective processing in visual cortical areas V1, V2, and V4 in the presence of competing stimuli. J Neurophysiol 70:909–919
Muller JR, Philiastides MG, Newsome WT (2005) Microstimulation of the superior colliculus focuses attention without moving the eyes. Proc Natl Acad Sci 102:524–529
Munoz DP, Everling S (2004) Look away: The anti-saccade task and the voluntary control of eye movement. Nat Rev Neurosci 5:218–228
Munoz DP, Istvan PJ (1998) Lateral inhibitory interactions in the intermediate layers of the monkey superior colliculus. J Neurophysiol 79:1193–1209
Munoz DP, Wurtz RH (1995) Saccade-related activity in monkey superior colliculus. I. characteristics of burst and buildup cells. J Neurophysiol 73:2313–2333
Munoz DP, Dorris MC, Paré M, Everling S (2000) On your mark, get set: Brainstem circuitry underlying saccadic initiation. Can J Physiol Pharmacol 78:934–944
Mysore SP, Knudsen EI (2011) The role of a midbrain network in competitive stimulus selection. Curr Opin Neurobiol 21:653–660
Mysore SP, Asadollahi A, Knudsen EI (2011) Signaling of the strongest stimulus in the owl optic tectum. J Neurosci 31:5186–5196

Noudoost B, Moore T (2011) Control of visual cortical signals by prefrontal dopamine. Nature 474:372–375
Ogawa T, Komatsu H (2004) Target selection in area V4 during a multidimensional visual search task. J Neurosci 24:6371–6382
Ottes FP, Van Gisbergen JA, Eggermont JJ (1986) Visuomotor fields of the superior colliculus: A quantitative model. Vis Res 26:857–873
Paré M, Wurtz RH (1997) Monkey posterior parietal cortex neurons antidromically activated from superior colliculus. J Neurophysiol 78:3493–3497
Paré M, Dorris MC (2011) The role of posterior parietal cortex in the regulation of saccadic eye movements. In: Liversedge S, Gilchrist I, Everling S (eds) Oxford handbook of eye movements. Oxford University Press, Oxford, pp 257–278
Phongphanphanee P, Marino RA, Kaneda K, Yanagawa Y, Munoz DP, Isa T (2014) Distinct local circuit properties of the superficial and intermediate layers of the rodent superior colliculus. Eur J Neurosci 40:2329–2343
Posner MI, Snyder CR, Davidson BJ (1980) Attention and the detection of signals. J Exp Psychol 109:160–174
Purcell BA, Weigand PK, Schall JD (2012a) Supplementary eye field during visual search: Salience, cognitive control, and performance monitoring. J Neurosci 32:10273–10285
Purcell BA, Schall JD, Logan GD, Palmeri TJ (2012b) From salience to saccades: multiple-alternative gated stochastic accumulator model of visual search. J Neurosci 32:3433–3446
Purcell BA, Heitz RP, Cohen JY, Schall JD, Logan GD, Palmeri TJ (2010) Neurally constrained modeling of perceptual decision making. Psychol Rev 117:1113–1143
Redgrave P, Gurney K (2006) The short-latency dopamine signal: A role in discovering novel actions? Nat Rev Neurosci 7:967–975
Redgrave P, Coizet V, Comoli E, McHaffie JG, Leriche M, Vautrelle N, Hayes LM, Overton P (2010) Interactions between the midbrain superior colliculus and the basal ganglia. Front Neuroanat 4:132
Reynolds JH, Chelazzi L (2004) Attentional modulation of visual processing. Annu Rev Neurosci 27:611–647
Reynolds JH, Pasternak T, Desimone R (2000) Attention increases sensitivity of V4 neurons. Neuron 26:703–714
Robinson DA (1972) Eye movements evoked by collicular stimulation in the alert monkey. Vis Res 12:1795–1808
Rodgers CK, Munoz DP, Scott SH, Paré M (2006) Discharge properties of monkey tectoreticular neurons. J Neurophysiol 95:3502–3511
Saito Y, Isa T (2005) Organization of interlaminar interactions in the rat superior colliculus. J Neurophysiol 93:2898–2907
Sato T, Murthy A, Thompson KG, Schall JD (2001) Search efficiency but not response interference affects visual selection in frontal eye field. Neuron 30:583–591
Schall JD, Cohen JY (2011) The neural basis of saccade target selection. In: Liversedge S, Gilchrist I, Everling S (eds) Oxford handbook of eye movements. Oxford University Press, Oxford, pp 357–381
Schall JD, Thompson KG (1999) Neural selection and control of visually guided eye movements. Annu Rev Neurosci 22:241–259
Schall JD, Hanes DP (1993) Neural basis of saccade target selection in frontal eye field during visual search. Nature 366:467–469
Schall JD, Morel A, King DJ, Bullier J (1995) Topography of visual cortex connections with frontal eye field in macaque: Convergence and segregation of processing streams. J Neurosci 15:4464–4487
Schall JD, Purcell BA, Heitz RP, Logan GD, Palmeri TJ (2011) Neural mechanisms of saccade target selection: Gated accumulator model of the visual-motor cascade. Eur J Neurosci 33:1991–2002

Schiller PH, Sandell JH (1983) Interactions between visually and electrically elicited saccades before and after superior colliculus and frontal eye field ablations in the rhesus monkey. Exp Brain Res 49:381–392
Schiller PH, True SD, Conway JL (1980) Deficits in eye movements following frontal eye-field and superior colliculus ablations. J Neurophysiol 44:1175–1189
Schlag J, Schlag-Rey M (1987) Evidence for a supplementary eye field. J Neurophysiol 57:179–200
Schmolesky MT, Wang Y, Hanes DP, Thompson KG, Leutgeb S, Schall JD, Leventhal AG (1998) Signal timing across the macaque visual system. J Neurophysiol 79:3272–3278
Scudder CA, Kaneko CS, Fuchs AF (2002) The brainstem burst generator for saccadic eye movements: A modern synthesis. Exp Brain Res 142:439–462
Segraves MA (1992) Activity of monkey frontal eye field neurons projecting to oculomotor regions of the pons. J Neurophysiol 68:1967–1985
Segraves MA, Goldberg ME (1987) Functional properties of corticotectal neurons in the monkey's frontal eye field. J Neurophysiol 58:1387–1419
Serences JT, Yantis S (2006) Selective visual attention and perceptual coherence. Trends Cogn Sci 10:38–45
Shen K, Paré M (2007) Neuronal activity in superior colliculus signals both stimulus identity and saccade goals during visual conjunction search. J Vis 7(15):1–13
Shen K, Valero J, Day GS, Paré M (2011) Investigating the role of the superior colliculus in active vision with the visual search paradigm. Eur J Neurosci 33:2003–2016
Shires J, Joshi S, Basso MA (2010) Shedding new light on the role of the basal ganglia-superior colliculus pathway in eye movements. Curr Opin Neurobiol 20:717–725
Shook BL, Schlag-Rey M, Schlag J (1990) Primate supplementary eye field: I. comparative aspects of mesencephalic and pontine connections. J Comp Neurol 301:618–642
Solomon SG, Lennie P (2007) The machinery of colour vision. Nat Rev Neurosci 8:276–286
Sparks DL (2002) The brainstem control of saccadic eye movements. Nat Rev Neurosci 3:952–964
Sparks DL, Mays LE (1980) Movement fields of saccade-related burst neurons in the monkey superior colliculus. Brain Res 190:39–50
Squire RF, Noudoost B, Schafer RJ, Moore T (2013) Prefrontal contributions to visual selective attention. Annu Rev Neurosci 36:451–466
Stanton GB, Goldberg ME, Bruce CJ (1988) Frontal eye field efferents in the macaque monkey: II. topography of terminal fields in midbrain and pons. J Comp Neurol 271:493–506
Stepniewska I, Ql HX, Kaas JH (2000) Projections of the superior colliculus to subdivisions of the inferior pulvinar in new world and old world monkeys. Vis Neurosci 17:529–549
Stuphorn V, Schall JD (2006) Executive control of countermanding saccades by the supplementary eye field. Nat Neurosci 9:925–931
Stuphorn V, Taylor TL, Schall JD (2000) Performance monitoring by the supplementary eye field. Nature 408:857–860
Thevarajah D, Mikulic A, Dorris MC (2009) Role of the superior colliculus in choosing mixed-strategy saccades. J Neurosci 29:1998–2008
Thomas NW, Paré M (2007) Temporal processing of saccade targets in parietal cortex area LIP during visual search. J Neurophysiol 97:942–947
Thompson KG, Bichot NP (2005) A visual salience map in the primate frontal eye field. Prog Brain Res 147:251–262
Thompson KG, Bichot NP, Schall JD (1997) Dissociation of visual discrimination from saccade programming in macaque frontal eye field. J Neurophysiol 77:1046–1050
Thompson KG, Hanes DP, Bichot NP, Schall JD (1996) Perceptual and motor processing stages identified in the activity of macaque frontal eye field neurons during visual search. J Neurophysiol 76:4040–4055
Tigges J, Tigges M (1981) Distribution of retinofugal and corticofugal axon terminals in the superior colliculus of squirrel monkey. Invest Ophthalmol Vis Sci 20:149–158
Treisman AM (1969) Strategies and models of selective attention. Psychol Rev 76:282–299

Treue S (2003) Visual attention: The where, what, how and why of saliency. Curr Opin Neurobiol 13:428–432
Treue S, Maunsell JH (1996) Attentional modulation of visual motion processing in cortical areas MT and MST. Nature 382:539–541
Ungerleider JT, Mishkin M (1982) Two cortical visual systems. In: Ingle DJ, Mansfield RJW, Goodale MS (eds) The analysis of visual behaviour. MIT Press, Cambridge, MA, pp 549–586
Van Gompel RPG, Fischer MH, Murray WS, Hill RL (2007) Eye movements: A window on mind and brain
Wang CA, Boehnke SE, White BJ, Munoz DP (2012) Microstimulation of the monkey superior colliculus induces pupil dilation without evoking saccades. J Neurosci 32:3629–3636
Wang CA, Munoz DP (2015) A circuit for pupil orienting responses: Implications for cognitive modulation of pupil size. Curr Opin Neurobiol 33:134–140
Wang Y, Luksch H, Brecha NC, Karten HJ (2006) Columnar projections from the cholinergic nucleus isthmi to the optic tectum in chicks (gallus gallus): A possible substrate for synchronizing tectal channels. J Comp Neurol 494:7–35
Wardak C, Ibos G, Duhamel JR, Olivier E (2006) Contribution of the monkey frontal eye field to covert visual attention. J Neurosci 26:4228–4235
Watanabe M, Munoz DP (2011) Probing basal ganglia functions by saccade eye movements. Eur J Neurosci 33:2070–2090
Werner JS, Chalupa LM (2014) The new visual neurosciences. MIT Press, MA
White BJ, Munoz DP (2011a) The superior colliculus. In: Liversedge S, Gilchrist I, Everling S (eds) Oxford handbook of eye movements. Oxford University Press, Oxford, pp 195–213
White BJ, Munoz DP (2011b) Separate visual signals for saccade initiation during target selection in the primate superior colliculus. J Neurosci 31:1570–1578
White BJ, Theeuwes J, Munoz DP (2012) Interaction between visual- and goal-related neuronal signals on the trajectories of saccadic eye movements. J Cogn Neurosci 24:707–717
White BJ, Kerzel D, Gegenfurtner KR (2006) Visually guided movements to color targets. Exp Brain Res 175:110–126
White BJ, Berg D, Itti L, Munoz DP (2014) Visual coding in the superior colliculus during free-viewing of natural dynamic stimuli. Soc Neurosci Abstr 288:11
White BJ, Boehnke SE, Marino RA, Itti L, Munoz DP (2009) Color-related signals in the primate superior colliculus. J Neurosci 29:12159–12166
White BJ, Marino RA, Boehnke SE, Itti L, Theeuwes J, Munoz DP (2013) Competitive integration of visual and goal-related signals on neuronal accumulation rate: A correlate of oculomotor capture in the superior colliculus. J Cogn Neurosci 25:1754–1768
Womelsdorf T, Fries P (2007) The role of neuronal synchronization in selective attention. Curr Opin Neurobiol 17:154–160
Wurtz RH, Mohler CW (1976) Organization of monkey superior colliculus: Enhanced visual response of superficial layer cells. J Neurophysiol 39:745–765
Wurtz RH, Goldberg ME (1972) Activity of superior colliculus in behaving monkey. 3. cells discharging before eye movements. J Neurophysiol 35:575–586
Zenon A, Krauzlis R (2014) Superior colliculus as a subcortical center for visual selection. Med Sci 30:637–643
Zenon A, Krauzlis RJ (2012) Attention deficits without cortical neuronal deficits. Nature 489:434–437
Zhang X, Zhaoping L, Zhou T, Fang F (2012) Neural activities in v1 create a bottom-up saliency map. Neuron 73:183–192
Zhaoping L (2008) Attention capture by eye of origin singletons even without awareness–a hallmark of a bottom-up saliency map in the primary visual cortex. J Vis 8:1.1–18
Zhou H, Desimone R (2011) Feature-based attention in the frontal eye field and area V4 during visual search. Neuron 70:1205–1217

Insights on Vision Derived from Studying Human Single Neurons

Jan Kamiński and Ueli Rutishauser

Investigating the living brain, and in particular relating its activity to behavior is one of the most important challenges in neuroscience. Researchers use many different techniques to explore this relationship. Careful observation of patients with brain lesions or neuroimaging methods such as functional magnetic resonance imaging (fMRI), electroencephalography (EEG), or near infra-red spectroscopy (NIRS) are examples of procedures which allow researchers to make inferences about brain activity in a non-invasive way. However, the most widely used such tool, fMRI, measures a blood oxygen dependent (BOLD) signal that is only indirectly related to the signal of interest, neural activity. Consequently, the BOLD contrast reflects hemodynamic changes rather than neuronal activity (Kim and Ogawa 2012). The relationship between BOLD and neural activity is complex and with the exception of primary visual cortex (V1) poorly understood (Logothetis et al. 2001; Logothetis and Wandell 2004). In contrast, the EEG signal reflects directly the (although averaged) activity of neurons, but it is highly filtered due to the large distance between the neuronal sources and the electrodes on the scalp. Thus, despite their undisputed utility in expanding our knowledge about the human brain, current non-invasive techniques do not permit direct study of activity such as local field potentials (LFPs) and spiking activity in individual neurons.

In rare clinical situations, invasive intracranial recordings are a necessary step in the diagnosis or treatment of illness, e.g. in patients suffering from drug resistant epilepsy or Parkinson's disease. These situations represent a rare opportunity for researchers to record electrophysiological signals directly from inside the human brain much like those usually restricted to animal models.

J. Kamiński · U. Rutishauser
Department of Neurosurgery, Cedars-Sinai Medical Center, Pasadena, USA

J. Kamiński (✉) · U. Rutishauser
Division of Biology and Biological Engineering, California Institute of Technology, Pasadena, USA
e-mail: jan.kaminski@cshs.org

Q. Zhao (ed.), *Computational and Cognitive Neuroscience of Vision*,
Cognitive Science and Technology, DOI 10.1007/978-981-10-0213-7_2

So far, most research utilizing single unit recordings in humans has focus on areas of the medial temporal lobe (MTL), in particular the hippocampus, amygdala, entorhinal cortex, and parahippocampal cortex. This is because the MTL is a routine target for depth electrode studies in patients suffering from drug resistant epilepsy. In this review, we will summarize key findings on how individual neurons in the MTL respond to visual stimuli and what has been discovered about the human visual system based on such recordings.

It is important to keep in mind the limitations of such invasive work conducted in human subjects. Subjects suffer from brain pathology, which means that these studies are never carried out on data from a healthy brain. The data we discuss all originates from epilepsy patients who are being investigated for possible surgical treatment of their drug resistant seizures. This raises the question to what extend these results generalize to non-epileptic subjects. For example, studies showed that some abnormalities in brain activity can be observed even in cortical areas not deemed epileptic (Engel et al. 1982; Bettus et al. 2011). Despite these limitations, it is remarkable that the type of findings we review in this chapter have been highly reproducible across a wide variety of patients with different kinds of epilepsy. This makes it very unlikely that these findings are abnormalities caused by epilepsy, because these would be expected to vary considerably according to their exact pathology. Also, since the location of the epileptic focus is not known in advance, electrodes are implanted in many areas which after the fact are deemed non-epileptic. Thus, eventually many of the recordings are made in healthy tissue unrelated to the seizure focus or network. Comparing data recorded within and outside of the seizure focus is a valuable tool to assess the sensitivity of a finding to epilepsy. Such comparisons have generally revealed little difference, with some notable exceptions that provide insights into specific deficits attributable to epilepsy (see Fig. SII in Rutishauser et al. 2008). Lastly, in spite of these limiting factors, human intracranial recodings have reproduced many key discoveries previously made in animal models, such as highly invariant and tuned visual neurons described in this chapter.

1 Latency

An important metric which can be quantified directly using human single neuron studies is the latency of neuronal responses. The most comprehensive knowledge about latencies is available for the different parts of the MTL. For example, Kreiman et al. quantified latencies in response to visual stimuli (pictures on a screen) for neurons in the amygdala, entorhinal cortex and hippocampus (Kreiman et al. 2000a). Onset latencies (defined as a significant increase of firing rate above baseline) for the amygdala were 240 ms, for entorhinal cortex 209 ms and for hippocampus 239 ms. Interestingly, latencies were not significantly different between the different areas. In a more recent study by Mormann et al. (2008) another method—analyzing distribution of interspike intervals (Hanes et al. 1995)

—was used to estimate the latency of single unit responses. Similarly to the earlier study, latencies were long (~400 ms) and there was no significant difference between amygdala, entorhinal cortex and hippocampal units. Noteable, the latency of neurons recorded from the parahippocampal cortex had a significantly shorter latency of 270 ms. Anatomically, one would expect this difference because the parahippocampal region is the major input to the entorhinal cortex (Suzuki and Amaral 1994). More puzzling is the lack of latency differences between the other areas. A potential explanation are the monosynaptic connections between amygdala, entorhinal cortex, and hippocampus (Suzuki 1996), but it remains unclear why other connections that are also monosynaptic result in a large latency difference. The most interesting, and puzzling, result is the ~100 ms gap between onset latencies of the parahippocampal cortex and other regions. It has been suggested that this is due to a recurrent process within parahippocampal region (Quian Quiroga 2012).

Latencies of neuronal responses from areas other than MTL have also been reported. In one study, single neuron responses to action execution and observation were studied in areas MTL and supplementary motor cortex (SMA) (Mukamel et al. 2010). Neurons in SMA had similar latencies as neurons in the MTL during action observation, but responded significantly faster than neurons in the MTL during action execution. Interestingly, neuronal latencies of units recorded in ventral prefrontal cortex were significantly shorter (faster) than those observed in the MTL (Kawasaki et al. 2005): latencies as short as 120–170 ms were observed in response to presentation of emotional visual stimuli. This fast response could indicate that prefrontal cortex modulates visual processing in other areas, including temporal cortex and the amygdala, via feedback projections (Sugase et al. 1999; Adolphs et al. 2014).

2 Visual Selectivity of Neurons in the Human MTL

Along the visual stream, selectivity becomes more complex (Gross 1994; Logothetis and Sheinberg 1996; Tanaka 1996; Grill-Spector and Malach 2004). In early stages of visual processing, such as in V1, neurons are tuned to simple visual features such as oriented bars. In contrast, neurons in higher visual areas such as V4/V5 are selective for directions of motion and neurons in inferotemporal cortex (IT) are selective for even more complex stimuli such as hands and faces (Gross 1994; Logothetis and Sheinberg 1996; Tanaka 1996). In even higher level visual areas of the MTL, such as the hippocampus and amygdala, visual selectivity is even more specific such as for particular aspects of faces such as eye contact (Leonard et al. 1985; Miyashita et al. 1989; Mosher et al. 2014). In humans, this area is where most is known about the selectivity of individual neurons. In 1997, Fried et al. (Fried et al. 1997) published one of the first comprehensive studies on the highly specific selectivity of human MTL neurons to complex visual stimuli, such as Ekman faces (Ekman and Friesen 1976). A key result was that single units in the

MTL could discriminate faces from inanimate objects. Following up on this initial discovery, Kreiman et al. (2000a) characterized the responses of single human MTL units for sets of pictures grouped into different semantic categories such as objects, animals, cars, and faces. The key finding was that most visually selective cells responded to all stimuli of a category, even if the different stimuli (pictures) belonging to this category looked visually very different. This indicated that these neurons are highly invariant to visual variation. While such highly invariant tuning has been found in numerous areas of the human MTL, there are some interesting inter-areal specializations that have been described. For example, Mormann et al. (2011) has demonstrated that neurons in the right but not left amygdala are significantly more likely to be tuned to pictures of animals. Importantly, the selectivity of such neurons can extend much beyond that of semantic categories. In a seminal study by Quiroga et al., neurons were identified that responded only when a specific individual, object or landmark were shown in an image (Quian Quiroga et al. 2005). This study was composed of two parts. In the first part of the experiment, around 90 images of people and objects familiar to the subject were presented in a screening session to identify candidates for selective units. Next, variations of the images which elicited a strong response in the screening session were used in a later session. Results from the second part of this experiment showed that MTL neurons where highly selective for pictures representing a particular person, object or landmark but not for other pictures. Figure 1 shows an example of a unit exhibiting this remarkable property.

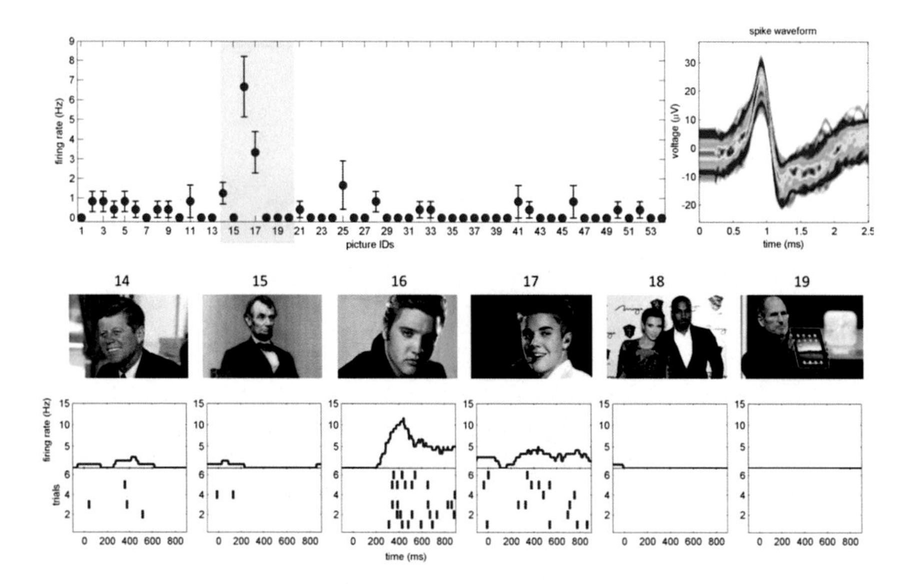

Fig. 1 Example of a highly selective response of an individual amygdala neuron. 54 different images were presented ("picture ID"), to most of which the neuron did not respond. However, to some images (such as ID 16) this unit responded very strongly (ID 16 is a picture showing Elvis Presley). Unpublished data

In another study the selectivity of putative pyramidal cells and interneurons in the MTL was assessed (Ison et al. 2011). Units were classified as putative pyramidal cells and interneurons based on their baseline firing rate and features of the extracellular waveform. Putative pyramidal cells were found to be much more selective than interneurons, a finding which was valid across different regions of the MTL, although the highest selectivity was observed in the hippocampus as compared to amygdala, entorhinal cortex and parahippocampal cortex. This gradient corresponds to the major signal pathways through the MTL (Suzuki and Amaral 1994), indicating that neuronal selectivity for visual stimuli increases successively at each stage in visual processing. Different analysis methods have supported this hierarchical gradient of selectivity (Mormann et al. 2008): parahippocampal neurons, on average, responded to five different stimuli whereas neurons in entorhinal cortex and hippocampus responded, on average, to only two stimuli. Mormann et al. (2008) also revealed a positive correlation between selectivity and latency, which further supports the concept of a mechanism of hierarchical processing in which in every step of processing of visual information the representation of a stimulus becomes more precise and selective (Grill-Spector and Malach 2004). The studies summarized so far show that MTL neurons respond to only small numbers of all possible stimuli. This indicates that each stimulus is encoded by a relatively small neuronal assembly (Quian Quiroga et al. 2008) and this data therefore support a sparse coding hypothesis (Olshausen and Field 2004; Barlow 2009). This theory proposes that a given percept in the brain is represented by the activity of a small population of specialized neurons. In contrast, an alternative view is distributed population coding (Decharms and Zador 2000), which suggests that a percept is represented by the activity of large numbers of broadly tuned neurons.

3 Invariance

Another remarkable feature of MTL neurons is their invariance. In their 2005 study, Quian Quiroga and collegues (2005) presented to subjects not only different images of the same object or person but they also used pencil sketches, caricatures of this person or object or even letter strings with their names. What they discovered was that even if those pictures had nothing in common visually (but were linked semantically), specific and selective neurons still showed an increase firing rate when presented with these stimuli. This result suggested that neurons responded to an abstract representation of the meaning of the object or individual rather than to any visual property of the picture. This phenomena has also been confirmed using a decoding analysis (Quian Quiroga et al. 2007) that demonstrated that it is possible to decode the identity of the person presented in a picture. In contrast, it was not possible to estimate which specific picture of the person was presented. Noticeable, the selectivity for the identity of a person is not limited to the visual modality but extends to auditory input as well. In one of the experiments, pictures and an audio playback of a voice reading the names of this images were used (Quian Quiroga et al. 2009). The results showed that there was no significant difference in selectivity of cells in MTL when comparing visual and auditory input. The strength of this multimodal invariance was different among different parts of the MTL. In the parahippocampal cortex, half of the responsive neurons showed visual invariance but none of this unit showed multimodal invariance. In entorhinal cortex, 70 % of all visually responsive neurons were visually invariance and 35 % were, in addition, multimodally invariant. In the hippocampus, even more neurons were invariant: 86 % showed visual invariance and 38 % multimodal invariance.

This systematic variability of multimodal invariance is compatible with the hierarchical structure of the MTL (Mormann et al. 2008; Ison et al. 2011).

4 Grandmother Cells

The highly invariant, sparse and specific responses of neurons in the MTL triggered a renewed discussion about the concept of so called "grandmother cells" (Ison and Quiroga 2008; Quian Quiroga et al. 2008; Quian Quiroga 2013). A grandmother cell is a hypothesized neuron that responds only to a complex, specific stimulus such as the image of one's grandmother. The concept of grandmother cells was originally proposed by Jerry Lettvin (Gross 2002), although unknown to Lettvin a few years earlier the same concept was introduced by the polish neurophysiologist Jerzy Konorski who used the term "gnostic" cell due to the Greek term "gnosis" for knowledge (Konorski 1967). The grandmother cell hypothesis is an extreme view of the sparse coding hypothesis (Olshausen and Field 2004; Barlow 2009), which proposed that a given percept in the brain is represented by the activity of a small neuronal population and in the extreme case only a few or even only one neuron. As summarized above, the results obtained from single units recordings in the human MTL support the sparse coding hypothesis. This raises the question of whether it is possible that ultimately, on the very last stages of visual processing, a percept could be represented by the highly specific activity of only a few neurons?

This question is difficult to answer with the current experimental setups used in research involving epilepsy patients. Because of time and stability constraints, only a limited number of pictures can be tested (typically less than 200). Consequently, it is impossible to dismiss the possibility that a given neuron that responded only to a particular picture during the experiment would not also respond to a different, but never shown, image. Nevertheless, there are several arguments against the grandmother cell hypothesis. Using Bayesian probabilistic analysis of a data set consisting of 1425 MTL units from 34 sessions Waydo and colleagues (Waydo et al. 2006) estimated that out of the approximately 10^9 (Harding et al. 1998; Henze et al. 2000; Schumann et al. 2004) neurons in the human MTL, approximately 2 million are involved in representing each individual concept. This estimate is far larger than that predicted by the grandmother cell hypothesis. Furthermore, estimating that average an person can distinguish between 10000 and 30000 objects (Biederman and Bederman 1987), we can assess that one neuron should respond to ~50–150

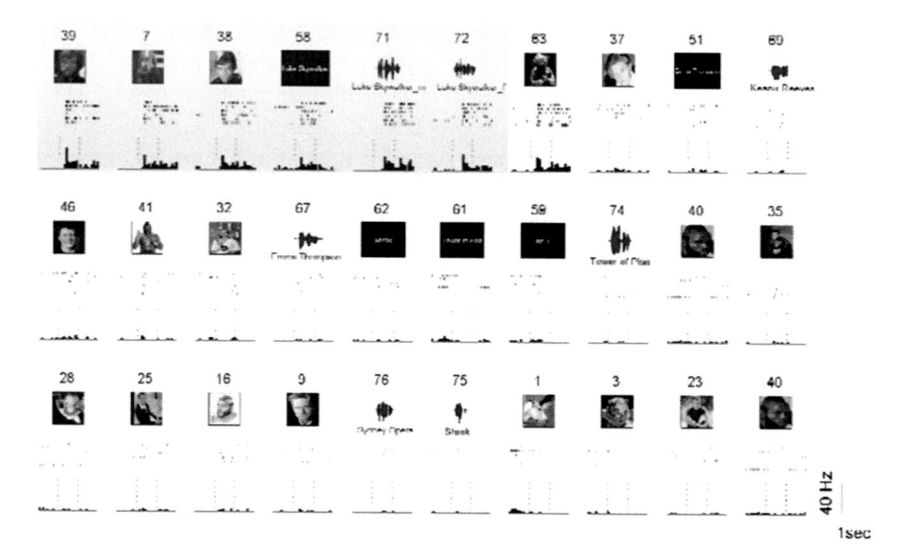

Fig. 2 Example single Unit recorded in the entorhinal cortex that responded selectively to pictures of Luke Skywalker, as well as the written and spoken name. The same unit showed also selectivity to Yoda—other character from movie "Star Wars". Modified from (Quian Quiroga et al. 2009)

distinct objects (Waydo et al. 2006; Quian Quiroga et al. 2008). It was also shown that some of the units indeed responded to very different objects. For example, a unit in entorhinal cortex that responded to Luke Skywalker from the Star Wars movie series was also activated by a picture of Yoda, another character appearing in the same movie (Quian Quiroga et al. 2009) (Fig. 2).

Similarly, another neuron responding to Jennifer Anistion also responded to Lisa Kudrow, both actresses in the TV series "Friends" (Quian Quiroga et al. 2005). In conclusion, it thus appears that these units responded to related but distinct concepts rather than to one specific object. This feature of MTL neurons led to the proposed term "concept cell" (Quian Quiroga et al. 2005; Quian Quiroga 2012) to appropriately describe these highly selective neurons.

5 Topography of Tuning

Another import piece of information which was obtained from individual human neurons concerns the topography of tuning properties of neurons. In many sensory systems in the brain, a topographic organization of neurons can be observed at least in lower levels (Hubel and Wiesel 1962; Simons and Woolsey 1979; Shou et al. 1986; Talavage et al. 2000). For example, in the visual system nearby neurons in the thalamus, as well as in the early visual cortices receive inputs from neighbouring parts of the visual field (Hubel and Wiesel 1962; Shou et al. 1986). This topographic organization can still be observed in higher order visual areas (Tanaka 1996; Kreiman et al. 2006; Tsao et al. 2008). In contrast, human MTL neurons do not show any apparent topographical organization. Neighboring neurons, such as those recorded on the same microwire, are typically tuned to unrelated concepts (Quian Quiroga et al. 2009; Mormann et al. 2011). This phenomenon is similar to that shown for place cells in the rodent hippocampus, where nearby units can encode totally different and non-overlapping place fields (O'Keefe and Nadel 1978).

6 Internally Generated Responses and Consciousness

Human recordings from MTL neurons are an ideal setup for studying internally (i.e. without sensory stimulation) generated responses because neurons are tuned to very specific high-level concepts and human subjects can easily follow instructions to produce certain thoughts such as "think about Bill Clinton". Similarly, concept cells represent a great opportunity to study awareness due to their tuning for specific abstract representations. Indeed there are many experiments studying these phenomena in patients. In one of these studies subjects were asked to imagine previously viewed images (Kreiman et al. 2000b). Results of this experiment showed that the selectivity and tuning for internally imagined pictures were very similar to selectivity during sensory stimulation: 88 % of neurons which were selective during perception and imagery had the same tuning in both situations. In another study subjects viewed short (5–10 s) video clips and were later asked to freely recall these clips and immediately report when a specific clip "came to mind" (Gelbard-Sagiv et al. 2008). Again, neurons in the MTL showed similar tuning during perception and during free recall (imaginary). This phenomena was subsequently utilized in an experiment in which subjects attempted to use knowledge of their own neuronal activity to alter pictures presented on a screen (Cerf et al. 2010). In this biofeedback study, subjects were presented with hybrid "morphs" of two images. For each of the two pictures at least one selective neuron was previously identified. The subject's task was to enhance one of the pictures by focusing their thoughts on it. A real-time decoder used the firing rates of MTL neurons to control the visibility of each image. Without any prior training, subjects succeeded in driving the visual stimulus towards one of the two images in over 70 % of trials entirely by focusing their thoughts on a given image. This result shows that internally generated responses can overwrite responses triggered by sensory input.

In an early study of visual awareness utilizing recordings from the MTL (Kreiman et al. 2002), the flash suppression method (Wolfe 1984; Sheinberg and Logothetis 1997)was used to alter perception of a visual input. With this method, the percept of an image presented to one eye only can be removed from awareness by the rapid presentation ("flash") of a different image to the other eye. The results of this study demonstrated that 70 % of the visually selective neurons followed the perceptual alternation. Strikingly, none of the selective neurons responded to the suppressed percept. This phenomena has also been studied using a backward masking paradigm. This revealed that visually tuned neurons fired whenever a picture was correctly recognized. In contrast, the same neurons remained at baseline when the picture was not recognized. Additionally, there was no difference in the evoked firing rate of visually selective cells regardless of whether the stimulus was presented for 33, 66 or 132 ms. All that was necessary is that the subject recognized the image, regardless of presentation duration. More recently, a face adaptation paradigm was used to study conscious perception (Quian Quiroga et al. 2014). In this paradigm, perception of an ambiguous hybrid face is biased by a preceding presentation of one of the faces that make up the hybrid (morphed) face. This preceding face (the "adaptor") biases perception toward the other, non-adapted, face. The hybrid faces were constructed from two familiar faces, to at least one of which a selectively tuned neuron was previously identified. Analysis of the neuronal firing patterns again demonstrated that the neuronal response followed the subjective percept, even if the physical input was exactly identical for two different percepts (Fig. 3).

A similar phenomena has also been observed during the perception of emotional faces (Wang et al. 2014). In this experiment, neurons in the amygdala were identified which signaled the presence of the emotion in a face (such as happy or fearful). Remarkably, these same neurons signaled the subjective judgment of emotions shown by the subjects rather than the stimulus features. For example,

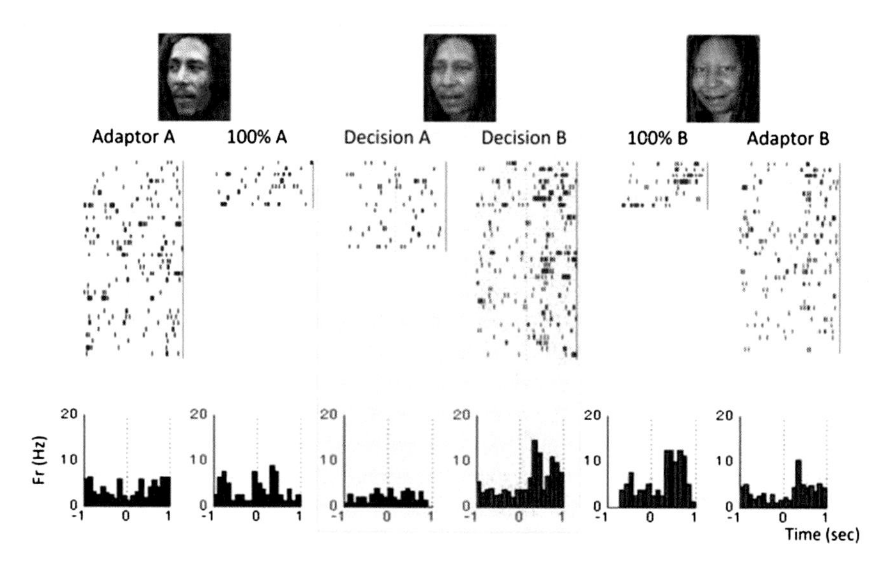

Fig. 3 Responses of a neuron recoded in the hippocampus that is tuned to an image of Whoopi Goldberg (100 % B) but not to Bob Marley (100 % A). The response was larger when the hybrid picture was recognized as Whoopi Goldberg (Decision B). Adapted from (Quian Quiroga et al. 2014)

when a patient thought a happy face was fearful, the neuronal response followed the subjective percept of fear.

Together, these results support the argument that neurons in the human MTL might play a role in forming conscious percepts. However, lesion studies neither report perceptual deficits nor impairment of moment-to-moment awareness resulting from MTL lesions. In contrast, such studies of patients with MTL lesions show major deficits in memory formation and consolidation (Scoville and Milner 1957; Milner et al. 1968; Squire et al. 2004). For example, patient H.M exhibits superior performance compared to age match control group in the Mooney face perception task, in which subjects need to recognize faces in chaotic-looking black and white patterns (Milner et al. 1968). Similarly, patients with bilateral lesions of the amygdala exhibit deficits in the recognition of emotions but not in other visual-perceptual discrimination tasks nor in consciousness processes (Tranel and Hyman 1990; Adolphs et al. 1994, 2005). This data thus suggest that activity of cells in the MTL is not required to create conscious percepts (Quian Quiroga 2012). So what other function could those neurons have?

7 Memory

As lesions of the MTL lead to significant impairments in acquiring new memories (Scoville and Milner 1957; Milner et al. 1968; Squire et al. 2004), it was hypothesized that concept cells could play an important role in the process of acquiring new memories (Quian Quiroga 2012). Some of the characteristic of these neurons supports this hypothesis (Quian Quiroga 2012). For example, the high degree of invariance of these neurons corresponds to the fact that humans have a tendency to remember concepts and forget details (Quian Quiroga et al. 2005). They also tend to be activated to relevant concepts for a given subject such as pictures of their family, themselves (Viskontas et al. 2009). Additionally, these cells respond to multimodal stimuli and they are also activated by internal processes. This indicates that their function may lay beyond perception (Kreiman et al. 2000b; Gelbard-Sagiv et al. 2008; Quian Quiroga et al. 2009; Cerf et al. 2010). Indeed, research concerning long term memory utilizing recordings of single unit MTL supports the crucial role of MTL neurons in forming new memories (Rutishauser et al. 2006, 2008, 2010). While this research has been reviewed extensively elsewhere (Rutishauser et al. 2014), we briefly summarize a few key findings relevant to the present discussion. In a series of experiments, subjects were shown a sequence of unfamiliar (novel) images. After a delay (10–30 min), during which subjects performed a distraction task, another sequence of images was presented. This second set contained both the images shown previously as well as unfamiliar (novel) images. The task of the subjects was to indicate for each image whether it has previously been seen before (old) or not (new). This is thus a typical declarative

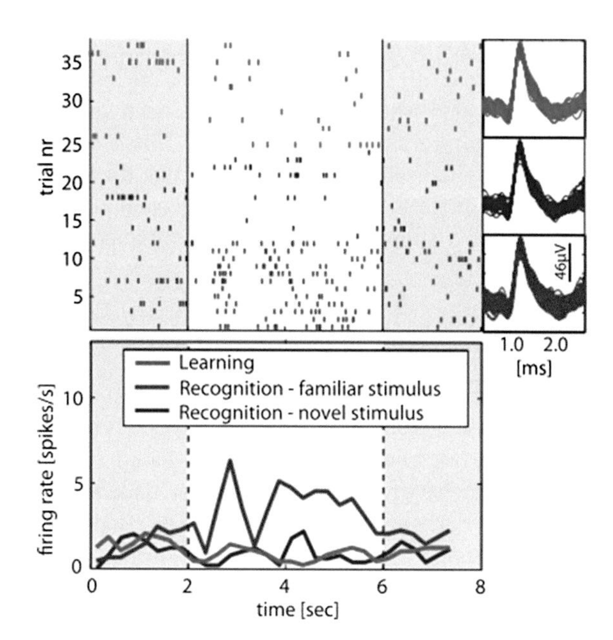

Fig. 4 Hippocampal neuron which increases its firing rate to familiar stimuli. *Green trials*—shows response during learning phase, *Blue trials*—shows response to novel stimuli during test recognition session, *Red trials*—shows response during recognition session to stimuli already presented during learning phase. Modified from Rutishauser et al. (2006)

memory task, requiring an intact MTL. Analysis of units in the amygdala and hippocampus revealed two subpopulations of neurons (Rutishauser et al. 2006, 2008). The first subpopulation increased their firing rate to pictures which had previously been presented (Fig. 4) whereas a second subpopulation responded only for novel pictures.

These differences could be observed as fast as 10 min after initial encoding and lasted at least 24 h (the shortest and longest period tested). In one of the variants of the experiment, subjects saw pictures in one of four quarters of the screen during the learning session and were ask to memorized not only which pictures they saw but also their position on the screen (Rutishauser et al. 2008). The group of neurons that increased their firing rate for familiar pictures increased their firing rate even more when the spatial position was correctly remembered by the subject. In contrast, the firing rate increase was of reduced magnitude if the subject failed to remember (recall) the spatial position. This data demonstrates a role of these neurons in the declarative memory process and suggests that these neurons encode a continuous strength-of-memory-gradient. Interestingly, it was possible to predict which pictures will later be remembered and which will be forgotten based on the response of neurons during the encoding phase (Rutishauser et al. 2010). The best predictive feature was the extent of phase locking to theta-band (3–8 Hz) oscillations of a subset of neurons. Thus, if during encoding of a particular picture theta phase locking was strong, the image was likely encoded and thus later remembered. In contrast, weak phase locking predicted that the image would be forgotten. This result shows the first direct demonstration of the crucial interplay between neuronal activity and ongoing oscillation in the process of forming new memories. Another interesting discovery made by Rutishauser et al. was that novelty/familiar responses were not visually tuned (Rutishauser et al. 2008, 2015). Instead, an individual novelty/familiarity coding neuron signaled whether a stimulus has been seen before or not regardless of the visual category of the presented picture. This is in stark contrast to the highly tuned neurons summarized earlier, indicating that these neurons are different. Thus, this neuronal response could be utilized as general novelty detector that signals the significance of a stimulus and facilitate its memorization (Lisman and Otmakhova 2001).

8 Closing Remarks

During the last two decades, research based on recoding single neurons in the human brain have greatly advanced our understating of cognition, memory, and perception. Early experiments largely replicated and extended to humans findings previously made only in animal models, in particular macaques. This is important because it validates the animal model. More recently, fundamental new discoveries only possible in humans have been made. Among those are the crucial new insights provided on the exquisite tuning, invariance and sparseness of visually responsive neurons summarized in this chapter. Unfortunately, little is known about the response of neurons in the human visual cortex due to the inability to routinely record from these areas in humans. Nevertheless, recordings from higher-level areas in the MTL have revealed a surprising number of interesting insights for vision and awareness and are thus highly valuable. Lesion studies of the MTL only have a limited impact on perception and visual awareness, but result in major memory impairments. Similarly, single unit research summarized in this chapter has revealed an important role for MTL neurons in the memory formation process. Together, this data supports the hypothesis that the highly specific visual responses observed in the MTL subserve memory formation or retrieval rather than perception as such. However, it should also be noted that the perceptual studies performed with MTL-lesioned subjects are rather crude and one can thus not exclude the possibility that the right experiments have not yet been done to uncover perceptual or perhaps awareness deficits in patients with MTL lesions.

References

Adolphs R, Gosselin F, Buchanan TW, Tranel D, Schyns P, Damasio AR (2005) A mechanism for impaired fear recognition after amygdala damage. Nature 433:68–72

Adolphs R, Kawasaki H, Tudusciuc O, Howard MA, Heller AC, Sutherling WW, Philpott L, Ross IB, Mamelak AN, Rutishauser U (2014) Electrophysiological responses to faces in the human amygdala. In: Fried I, Rutishauser U, Cref U, Kreiman G (eds) Single neuron studies of the human brain. MIT Press, Boston, pp 229–247

Adolphs R, Tranel D, Damasio H, Damasio A (1994) Impaired recognition of emotion in facial expressions following bilateral damage to the human amygdala. Nature 372:669–672
Barlow HB (2009) Single units and sensation: a neuron doctrine for perceptual psychology? Perception 38:371–394
Bettus G, Ranjeva JP, Wendling F, Bénar CG, Confort-Gouny S, Régis J, Chauvel P, Cozzone PJ, Lemieux L, Bartolomei F, Guye M (2011) Interictal functional connectivity of human epileptic networks assessed by intracerebral EEG and BOLD signal fluctuations. PLoS ONE 6:e20071
Biederman I, Bederman I (1987) Recognition-by-components: a theory of human image understanding. Psychol Rev 94:115–147
Cerf M, Thiruvengadam N, Mormann F, Kraskov A, Quiroga RQ, Koch C, Fried I (2010) On-line, voluntary control of human temporal lobe neurons. Nature 467:1104–1108
Decharms R, Zador A (2000) Neural representation and the cortical code. Annu Rev Neurosci:613–647
Ekman P, Friesen WV (1976) Pictures of Facial Affect. Consulting Psychologists Press, Palo Alto, CA
Engel J, Kuhl DE, Phelps ME, Mazziotta JC (1982) Interictal cerebral glucose metabolism in partial epilepsy and its relation to EEG changes. Ann Neurol 12:510–517
Fried I, MacDonald KA, Wilson CL (1997) Single neuron activity in human hippocampus and amygdala during recognition of faces and objects. Neuron 18:753–765
Gelbard-Sagiv H, Mukamel R, Harel M, Malach R, Fried I (2008) Internally generated reactivation of single neurons in human hippocampus during free recall. Science 322:96–101
Grill-Spector K, Malach R (2004) The human visual cortex. Annu Rev Neurosci 27:649–677
Gross CG (1994) How inferior temporal cortex became a visual area. Cereb Cortex 4:455–469
Gross CG (2002) Genealogy of the “grandmother cell”. Neuroscientist 8:512–518
Hanes DP, Thompson KG, Schall JD (1995) Relationship of presaccadic activity in frontal eye field and supplementary eye field to saccade initiation in macaque: Poisson spike train analysis. Exp Brain Res 103:85–96
Harding AJ, Halliday GM, Kril JJ (1998) Variation in hippocampal neuron number with age and brain volume 8:710–718
Henze DA, Borhegyi Z, Csicsvari J, Mamiya A, Harris KD, Buzsáki G (2000) Intracellular features predicted by extracellular recordings in the hippocampus in vivo. J Neurophysiol 84:390–400
Hubel D, Wiesel T (1962) Receptive fields, binocular interaction and functional architecture in the cat’s visual cortex. J Physiol:106–154
Ison M, Quiroga R (2008) Selectivity and invariance for visual object perception. Front Biosci:4889–4903
Ison MJ, Mormann F, Cerf M, Koch C, Fried I, Quiroga RQ (2011) Selectivity of pyramidal cells and interneurons in the human medial temporal lobe. J Neurophysiol 106:1713–1721
Kawasaki H, Adolphs R, Oya H, Kovach C, Damasio H, Kaufman O, Howard M (2005) Analysis of single-unit responses to emotional scenes in human ventromedial prefrontal cortex. J Cogn Neurosci 17:1509–1518
Kim S-G, Ogawa S (2012) Biophysical and physiological origins of blood oxygenation level-dependent fMRI signals. J Cereb Blood Flow Metab 32:1188–1206
Konorski J (1967) Integrative activity of the brain; an interdisciplinary approach. Chicago University, Chicago
Kreiman G, Fried I, Koch C (2002) Single-neuron correlates of subjective vision in the human medial temporal lobe. Proc Natl Acad Sci USA 99:8378–8383
Kreiman G, Hung CP, Kraskov A, Quiroga RQ, Poggio T, DiCarlo JJ (2006) Object selectivity of local field potentials and spikes in the macaque inferior temporal cortex. Neuron 49:433–445
Kreiman G, Koch C, Fried I (2000a) Category-specific visual responses of single neurons in the human medial temporal lobe. Nat Neurosci 3:946–953
Kreiman G, Koch C, Fried I (2000b) Imagery neurons in the human brain. Nature 408:357–361
Leonard CM, Rolls ET, Wilson FAW, Baylis GC (1985) Neurons in the amygdala of the monkey with responses selective for faces. Behav Brain Res 15:159–176

Lisman JE, Otmakhova NA (2001) Storage, recall, and novelty detection of sequences by the hippocampus: Elaborating on the SOCRATIC model to account for normal and aberrant effects of dopamine. Hippocampus 11:551–568
Logothetis NK, Pauls J, Augath M, Trinath T, Oeltermann A (2001) Neurophysiological investigation of the basis of the fMRI signal. Nature 412:150–157
Logothetis NK, Sheinberg DL (1996) Visual object recognition. Annu Rev Neurosci 19:577–621
Logothetis NK, Wandell BA (2004) Interpreting the BOLD signal. Annu Rev Physiol 66:735–769
Milner B, Corkin S, Teuber H-L (1968) Further analysis of the hippocampal amnesic syndrome: 14-year follow-up study of H.M. Neuropsychologia 6:215–234
Miyashita Y, Rolls ET, Cahusac PM, Niki H, Feigenbaum JD (1989) Activity of hippocampal formation neurons in the monkey related to a conditional spatial response task. J Neurophysiol 61:669–678
Mormann F, Dubois J, Kornblith S, Milosavljevic M, Cerf M, Ison M, Tsuchiya N, Kraskov A, Quiroga RQ, Adolphs R, Fried I, Koch C (2011) A category-specific response to animals in the right human amygdala. Nat Neurosci 14:1247–1249
Mormann F, Kornblith S, Quiroga RQ, Kraskov A, Cerf M, Fried I, Koch C (2008) Latency and selectivity of single neurons indicate hierarchical processing in the human medial temporal lobe. J Neurosci 28:8865–8872
Mosher CP, Zimmerman PE, Gothard KM (2014) Neurons in the monkey amygdala detect eye contact during naturalistic social interactions. Curr Biol 24:2459–2464
Mukamel R, Ekstrom AD, Kaplan J, Iacoboni M, Fried I (2010) Single-neuron responses in humans during execution and observation of actions. Curr Biol 20:750–756
O'Keefe J, Nadel L (1978) The hippocampus as a cognitive map. Oxford University Press
Olshausen BA, Field DJ (2004) Sparse coding of sensory inputs. Curr Opin Neurobiol 14:481–487
Quian Quiroga R (2012) Concept cells: the building blocks of declarative memory functions. Nat Rev Neurosci 13:587–597
Quian Quiroga R (2013) Gnostic cells in the 21st century. Acta Neurobiol Exp (Wars) 73:463–471
Quian Quiroga R, Kraskov A, Koch C, Fried I (2009) Explicit encoding of multimodal percepts by single neurons in the human brain. Curr Biol 19:1308–1313
Quian Quiroga R, Kreiman G, Koch C, Fried I (2008) Sparse but not "grandmother-cell" coding in the medial temporal lobe. Trends Cogn Sci 12:87–91
Quian Quiroga R, Reddy L, Koch C, Fried I (2007) Decoding visual inputs from multiple neurons in the human temporal lobe. J Neurophysiol 98:1997–2007
Quian Quiroga R, Reddy L, Kreiman G, Koch C, Fried I (2005) Invariant visual representation by single neurons in the human brain. Nature 435:1102–1107
Quian Quiroga R, Kraskov A, Mormann F, Fried I, Koch C (2014) Single-cell responses to face adaptation in the human medial temporal lobe. Neuron 84:363–369
Rutishauser U, Mamelak AN, Schuman EM (2006) Single-trial learning of novel stimuli by individual neurons of the human hippocampus-amygdala complex. Neuron 49:805–813
Rutishauser U, Ross IB, Mamelak AN, Schuman EM (2010) Human memory strength is predicted by theta-frequency phase-locking of single neurons. Nature 464:903–907
Rutishauser U, Schuman EM, Mamelak A (2014) Single neuron correlates of declarative memory formation and retrieval in the human medial temporal lobe. In: Fried I, Rutishauser U, Cref U, Kreiman G (eds) Single neuron studies of the human brain. MIT Press, Boston
Rutishauser U, Schuman EM, Mamelak AN (2008) Activity of human hippocampal and amygdala neurons during retrieval of declarative memories. Proc Natl Acad Sci USA 105:329–334
Rutishauser U, Ye S, Koroma M, Tudusciuc O, Ross IB, Chung JM, Mamelak AN (2015) Representation of retrieval confidence by single neurons in the human medial temporal lobe. Nat Neurosci 18:1–12
Schumann CM, Hamstra J, Goodlin-Jones BL, Lotspeich LJ, Kwon H, Buonocore MH, Lammers CR, Reiss AL, Amaral DG (2004) The amygdala is enlarged in children but not adolescents with autism; the hippocampus is enlarged at all ages. J Neurosci 24:6392–6401
Scoville WB, Milner B (1957) Loss of recent memory after bilateral hippocampal lesions. J Neurol Neurosurg Psychiatry 20:11–21

Sheinberg DL, Logothetis NK (1997) The role of temporal cortical areas in perceptual organization. Proc Natl Acad Sci USA 94:3408–3413
Shou T, Ruan D, Zhou Y (1986) The orientation bias of LGN neurons shows topographic relation to area centralis in the cat retina. Exp Brain Res 64:233–236
Simons DJ, Woolsey TA (1979) Functional organization in mouse barrel cortex. Brain Res 165:327–332
Squire LR, Stark CEL, Clark RE (2004) The medial temporal lobe. Annu Rev Neurosci 27: 279–306
Sugase Y, Yamane S, Ueno S, Kawano K (1999) Global and fine information coded by single neurons in the temporal visual cortex. Nature 400:869–873
Suzuki WA (1996) Neuroanatomy of the monkey entorhinal, perirhinal and parahippocampal cortices: organization of cortical inputs and interconnections with amygdala and striatum. Semin Neurosci 8:3–12
Suzuki WA, Amaral DG (1994) Perirhinal and parahippocampal cortices of the macaque monkey: cortical afferents. J Comp Neurol 350:497–533
Talavage TM, Ledden PJ, Benson RR, Rosen BR, Melcher JR (2000) Frequency-dependent responses exhibited by multiple regions in human auditory cortex. Hear Res 150:225–244
Tanaka K (1996) Inferotemporal cortex and object vision. Annu Rev Neurosci 19:109–139
Tranel D, Hyman BT (1990) Neuropsychological correlates of bilateral amygdala damage. Arch Neurol 47:349–355
Tsao DY, Schweers N, Moeller S, Freiwald WA (2008) Patches of face-selective cortex in the macaque frontal lobe. Nat Neurosci 11:877–879
Viskontas IV, Quiroga RQ, Fried I (2009) Human medial temporal lobe neurons respond preferentially to personally relevant images. Proc Natl Acad Sci USA 106:21329–21334
Wang S, Tudusciuc O, Mamelak AN, Ross IB, Adolphs R, Rutishauser U (2014) Neurons in the human amygdala selective for perceived emotion. Proc Natl Acad Sci USA 111:E3110–E3119
Waydo S, Kraskov A, Quian Quiroga R, Fried I, Koch C (2006) Sparse representation in the human medial temporal lobe. J Neurosci 26:10232–10234
Wolfe JM (1984) Reversing ocular dominance and suppression in a single flash. Vision Res 24:471–478

Recognition of Occluded Objects

Hanlin Tang and Gabriel Kreiman

Pattern recognition involves building a mental model to interpret incoming inputs. This incoming information is often incomplete and the mental model must extrapolate to complete the patterns, a process that is constrained by the statistical regularities in nature. Examples of pattern completion involve identification of objects presented under unfavorable luminance or interpretation of speech corrupted by acoustic noise. Pattern completion is also at the heart of other high-level cognitive phenomena including our ability to discern actions from still images or to predict behavioral patterns from observations.

Pattern completion constitutes a ubiquitous challenge during natural vision. Stimuli are often partially occluded, or degraded by changes in illumination and contrast. While much progress has been made towards understanding the mechanisms underlying recognition of complete objects, the neural computations underlying more challenging recognition problems such as identifying occluded objects remain poorly understood. Here we generically refer to object occlusion to include any transformation where only partial information about an object is accessible (such as the multiple examples illustrated in Fig. 1), and refer to object completion as the ability to infer object identity from partial information (without necessarily implying that subjects perceptually fill in the missing information).

Understanding how the neural representations of visual signals are modified when objects are occluded is critical to developing biologically constrained computational models of occluded object recognition and may also shed light on how to solve manifestations of pattern completion in other domains. The development of feed-forward models for visual recognition of whole objects has been driven by behavioral and physiological experiments establishing the hierarchy of feature tuning and robustness to image transformations. Similarly, systematically examining when and where neural representations that are robust to occlusion

H. Tang · G. Kreiman (✉)
Harvard Medical School, 1 Blackfan Circle, Karp 11217, Boston, MA 02115, USA
e-mail: gabriel.kreiman@tch.harvard.edu

Q. Zhao (ed.), *Computational and Cognitive Neuroscience of Vision*,
Cognitive Science and Technology, DOI 10.1007/978-981-10-0213-7_3

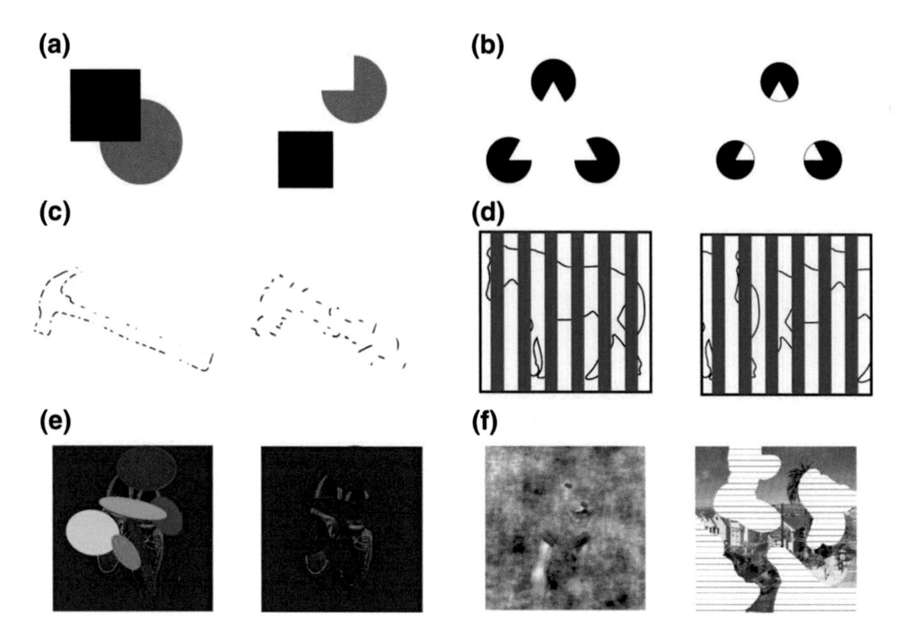

Fig. 1 Object completion examples. **a** Occluded geometric shape (*left*) and its mosaic counterpart (*right*) (e.g. Murray 2004). **b** Example of modal completion inducing an illusory triangle (*left*). This percept is disrupted by adding edges to the inducers (*right*). **c** Line drawing of an object defined by disconnected segments and its fragmented counterpart (e.g. Doniger et al. 2000; Sehatpour et al. 2008). **d** Line drawing of an occluded object and its scrambled counterpart (e.g. Lerner et al. 2004, 2002). **e** Occluded object and its 'deleted' counterpart (e.g. Johnson and Olshausen 2005). **f** (*left*) Example partial image of an object seen through bubbles with a phase-scrambled background to equalize contrast (e.g. Tang et al. 2014), (*right*) Example partial image of a scene (e.g. Nielsen et al. 2006a, b)

emerge can help extend our theoretical understanding of vision and develop the next generation of computational models in vision. Understanding pattern completion and recognition of occluded objects is a challenging task: performance depends on the stimulus complexity, the type and amount of occlusion and the task itself. In this chapter, we summarize recent efforts to examine the mechanisms underlying recognition of partially occluded objects and discuss several avenues for future work towards a theory of object completion.

1 Visual System Hierarchy

Object recognition is orchestrated by a semi-hierarchical series of processing areas along ventral visual cortex (Connor et al. 2007; DiCarlo et al. 2012; Felleman and Van Essen 1991; Logothetis and Sheinberg 1996; Riesenhuber and Poggio 1999; Schmolesky et al. 1998; Tanaka 1996). At each step in this hierarchy, the feature specificity of the neurons increases in complexity. For example, neurons in primary visual cortex (V1), respond selectively to bars of a particular orientation (Hubel and Wiesel 1959) whereas neurons in inferior temporal cortex respond preferentially to complex shapes including faces and other objects (Desimone et al. 1984; Gross et al. 1969; Perrett et al. 1992; Richmond et al. 1983; Rolls 1991). In addition to this increase in feature complexity, there is a concomitant progression in the degree of tolerance to object transformations such as changes in object position or scale (Hung et al. 2005; Ito et al. 1995; Logothetis et al. 1995). The selective and tolerant physiological responses characterized in the macaque inferior temporal cortex have also been observed in the human inferior temporal cortex (Allison et al. 1999; Liu et al. 2009). The timing of these neural responses provides important constraints on the number of possible computations involved in visual recognition. Multiple lines of evidence from human psychophysical measurements (Potter and Levy 1969; Thorpe et al. 1996), macaque single unit recordings (Hung et al. 2005; Keysers et al. 2001), human EEG (Thorpe et al. 1996) and human intracranial recordings (Allison et al. 1999; Liu et al. 2009) have established that selective responses to whole objects emerge within 100–150 ms of stimulus onset in the highest echelons of the ventral visual stream.

Research over the last several decades characterizing the spatiotemporal dynamics involved in the neural representation of objects in these successive areas has led to the development of a theoretical framework to explain the mechanisms underlying object recognition. An influential theoretical framework suggests that, to a first approximation, processing of visual information traverses through the ventral stream in a feed-forward fashion, without significant contributions from long top-down feedback loops or within-area recurrent computations (Deco and Rolls 2004; Fukushima 1980; LeCun et al. 1998; Mel 1997; Olshausen et al. 1993; Riesenhuber and Poggio 1999; Wallis and Rolls 1997). Consistent with this notion, computational models of object recognition instantiating feed-forward processing provide a parsimonious explanation for the selectivity and tolerances observed experimentally (Serre et al. 2007a, b). The activity of these computational units at various stages of processing captures the variability in the neural representation from macaque single unit recordings along the visual hierarchy (Cadieu et al. 2014; Yamins et al. 2014). These feed-forward computational models have inspired the development of deep convolutional networks that demonstrate a significant degree of success in a variety of computer vision approaches to object recognition (e.g. Hinton and Salakhutdinov 2006; LeCun et al. 1998; Russakovsky et al. 2015; Sun et al. 2014; Taigman et al. 2014).

These purely feedforward architectures do not incorporate any feedback or recurrent connections. However, at the anatomical level, feedback and recurrent connections figure prominently throughout the visual system (Felleman and Van Essen 1991). In fact, quantitative anatomical studies have suggested that feedback and recurrent connections significantly outnumber feedforward ones (Callaway 2004; Douglas and Martin 2004). The computational contributions of these feedback and recurrent projections are largely underexplored in existing computational models of visual recognition because their underlying roles remain unclear. Several investigators have suggested that these feedback and recurrent projections could play an important role during object recognition under conditions where the visual cues are impoverished (e.g. poor illumination, low contrast) or even partially missing (e.g. visual occlusion) (Carpenter and Grossberg 2002; Hopfield 1982; Mumford 1992; Tang et al. 2014; Wyatte et al. 2012a).

Following the approach suggested by Marr in his classic book on vision (Marr 1982), we subdivide our discussion of recognition of occluded objects into three parts: (i) definition of the *computational problem* by describing behavioral performance during recognition of occluded objects, (ii) characterization of the *implementation at the physical level* describing the neural responses to occluded objects and (iii) initial sketches of *theoretical ideas instantiated into computational models* that aim to recognize occluded objects.

2 The Computational Problem of Object Completion

Figure 1 shows examples of several images that induce object completion. In the natural world, objects can be partially occluded in multiple different ways due to the presence of explicit occluders, shadows, camouflage and differential illumination. Object completion is an ill-posed problem: in general, there are infinite ways of completing contours from partial information. The visual system must be able to infer what the object is despite the existence of all of these possible solutions consistent with the visual input.

2.1 Amodal Completion

Occluded shapes can be perceived as whole (Fig. 1a, compare left and right panels). Object completion can be *amodal* when there is an explicit occluder and the subject cannot see the contours behind the occluder despite being aware of the overall shape (Singh 2004). In contrast, in the famous illusory triangle example (Fig. 1b), Kanizsa describes the phenomenon known as *modal completion* whereby the object is completed by inducing illusory contours that are perceived by the observer (Kanizsa 1979). Because these illusory inducers are rare in natural vision, in this chapter we focus on amodal completion. Even though occluded or partial objects such as the ones shown in Fig. 1c, d are segmented, observers view the object as a single percept, not as disjointed segments. Amodal completion is also important for achieving this single ‘gestalt’. Investigators have used a variety of different stimuli to probe the workings of object completion, ranging from simple lines and geometric shapes to naturalistic objects such as the ones shown in Fig. 1e, f.

Psychophysical studies of amodal completion have provided many clues to the underlying computations (Kellman et al. 2001; Sekuler and Murray 2001). Amodal completion relies on an inferred depth between the occluder shape and the occluded object, which in turns generates a surface-based representation of the scene (Nakayama et al. 1995). In fact, presence of the occluder aids in identifying the occluded object, as powerfully illustrated by the Bregman's occluded B letters (Bregman 1981). Grouping of different parts into a whole, and the 'completion' of missing lines and contours, represent important components of object recognition. There are infinite possible ways of completing occluded objects. The ambiguities arise from the many combinations with which occluded edges, called 'inducers' can be paired together, as well as the infinite possible contours between two pairs of inducers (Kellman et al. 2001; Nakayama et al. 1995; Ullman 1976). Despite the many possible solutions, the visual system typically arrives at a single (and correct) interpretation of the image.

The temporal dynamics of shape completion can constrain the computational steps involved in processing occluded images. Psychophysics experiments have measured the time course of amodal completion with a diverse array of experimental paradigms. The most common method is a contrast of an occluded shape against its mosaic parts (e.g. Figure 1a). For example, in the prime matching paradigm, subjects are first primed with a stimulus, and then asked to judge whether a pair of test stimuli represent the same or a different shape. Subjects are faster to correctly respond 'same' when the primed shape is the same as the test stimuli. When partly occluded objects are used as the prime, this priming effect depends on the exposure time (Sekuler and Palmer 1992). At short durations (50 ms), occluded objects primed subject's responses towards mosaic shapes, suggesting that 50 ms is not enough time for amodal completion of the prime stimulus. At longer durations (100 ms or more), the priming effect switched to favor whole shapes. Therefore, the authors estimate that amodal completion for simple geometric shapes takes between 100 and 200 ms, depending on the amount of occlusion (Sekuler et al. 1994). A different set of behavioral experiments suggests approximately the same time scales for amodal completion: in several studies, subjects are asked to discriminate shapes in a timed forced-choice task. Response times to occluded shapes lagged those to whole shapes by about 75–150 ms (Murray et al. 2001; Shore and Enns 1997). For naturalistic objects such as faces, however, Chen et al. report that amodal completion takes longer than 200 ms (Chen et al. 2009), well beyond the previous estimates based on simpler shapes and tasks.

2.2 From Amodal Completion to Recognition of Occluded Objects

Many studies of amodal completion have used simple shapes and contours, as outlined in the previous section. Psychophysical studies using simple shapes and neurophysiological studies describing the linear filters in primary visual areas led to the basic building blocks to develop deep models of visual recognition capable of detecting complex shapes. Inspired by the success of this approach, we assume that

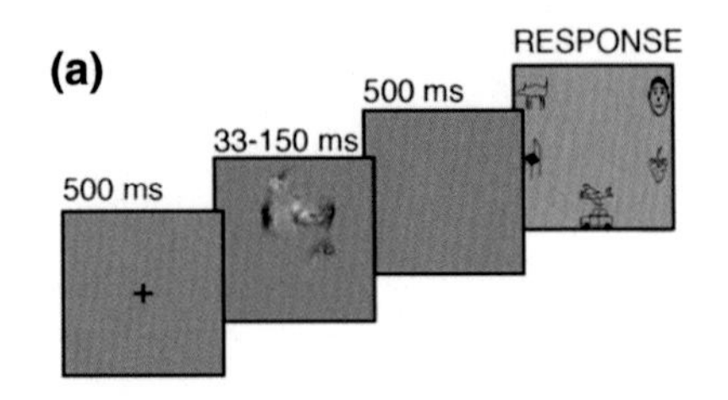

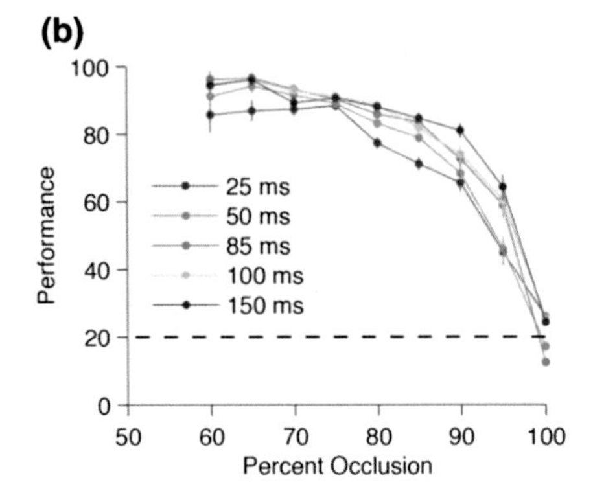

Fig. 2 Robust object categorization despite strong occlusion. **a** Experiment timeline. Partial images similar to the ones shown in Fig. 1f were presented during exposure times ranging from 33 to 150 ms. Subjects performed a five alternative forced-choice categorization task determining whether the image contained animals, chairs, faces, fruits, or vehicles. **b** Performance as a function of percentage occlusion across n = 14 subjects for various exposure times. Error bars denote SEM. Dashed line indicates chance performance

contour completion is one of the initial steps involved in interpreting complex objects that are partially occluded.

Recognition remains robust to partial occlusion for complex objects. We used naturalistic objects that were occluded by presenting information through "bubbles" (Gosselin and Schyns 2001) (Fig. 1f). After a variable exposure time from 33 to 150 ms, subjects performed a five alternative forced-choice categorization task. Recognition was robust even when 80–90 % of the object was occluded across the various exposure times (Fig. 2). Similar results were obtained by Wyatte et al. (2012a). As illustrated by Bregman (1981), the presence of an occluder during object completion aids recognition performance. For example, one study presented natural objects that were either occluded (Fig. 1e, left) or where the same object parts were deleted (Fig. 1e, right). Recognition was significantly impaired in the deleted part case compared to the occluded part case, but only when using high percentage of occlusion (>75 % missing pixels) (Johnson and Olshausen 2005).

Intuitively, we would expect successful recognition to depend on the exact features shown. This intuition was quantitatively measured by Gosselin and Schyns (2001): the facial features critical to recognition varied depending on the task (e.g. the eyes and eyebrows were more relevant for gender discrimination and the mouth provided more information when evaluating expressiveness). Similar conclusions were reached when using the same paradigm to evaluate recognition in monkeys (Nielsen et al. 2006b).

3 Neural Representation of Occluded Objects

A series of scalp electroencephalography (EEG) studies have measured the latency at which responses differ between occluded objects and suitable control images. Using simple geometric stimuli, differences between occluded shapes and notched shapes emerged at 140–240 ms (Murray 2004). Using more naturalistic stimuli (e.g. Figure 1e), other investigators report differential activity in the 130–220 ms (Chen et al. 2010) and 150–200 ms (Johnson and Olshausen 2005) ranges. In a more difficult task with fragmented line drawings that are progressively completed, Doniger et al. reported that differences are only observed in the 200–250 ms response window (2000). Even though these studies use different stimuli and make different comparisons, they consistently conclude that amodal completion effects manifest within 130–250 ms.

Several neuroimaging (Hegde et al. 2008; Komatsu 2006; Lerner et al. 2004, 2002; Olson et al. 2004; Rauschenberger et al. 2004) and scalp EEG (Chen et al. 2010; Doniger et al. 2000; Johnson and Olshausen 2005) studies with more complex objects have contrasted activity changes between an occluded object and an appropriately scrambled counterpart (e.g. Fig. 1c, d). In these stimuli, the low-level features are maintained but disruption in their geometric arrangement destroys the percept. For example, investigators have reported differential activity in the lateral-occipital complex between occluded line drawings and their scrambled counterparts (Lerner et al. 2002). The authors reason that, since the occluded images elicit a larger response in the lateral-occipital complex (LOC) than scrambled images, the LOC could be involved in object completion. It should be noted that LOC also demonstrates increased activity to whole objects compared to scrambled versions of those objects without any occlusion (Grill-Spector et al. 2001). Thus, the increased responses to whole objects may not be necessarily related to object completion mechanisms per se, but rather neural activity related to perceptual recognition.

Similarly, EEG and intracranial studies compared line drawings against their fragmented counterparts (Doniger et al. 2000; Sehatpour et al. 2008) to measure the timing and brain regions involved in object completion. Sehatpour et al. worked with epilepsy patients who have intracranial electrodes implanted for clinical purposes. This approach allows a rare opportunity to record directly from the human brain. The authors take advantage of simultaneous recordings from multiple brain regions to show that line fragments elicited greater coherence in the LOC-Prefrontal Cortex-Hippocampus network compared to scrambled line fragments. They suggest that this network synchrony is responsible for the perceptual line closure of objects.

In order to understand the neural mechanisms orchestrating object completion, it is also critical to examine the neural architectures that could implement the computational solutions suggested in the previous section. Essential aspects of object completion can be traced back to the earliest stages in visual processing. An early study demonstrated that neurons in area V2 show selective responses to illusory contours (Peterhans and von der Heydt 1991; von der Heydt et al. 1984). Other work has demonstrated that even V1 neurons can respond to occluded shapes. One study recorded single cells in macaque V1 when presented with occluded moving bars (Sugita 1999). Approximately 12 % of orientation-selective cells responded to the moving oriented bar even when it was occluded, thus potentially underlying the phenomenology of amodal completion. These cells responded strongly only when the occluder was presented in front of the moving bar (positive disparity), and not at zero or negative disparity. Notably, responses to the occluded bar were not different from those obtained when presenting the bar alone. These results have led to the suggestion that amodal completion is achieved by contextual modulation from outside the classical receptive field. While other studies have suggested that contextual modulation occurs with a delay of 50–70 ms with respect to the onset of the visually evoked responses (Bakin et al. 2000; Zipser et al. 1996), Sugita did not observe any latency delays for the amodally-completed response. The author suggests that these contextual modulations may come from lateral connections or feedback from proximal areas. In another study, responses to illusory contours in V1 were delayed by about 55 ms compared to the response to real contours (Lee and Nguyen 2001). Importantly, illusory contour responses appeared first in V2 before emerging in V1, suggesting that modal completion in V1 might require feedback modulation from V2. Complementing these studies, psychophysical studies on the effect of inferred depth and apparent motion on the perception of occluded surfaces also conclude that amodal completion effects manifests in early visual processing (Shimojo and Nakayama 1990a, b).

These neurophysiology studies have focused on the occlusion or inducing of linear contours, where the inducers are close in proximity to the classical receptive field. However in natural vision we complete curvilinear contours over distances much longer than the width of classical V1 receptive fields. Often in these cases, correct completion of an object depends on the global context in which the object is embedded. Future studies are needed to examine whether and when V1 neurons respond to completed contours of varying curvature, length, and context.

As outlined above, V1 neurons feed onto a cascade of semi-hierarchical processing steps through V2 and V4, culminating in the inferior temporal cortex (ITC) (Felleman and Van Essen 1991). How do these higher visual areas respond to occluded shapes? Few studies have examined the responses in intermediate visual areas to occluded shapes. A recent elegant study has begun to fill in this gap by characterizing how macaque V4 neurons respond to different curvatures when they are partially occluded by dots (Kosai et al. 2014). The authors report that neurons can maintain selectivity within a range of occlusion. While the response onset times of these neurons were not delayed by occlusion, the latency at which selectivity arose was delayed by hundreds of milliseconds.

Kovacs et al. found that visually selective responses to complex shapes in ITC were similar between whole shapes and occluded shapes defined by adding noise, occluders or deleting shape parts (Kovács et al. 1995). Although selectivity to complex shapes was retained despite up to ~50 % occlusion, the absolute magnitude of the responses was modulated linearly with the amount of occlusion. Contrary to what Kosai et al. found in V4, these authors observed delays of up to 50 ms in the response latency to occluded shapes. While it is tempting to attribute this discrepancy to differences in processing between V4 and IT, we note that the stimuli and occluding patterns used are different between the two studies.

Nielsen et al. extended this work by examining the responses of IT neurons to occluded objects embedded in naturalistic stimuli (Fig. 1f) (Nielsen et al. 2006a). Using the bubbles paradigm (Gosselin and Schyns 2001), the authors defined parts of an image that provided more diagnostic value (i.e. provided information that aided recognition) versus other non-diagnostic parts. The authors first demonstrated that monkeys and humans show striking behavioral similarities in terms of what object parts are considered diagnostic (Nielsen et al. 2006b). For occluded scenes containing diagnostic parts, both firing rates and local field potentials in ITC remained largely invariant to significant amounts of occlusion, in contrast to the findings of the Kovacs study with simpler stimuli (Kovacs et al. 1995). However, for scenes that contained only non-diagnostic parts, the results from the Kovacs study were reproduced—the firing rate varied linearly with the amount of occlusion. This comparison also serves as a cautionary tale against extrapolating results based on geometric shapes to the processing of more naturalistic stimuli because the details of which features are revealed can play a very important role in dictating the effects of occlusion. Issa et al. reached similar conclusions when demonstrating that ITC responses selective to faces were particularly sensitive to occlusion of certain parts (one eye) and that those parts could drive the responses almost as well as the whole face (Issa and Dicarlo 2012). These results suggest that the robustness of the neural representation to missing parts depends on the diagnosticity of the visible features.

Tang et al. used intracranial recordings to evaluate how and when visually selective responses to occluded objects emerge. Naturalistic objects were presented through the bubbles paradigm (Gosselin and Schyns 2001) in a task similar to the one illustrated in Fig. 2. The use of objects seen through bubbles evaluates the core ability to spatially integrate multiple parts to subserve recognition. Figure 3 (left) shows the responses from an example electrode in the fusiform gyrus that displayed a strong response to face stimuli. Remarkably, the electrode showed similar responses to images that displayed only ~11 % of the object content (Fig. 3, columns 2–6). Even in cases where the images shared essentially no common pixel (e.g. columns 3 and 5), the responses remained similar. Overall, the magnitude of the responses did not vary with the amount of visible pixels. Yet, the responses to occluded shapes were not identical to those obtained upon presenting whole images. A notable difference was that the responses to occluded objects were significantly delayed (compare the position of the arrows in Fig. 3). Selective neural responses emerged with a delay of ~100 ms. These delays were also apparent when using a machine learning approach to decode the category or identity of the objects from the physiological responses to the whole or occluded images (Fig. 4).

Image processing does not end with visual cortex. Information from visual cortex is conveyed to frontal cortex and to medial temporal lobe structures including the amygdala, hippocampus and entorhinal cortex. One group recorded single unit activity in the amygdala of human epilepsy patients, and found that

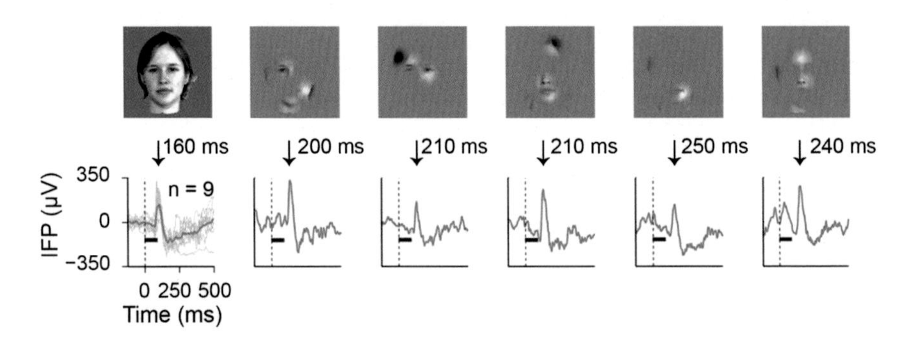

Fig. 3 Physiological responses in the human fusiform gyrus show tolerance to strong occlusion. Invasive intracranial field potential recordings from an electrode in the fusiform gyrus in a subject with epilepsy (modified from Tang et al. 2014). This electrode responded selectively to faces (*left*, *gray* = individual trials, *green* = average of 9 repetitions). The other panels show single trial responses to five partial images of the face. *Black bar* indicates stimulus presentation time (150 ms). Despite heavy occlusion (89 %), the neural responses were similar to those obtained when presenting the whole object. However, the responses to partial images were significantly delayed. The *arrow* indicates latency of the response peak with respect to image onset

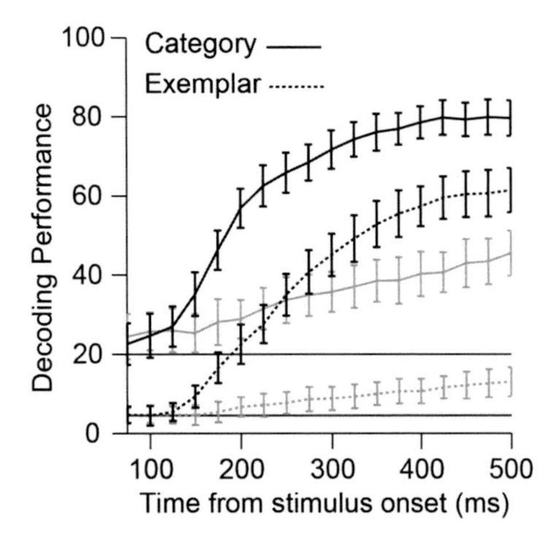

Fig. 4 Physiological responses to occluded objects were delayed compared to whole objects. Single trial decoding performance from a pseudopopulation of n = 60 electrodes for categorization (*black*, n = 5 categories) or identification (*gray*, n = 25 exemplars) (modified from (Tang et al. 2014)). The latency of selective information was delayed for occluded objects (*dashed line*) compared to whole objects (*solid line*)

neurons were surprisingly sensitive to even small degrees of occlusion (Rutishauser et al. 2011). Their firing rates varied non-linearly with the amount of occlusion and the responses to image parts did not necessarily bear a resemblance to the responses to the whole images. These non-visual medial temporal lobe neurons, in effect, 'lost' the robustness developed in the ventral visual stream, in that individual parts were not sufficient to drive the response to the level observed for whole faces.

In addition to the type of spatial integration demonstrated in the studies above, the visual system is able to integrate information over time. Temporal integration is particularly prominent in examples of action recognition. Yet, in some cases, different parts of the same object may appear in a dynamic fashion over time. At the behavioral level, presenting object parts asynchronously significantly disrupts object recognition performance, even when the temporal lag is as short as 16 ms (Singer and Kreiman 2014). This disruption by asynchronous presentation is also evident at the physiological level (Singer et al. 2015). Thus, the ability to spatially integrate parts into a whole for recognition is quite sensitive to deviations from the synchronous presentation of those parts.

In sum, early visual areas show evidence of contour completion in the presence of both occluded and illusory contours when the corresponding edges are in close spatial proximity. In some, but not all cases, these contour completion responses show a delay with respect to both responses to real contours and responses in higher visual areas. In higher visual areas responsible for object recognition, physiological signals show strong robustness to large degrees of occlusion, consistent with behavioral recognition performance, and these physiological signals also show a significant delay. The robustness in the physiological responses and the dynamic delays are consistent with the behavioral observations. These delays are interpreted as originating from the involvement of recurrent and/or top-down connections during the process of object completion. Given these initial steps in understanding plausible neural circuits underlying recognition of occluded objects, we turn our attention to describing the possible biological algorithms instantiated by these neural signals.

4 Computational Models of Occluded Object Recognition

There has been significant progress over the last decade in developing computational models of object recognition (Deco and Rolls 2004; DiCarlo et al. 2012; Kreiman 2013; Riesenhuber and Poggio 1999; Serre et al. 2007a). To a first approximation, these models propose a hierarchical sequence of linear filtering and non-linear max operations inspired by the basic principles giving rise to simple and complex cells in primary visual cortex (Hubel and Wiesel 1962). Concatenating multiple such operations gave rise to some of the initial models for object recognition (Fukushima 1980). Recently, these ideas have also seen wide adoption in the computer science literature in the form of deep convolutional neural networks (e.g. (Krizhevsky et al. 2012) among many others). Both biologically inspired models and deep convolutional neural networks (CNNs) geared towards performance on benchmark datasets share similar core architectures.

4.1 *Performance of Feed-Forward Models in Recognizing Occluded Objects*

The canonical steps in feed-forward computational models is inspired by the observation of simple and complex cells in primary visual cortex of anesthetized cats. In their seminal study, Hubel and Wiesel discovered 'simple' cells tuned to bars oriented at a particular orientation (Hubel and Wiesel 1959). They also described 'complex' cells, which were also tuned to a preferred orientation, but exhibited a degree of tolerance to spatial translation of the stimulus. They hypothesized that to generate this spatial invariance, the complex cells pool over simple cells whose receptive fields tile the visual space with a max-like operation. This complex cell would then respond to an oriented bar regardless of its spatial location. Both hierarchical models of biological vision such as HMAX (Riesenhuber and Poggio 1999; Serre et al. 2007a, b) and CNNs are composed of alternating layers of tuning and pooling with increasingly more complicated tuning functions as one ascends this hierarchy. Whereas biologically-inspired models such as HMAX have about 4 layers, state-of-the-art computer vision models have moved to complex topologies with tens of layers and different mixtures of tuning and pooling layers (e.g. Russakovsky et al. 2015). Performance of feed-forward models such as HMAX on object recognition datasets match the pattern of human performance (Serre et al. 2007b). Additionally, the activity of individual layers in deep learning networks can capture the variance of neurons in the corresponding layers in macaque cortex (Cadieu et al. 2014; Yamins et al. 2014).

While these feed-forward architectures are designed to build tolerance to image transformations such as position and scale changes, they are not necessarily robust to the removal or occlusion of object features. Indeed, Fig. 5 shows the performance of an HMAX-like architecture in recognizing the same occluded objects form Figs. 2, 3 and 4. Small amounts of occlusion do not impair performance. However, performance drops rapidly with increasing occlusion, much more rapidly than human performance (see "behavior" line in Fig. 5 and compare Fig. 2 versus Fig. 5. Experimental simulations with other similar models confirm that that both HMAX and CNN models are challenged by recognition of occluded objects (Pepik et al. 2015; Wyatte et al. 2012b). Unlike position or scale transformations, the underlying representation in these models is not robust to occlusion. Feed-forward networks do not have explicit mechanisms to compensate for the missing features of occluded objects. In addition, because these models do not distinguish between the occluder and the object, the occluder can introduce spurious features that, through spatial pooling, are mixed with the object features. With small amounts of occlusion, the remaining features may be sufficient to lead to successful classification. However, with increasing levels of occlusion, the lack of sufficient information and completion mechanisms lead to a significant impairment in performance.

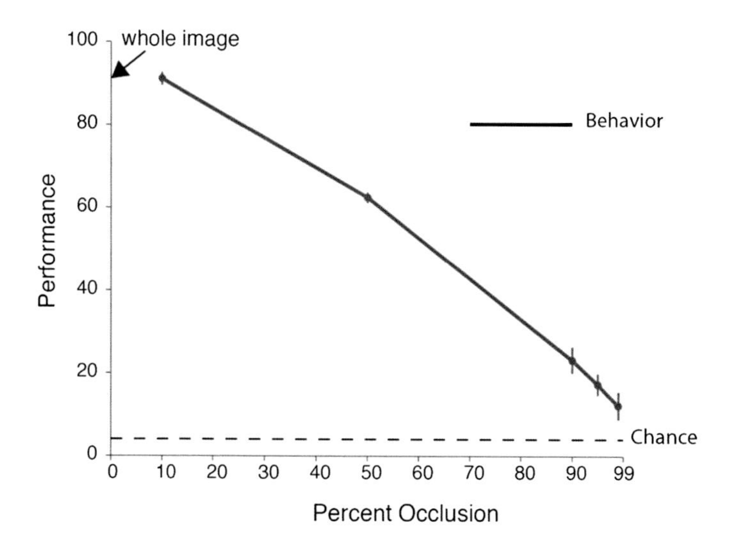

Fig. 5 Challenge to feed-forward computational models Performance of a hierarchical feed-forward model of biological vision (Serre et al. 2007a) on recognition of partial images similar to the one shown in Fig. 1f. *Dashed line* indicates chance level. Performance (*red line*) of the model is well below that of human subjects (*solid black line*) for heavy occlusion. Span of the *black line* indicates the range of difficulty tested for humans

4.2 Beyond Feed-Forward Models

Models that incorporate additional computations beyond feed-forward architectures can be subdivided into several categories. A group of computational models describes the process by which contours are amodally completed (Ullman 1976; Yuille and Kersten 2006). These models typically rely on an axiomatic set of desirable qualities that completed curves must satisfy (such as minimum total bending energy).

Another set of ideas argues that creating an understanding of the surfaces in a scene is critical to object recognition for occluded objects. Nakayama (1995) proposes a theory where surface representation is constructed via feedback in early visual cortex by learning associations between a viewed image and the underlying surface representations. The authors argue that this intermediate surface representation is vital for subsequent recognition for occluded objects and mediates many important functions in texture segregation and visual search.

A different model proposes that neural representations of surfaces are created in three stages based on the low-level features (Sajda and Finkel 1995). First, points belonging to the same contour are bound together, followed by a process that determines the surfaces, and finally the surfaces are ordered by depth. Both feed-forward and feedback connections subserve to communicate between these three stages.

Several theories based on the role of feedback connections emphasize inference, but these ideas have largely not been operationalized into object recognition models. Predictive coding models provide specific hypotheses for the roles of feedback connections during object recognition (Rao and Ballard 1999). In these models, higher visual areas send their predictions to lower levels, which then return only the mismatch between the predicted activity and the actual activity. This creates an efficient system where each layer only sends forwards signals that deviate from the receiving layer's predictions. The higher layers then attempt to generate the correct hypothesis of the image by reducing the incoming prediction errors. A related model proposes that visual cortex is essentially performing Bayesian inference where feed-forward inputs combine with top-down priors for recognition (Lee and Mumford 2003; Yuille and Kersten 2006).

As pointed out by Wyatte et al. (2014), predictive coding models would expect occluded images to lead to increased activity in visual cortex, since the first generated hypotheses would have very large prediction errors due to occlusion. Over time, we would expect this activity to subside as the system converged on an accurate hypothesis. Neurophysiological studies, however, find decreased or unchanged activity throughout visual cortex (Kovacs et al. 1995; Nielsen et al. 2006a; Tang et al. 2014).

Several models that more directly examine object recognition deal with recognition from partial information, and do not incorporate any of the amodal completion mechanisms previously described. These models take advantage of the extensive feedback and recurrent connections in visual cortex. While the role of these connections in attentional modulation has been extensively studied, their contribution to object recognition remains unclear. A particularly prominent and attractive class of models that can perform pattern completion is the all-to-all connectivity architectures such as Hopfield networks (Hopfield 1982). The Hopfield network generates attractors for previously learned patterns in such a way that if the network is initialized with partial information, it can dynamically evolve towards the right attractor. Interestingly, this type of dynamical convergence towards the attractor state could account for the type of delays observed in the behavioral and physiological experiments. This general principle is operationalized by a neural network model that combines bottom-up input with top-down signals carrying previously learned patterns to complete occluded objects (Fukushima 2005). For occluded patterns that are novel, this network attempts to interpolate from visible edges. The author applies this model to complete occluded letters of the alphabet. This concept has been extended to naturalistic objects with a feed-forward model that is augmented with recurrent feedback (Wyatte et al. 2012a). This recurrent feedback served to strengthen the feed-forward signals that were diminished from the occluded image.

Given that strong behavioral and neurophysiological evidence exists for amodal completion in the human brain, and that surface representations are important for organizing the visual scene, theories of object recognition would be remiss to exclude these features in favor of purely feature-based recognition. An important step towards models that fully capture natural biological vision would be to integrate traditional feed-forward models with feedback mechanisms, including amodal completion, surface generation, and top-down modulation based on priors and context. The challenge of recognizing occluded objects stands as the first test of these future integrative theories.

References

Allison T, Puce A, Spencer D, McCarthy G (1999) Electrophysiological studies of human face perception. I: Potentials generated in occipitotemporal cortex by face and non-face stimuli. Cereb Cortex 9:415–430

Bakin JS, Nakayama K, Gilbert CD (2000) Visual responses in monkey areas V1 and V2 to three-dimensional surface configurations. J Neurosci Off J Soc Neurosci 20:8188–8198

Bregman AS (1981) Asking the 'What for'question in auditory perception. In: Perceptual organization, pp 99–118

Cadieu CF, Hong H, Yamins DLK, Pinto N, Ardila D, Solomon EA, Majaj NJ, DiCarlo JJ (2014) Deep neural networks rival the representation of primate IT cortex for core visual object recognition. PLoS Comput Biol 10:e1003963

Callaway EM (2004) Feedforward, feedback and inhibitory connections in primate visual cortex. Neural Netw 17:625–632

Carpenter G, Grossberg S (2002) Adaptive resonance theory. In: The handbook of brain theory and neural networks. MIT Press, Cambridge

Chen J, Liu B, Chen B, Fang F (2009) Time course of amodal completion in face perception. Vis Res 49:752–758

Chen J, Zhou T, Yang H, Fang F (2010) Cortical dynamics underlying face completion in human visual system. J Neurosci Off J Soc Neurosci 30:16692–16698

Connor CE, Brincat SL, Pasupathy A (2007) Transformation of shape information in the ventral pathway. Curr Opin Neurobiol 17:140–147

Deco G, Rolls ET (2004) A neurodynamical cortical model of visual attention and invariant object recognition. Vis Res 44:621–642

Desimone R, Albright T, Gross C, Bruce C (1984) Stimulus-selective properties of inferior temporal neurons in the macaque. J Neurosci 4:2051–2062

DiCarlo JJ, Zoccolan D, Rust NC (2012) How does the brain solve visual object recognition? Neuron 73:415–434

Doniger GM, Foxe JJ, Murray MM, Higgins BA, Snodgrass JG, Schroeder CE, Javitt DC (2000) Activation timecourse of ventral visual stream object-recognition areas: high density electrical mapping of perceptual closure processes. J Cogn Neurosci 12:615–621

Douglas RJ, Martin KA (2004) Neuronal circuits of the neocortex. Annu Rev Neurosci 27:419–451

Felleman DJ, Van Essen DC (1991) Distributed hierarchical processing in the primate cerebral cortex. Cereb Cortex 1:1–47

Fukushima K (1980) Neocognitron: a self organizing neural network model for a mechanism of pattern recognition unaffected by shift in position. Biol Cybern 36:193–202

Fukushima K (2005) Restoring partly occluded patterns: a neural network model. Neural Netw 18:33–43

Gosselin F, Schyns PG (2001) Bubbles: a technique to reveal the use of information in recognition tasks. Vis Res 41:2261–2271

Grill-Spector K, Kourtzi Z, Kanwisher N (2001) The lateral occipital complex and its role in object recognition. Vis Res 41:1409–1422

Gross C, Bender D, Rocha-Miranda C (1969) Visual receptive fields of neurons in inferotemporal cortex of the monkey. Science 166:1303–1306

Hegde J, Fang F, Murray S, Kersten D (2008) Preferential responses to occluded objects in the human visual cortex. J Vis 8:1–16
Hinton GE, Salakhutdinov RR (2006) Reducing the dimensionality of data with neural networks. Science 313:504–507
Hopfield JJ (1982) Neural networks and physical systems with emergent collective computational abilities. PNAS 79:2554–2558
Hubel D, Wiesel T (1959) Receptive fields of single neurons in the cat's striate cortex. J Physiol (Lond) 148:574–591
Hubel DH, Wiesel TN (1962) Receptive fields, binocular interaction and functional architecture in the cat's visual cortex. J Physiol 160:106–154
Hung C, Kreiman G, Poggio T, DiCarlo J (2005) Fast read-out of object identity from macaque inferior temporal cortex. Science 310:863–866
Issa EB, Dicarlo JJ (2012) Precedence of the eye region in neural processing of faces. J Neurosci Off J Soc Neurosci 32:16666–16682
Ito M, Tamura H, Fujita I, Tanaka K (1995) Size and position invariance of neuronal responses in monkey inferotemporal cortex. J Neurophysiol 73:218–226
Johnson JS, Olshausen BA (2005) The recognition of partially visible natural objects in the presence and absence of their occluders. Vis Res 45:3262–3276
Kanizsa G (1979) Organization in vision: essays on gestalt perception. Praeger Publishers
Kellman PJ, Guttman SE, Wickens TD (2001) Geometric and neural models of object. In: From fragments to objects: segmentation and grouping in vision, vol 130, p 183
Keysers C, Xiao DK, Foldiak P, Perret DI (2001) The speed of sight. J Cogn Neurosci 13:90–101
Komatsu H (2006) The neural mechanisms of perceptual filling-in. Nat Rev Neurosci 7:220–231
Kosai Y, El-Shamayleh Y, Fyall AM, Pasupathy A (2014) The role of visual area V4 in the discrimination of partially occluded shapes. J Neurosci Off J Soc Neurosci 34:8570–8584
Kovács G, Vogels R, Orban GA (1995) Selectivity of macaque inferior temporal neurons for partially occluded shapes. J Neurosci Off J Soc Neurosci 15:1984–1997
Kreiman G (2013) Computational models of visual object recognition. In: Panzeri S, Quian Quiroga R (eds) Principles of neural coding. Taylor and Fracis Group
Krizhevsky A, Sutskever I, Hinton GE (2012) Imagenet classification with deep convolutional neural networks. In: Advances in neural information processing systems, pp 1097–1105
LeCun Y, Bottou L, Bengio Y, Haffner P (1998) Gradient-based learning applied to document recognition. Proc IEEE 86:2278–2324
Lee TS, Mumford D (2003) Hierarchical Bayesian inference in the visual cortex. J Opt Soc Am A Opt Image Sci Vis 20:1434–1448
Lee TS, Nguyen M (2001) Dynamics of subjective contour formation in the early visual cortex. Proc Natl Acad Sci USA 98:1907–1911
Lerner Y, Harel M, Malach R (2004) Rapid completion effects in human high-order visual areas. Neuroimage 21:516–526
Lerner Y, Hendler T, Malach R (2002) Object-completion effects in the human lateral occipital complex. Cereb Cortex 12:163–177
Liu H, Agam Y, Madsen JR, Kreiman G (2009) Timing, timing, timing: fast decoding of object information from intracranial field potentials in human visual cortex. Neuron 62:281–290
Logothetis NK, Pauls J, Poggio T (1995) Shape representation in the inferior temporal cortex of monkeys. Curr Biol 5:552–563
Logothetis NK, Sheinberg DL (1996) Visual object recognition. Annu Rev Neurosci 19:577–621
Marr D (1982) Vision. Freeman Publishers, San Francisco
Mel B (1997) SEEMORE: combining color, shape and texture histogramming in a neurally inspired approach to visual object recognition. Neural Comput 9:777
Mumford D (1992) On the computational architecture of the neocortex. II. The role of cortico-cortical loops. Biol Cybern 66:241–251
Murray MM (2004) Setting boundaries: brain dynamics of modal and amodal illusory shape completion in humans. J Neurosci 24:6898–6903

Murray RF, Sekuler AB, Bennett PJ (2001) Time course of amodal completion revealed by a shape discrimination task. Psychon Bull Rev 8:713–720

Nakayama K, He Z, Shimojo S (1995) Visual surface representation: a critical link between lower-level and higher-level vision. In: Kosslyn S, Osherson D (eds) Visual cognition. The MIT Press, Cambridge

Nielsen K, Logothetis N, Rainer G (2006a) Dissociation between LFP and spiking activity in macaque inferior temporal cortex reveals diagnostic parts-based encoding of complex objects. J Neurosci 26:9639–9645

Nielsen KJ, Logothetis NK, Rainer G (2006b) Discrimination strategies of humans and rhesus monkeys for complex visual displays. Curr Biol 16(8):814–820

Olshausen BA, Anderson CH, Van Essen DC (1993) A neurobiological model of visual attention and invariant pattern recognition based on dynamic routing of information. J Neurosci Off J Soc Neurosci 13:4700–4719

Olson IR, Gatenby JC, Leung HC, Skudlarski P, Gore JC (2004) Neuronal representation of occluded objects in the human brain. Neuropsychologia 42:95–104

Pepik B, Benenson R, Ritschel T, Schiele B (2015) What is holding back convnets for detection? arXiv:150802844

Perrett D, Hietanen J, Oeam M, Benson P (1992) Organization and functions of cells responsive to faces in the temporal cortex. Phil Trans Roy Soc 355:23–30

Peterhans E, von der Heydt R (1991) Subjective contours - bridging the gap between psychophysics and physiology. Trends Neurosci 14:112–119

Potter M, Levy E (1969) Recognition memory for a rapid sequence of pictures. J Exp Psychol 81:10–15

Rao RP, Ballard DH (1999) Predictive coding in the visual cortex: a functional interpretation of some extra-classical receptive-field effects. Nat Neurosci 2:79–87

Rauschenberger R, Peterson MA, Mosca F, Bruno N (2004) Amodal completion in visual search: preemption or context effects? Psychol Sci 15:351–355

Richmond B, Wurtz R, Sato T (1983) Visual responses in inferior temporal neurons in awake Rhesus monkey. J Neurophysiol 50:1415–1432

Riesenhuber M, Poggio T (1999) Hierarchical models of object recognition in cortex. Nat Neurosci 2:1019–1025

Rolls E (1991) Neural organization of higher visual functions. Curr Opin Neurobiol 1:274–278

Russakovsky O, Deng J, Su H, Krause J, Satheesh S, Ma S, Huang Z, Karpathy A, Khosla A, Bernstein M et al (2015). Imagenet large scale visual recognition challenge. Int J Comput Vis

Rutishauser U, Tudusciuc O, Neumann D, Mamelak AN, Heller AC, Ross IB, Philpott L, Sutherling WW, Adolphs R (2011) Single-unit responses selective for whole faces in the human amygdala. Curr Biol CB 21:1654–1660

Sajda P, Finkel LH (1995) Intermediate-level visual representations and the construction of surface perception. J Cogn Neurosci 7:267–291

Schmolesky M, Wang Y, Hanes D, Thompson K, Leutgeb S, Schall J, Leventhal A (1998) Signal timing across the macaque visual system. J Neurophysiol 79:3272–3278

Sehatpour P, Molholm S, Schwartz TH, Mahoney JR, Mehta AD, Javitt DC, Stanton PK, Foxe JJ (2008) A human intracranial study of long-range oscillatory coherence across a frontal-occipital-hippocampal brain network during visual object processing. Proc Natl Acad Sci USA 105:4399–4404

Sekuler AB, Murray RF (2001) Amodal completion: a case study in grouping. Advances in Psychology 130:265–293

Sekuler AB, Palmer SE (1992) Perception of partly occluded objects: a microgenetic analysis. J Exp Psychol Gen 121:95–111

Sekuler AB, Palmer SE, Flynn C (1994) Local and global processes in visual completion. Psychol Sci 5:260–267

Serre T, Kreiman G, Kouh M, Cadieu C, Knoblich U, Poggio T (2007a) A quantitative theory of immediate visual recognition. Prog Brain Res 165C:33–56. doi:10.1016/S0079-6123(06)65004-8

Serre T, Oliva A, Poggio T (2007b) Feedforward theories of visual cortex account for human performance in rapid categorization. PNAS 104:6424–6429
Shimojo S, Nakayama K (1990a) Amodal representation of occluded surfaces: role of invisible stimuli in apparent motion correspondence. Perception 19:285–299
Shimojo S, Nakayama K (1990b) Real world occlusion constraints and binocular rivalry. Vis Res 30:69–80
Shore DI, Enns JT (1997) Shape completion time depends on the size of the occluded region. J Exp Psychol Hum Percept Perform 23:980–998
Singer JM, Kreiman G (2014) Short temporal asynchrony disrupts visual object recognition. J Vis 14:7
Singer JM, Madsen JR, Anderson WS, Kreiman G (2015) Sensitivity to timing and order in human visual cortex. J Neurophysiol 113:1656–1669
Singh M (2004) Modal and amodal completion generate different shapes. Psychol Sci 15:454–459
Sugita Y (1999) Grouping of image fragments in primary visual cortex. Nature 401:269–272
Sun Y, Wang X, Tang X (2014) Deeply learned face representations are sparse, selective, and robust. arXiv:14121265
Taigman Y, Yang M, Ranzato MA, Wolf L (2014) Deepface: closing the gap to human-level performance in face verification. In: 2014 IEEE conference on computer vision and pattern recognition (CVPR), pp. 1701–1708. IEEE
Tanaka K (1996) Inferotemporal cortex and object vision. Annu Rev Neurosci 19:109–139
Tang H, Buia C, Madhavan R, Crone NE, Madsen JR, Anderson WS, Kreiman G (2014) Spatiotemporal dynamics underlying object completion in human ventral visual cortex. Neuron 83:736–748
Thorpe S, Fize D, Marlot C (1996) Speed of processing in the human visual system. Nature 381:520–522
Ullman S (1976) Filling-in the gaps: the shape of subjective contours and a model for their generation. Biol Cybern 25:1–6
von der Heydt R, Peterhans E, Baumgartner G (1984) Illusory contours and cortical neuron responses. Science 224:1260–1262
Wallis G, Rolls ET (1997) Invariant face and object recognition in the visual system. Prog Neurobiol 51:167–194
Wyatte D, Curran T, O'Reilly R (2012a) The limits of feedforward vision: recurrent processing promotes robust object recognition when objects are degraded. J Cogn Neurosci 24:2248–2261
Wyatte D, Jilk DJ, O'Reilly RC (2014) Early recurrent feedback facilitates visual object recognition under challenging conditions. Front Psychol 5:674
Wyatte D, Tang H, Buia C, Madsen J, O'Reilly R, Kreiman G (2012b) Object completion along the ventral visual stream: neural signatures and computational mechanisms. In: Computation and systems neuroscience, Salt Lake City, Utah
Yamins DLK, Hong H, Cadieu CF, Solomon EA, Seibert D, DiCarlo JJ (2014) Performance-optimized hierarchical models predict neural responses in higher visual cortex. Proc Natl Acad Sci USA 111:8619–8624
Yuille A, Kersten D (2006) Vision as Bayesian inference: analysis by synthesis? Trends Cogn Sci 10:301–308
Zipser K, Lamme VA, Schiller PH (1996) Contextual modulation in primary visual cortex. J Neurosci 16:7376–7389

Towards a Theory of Computation in the Visual Cortex

David A. Mély and Thomas Serre

Abstract One of the major goals in visual neuroscience is to understand how the cortex processes visual information (Marr 1982). A substantial effort has thus gone into characterizing input-output relationships across areas of the visual cortex (Dicarlo et al. 2012), which has yielded an array of computational models. These models have, however, typically focused on one or very few visual areas, modules (form, motion, depth, color) or functions (e.g., object recognition, boundary detection, action recognition, etc.), see (Poggio and Serre 2013) for a recent review. An integrated framework that would explain the computational mechanisms underlying *vision* beyond any specific visual area, module or function, while being at least consistent with the known anatomy and physiology of the visual cortex is still lacking. The goal of this review is to draft an initial integrated theory of visual processing in the cortex. We highlight the computational mechanisms that are shared across many successful models and derive a taxonomy of canonical computations. Such an enterprise is reductionist in nature as we break down the myriad of input-output functions found in the visual cortex into a basic set of computations. Identifying canonical computations that are repeated and combined across visual functions will pave the way for the identification of their cortical substrate (Carandini 2012).

Canonical microcircuits are a theorist's dream because their very existence provides evidence that different visual cortices indeed tackle one common set of computational problems with a shared toolbox of computations, and constitute the building blocks of the different information processing pathways throughout the visual cortex. While the identification of canonical circuits has proved elusive (Douglas and Martin 2007), recent technological advances hold new promises: from complete anatomical reconstructions of circuits in increasingly large volumes of brain tissue (Plaza et al. 2014) to the recording of increasingly large populations of neurons (Stevenson and

D.A. Mély · T. Serre (✉)
Cognitive, Linguistic & Psychological Sciences Department, Brown Institute for Brain Science, Brown University, Providence, RI 02912, USA
e-mail: thomas_serre@brown.edu

Q. Zhao (ed.), *Computational and Cognitive Neuroscience of Vision*, Cognitive Science and Technology, DOI 10.1007/978-981-10-0213-7_4

Kording 2011) combined with an ability to selectively modulate their activity (Fenno et al. 2011).

We start this review with an overview of cortical filter models, focusing on the two-dimensional Gabor function, a notorious example successfully used to characterize the receptive field of orientation-selective cells as found in the primary visual cortex. We use this model as a case in point for understanding how computational mechanisms for the processing of form extends to other domains including color, binocular disparity, and motion. We further demonstrate how these basic computations can be cascaded within a hierarchical architecture to account for information processing in extra-striate areas, examining the theoretical underpinnings of their effectiveness. Finally, we describe other possible canonical circuits and principles of design that hold promise for future research.

1 Cortical Filter Models of Form Processing

1.1 The Linear-Nonlinear (LN) Model

Cortical filter models, also referred to as "cortex transform" models (Watson 1987), have been widely used to describe the input-output transfer function of neurons across cortical areas and visual functions (see Landy and Movshon 1991 for an overview). Under this broad family of models, the output of a model cell (also called "unit") depends on the activity of input units that feed into it (the "afferent" units), which define its "receptive field". In turn, a particular unit may project onto a set of output units also called "projection" units. Any afferent unit has its own input units; in the case of vision, such cascades can be traced all the way back to the retina. Thus, by extension, the receptive field of a unit also designates the unique sub-region of the visual field that if properly stimulated may elicit a response from the unit.

Well before the advent of modern computational modeling, neurophysiologists had developed the tools to map out the input-output function of cortical cells in the primary visual cortex. One prominent experimental method, derived from systems theory, is known as "reverse correlation" (see Ringach 2004 for a review): a neuron is treated as a black box which transforms a visual input $\mathbf{x}$, i.e., the set of image elements (or pixels) $x_{i,j}$ for i, j in its receptive field, into an output response y. The neuron's input-output relationship is characterized as a linear function of its input given by the following equation:

$$y = \mathbf{w} \cdot \mathbf{x} = \sum_{i,j} w_{i,j} x_{i,j}. \tag{1}$$

The scalars $w_{i,j}$ correspond to the (synaptic) weights of a linear filter and are sometimes referred to as "the linear receptive field". These weights can be estimated empirically by presenting white noise as an input to the neuron while recording its response.

A good parametrization of such linear receptive field for the simple cells found in the primary visual cortex (Hubel and Wiesel 1962) is the Gabor function (Jones and Palmer 1987), which is given by the following equation:

$$w_{ij} = \exp\left(-\frac{(u^2 + \gamma^2 v^2)}{2\sigma^2}\right) \times \cos\left(\frac{2\pi}{\lambda}u + \phi\right) \tag{2}$$

$$\text{s.t. } u = i\cos\theta + j\sin\theta \text{ and } v = -i\sin\theta + j\cos\theta. \tag{3}$$

The five parameters, i.e., the orientation θ, the aspect ratio γ, the effective width σ, the phase ϕ and the wavelength λ, determine the properties of the spatial receptive field of the corresponding model simple cell. Figure 1 shows examples of model simple cells varying in orientation, spatial frequency and phase. The Gabor function thus describes a process that extends from the visual input falling onto the retina on

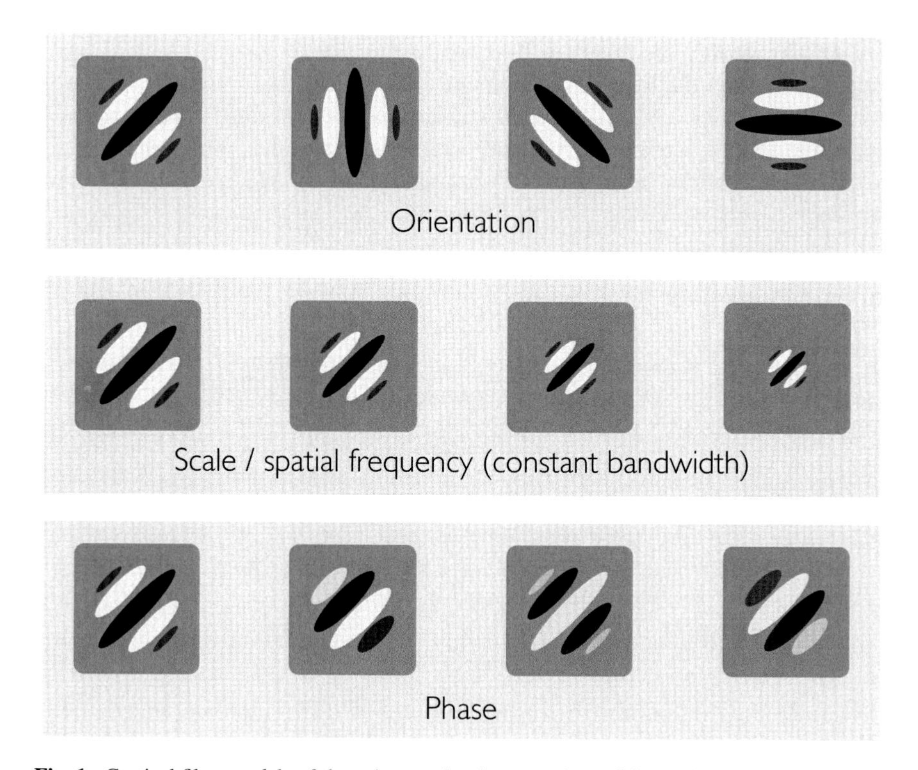

Fig. 1 Cortical filter models of the primary visual cortex. A good linear filter model that can be fitted well to simple cells is the Gabor filter (Jones and Palmer 1987). (Note that other parametrizations are also possible, see text.) Computational models of the primary visual cortex typically include a battery of such filters spanning a range of orientations, spatial frequencies and phases

the one end, to the output activity of a model simple cell in the primary visual cortex on the other end, including intermediate stages such as processing by the lateral geniculate nucleus (LGN).

Other parametrizations have been proposed for simple cells. These include Gaussian derivatives which have been shown to provide an excellent fit to cortical cells receptive fields both in the spatial (Young 1987a) and spatio-temporal domain (Young and Lesperance 2001). They were used in one of the first early-vision models of pre-attentive texture discrimination (Malik and Perona 1990).

However, biological neurons also behave in nonlinear ways; e.g., their output tends to saturate as their input grows stronger, instead of increasing indefinitely as a linear input-output function would predict. Thus, cortical filter models always include a nonlinear transfer function following the linear part, which is why they are also referred to as linear-nonlinear (LN) models. The above input-output function of a model unit then becomes:

$$y = f(\mathbf{w} \cdot \mathbf{x}), \tag{4}$$

where f is a nonlinear transfer function. Popular choices for the function f include (half-) linear rectification functions, exponential functions (square and square root), or the logistic and the hyperbolic functions (see Fig. 2). The LN model has been shown to account for a host of experimental data (Rieke et al. 1997) and it has been shown that in many cases, biophysically more realistic models of neurons (which include a spike generation process) can be reduced to a simple LN model (Ostojic and Brunel 2011).

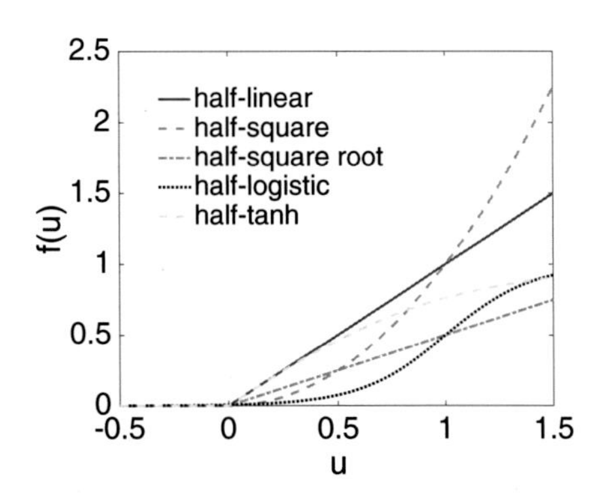

Fig. 2 Common nonlinear transfer functions used in cortical filter models of the primary visual cortex. See text for details

1.2 *Divisive Normalization*

An extension of the LN model includes the addition of a normalization stage:

$$y = \frac{f(\mathbf{w} \cdot \mathbf{x})}{k + \sum_{j \in J} g(\mathbf{v} \cdot \mathbf{u}^{\mathbf{j}})}, \tag{5}$$

where $k > 0$ is a constant to avoid division by zero. The pool of units used for normalization, indexed by $j \in J$, may correspond to the same set of input units ($\mathbf{u} = \mathbf{x}$) as in the *tuning circuits* (Kouh and Poggio 2008) or another set of output units as in the divisive normalization model (Heeger 1992). In addition, the spatial area covered by the normalization pool (whether input or output) may be limited to the (classical) receptive field of the unit $\mathbf{x}$, or possibly extend beyond to account for extra-classical receptive field effects (Series et al. 2003; see Fig. 3 for a common implementation of divisive normalization in the primary visual cortex). Such normalization circuits were originally proposed to explain the contrast response of cells in the primary visual cortex (Heeger 1992) and are now thought to operate throughout the visual system, and in many other sensory modalities and brain regions (see Carandini and Heeger 2012 for a recent review).

For instance, in the HMAX model (Riesenhuber and Poggio 1999; Serre et al. 2007; see also Sect. 3.3), two types of operations are being assumed: a bell-shape

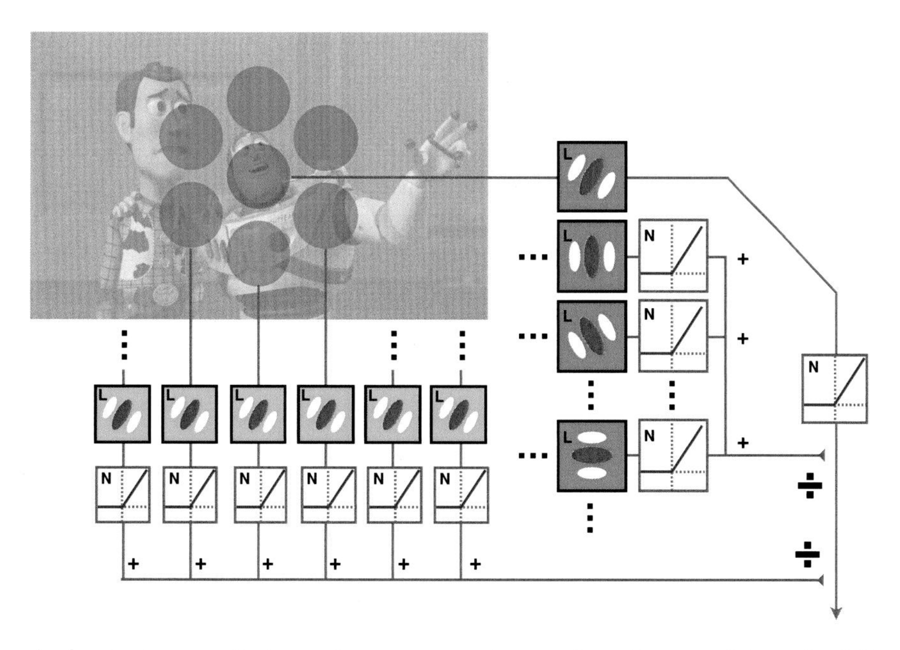

Fig. 3 Divisive normalization in the primary visual cortex. The reference unit (receptive field shown as a *red circle*) gets normalized by other units which share the same tuning preference but with receptive fields located outside its receptive field (*blue circles*; this region is called the "extra-classical receptive field" or "surround region"). This unit also gets inhibited by other units within the same cortical hypercolumn (in *red*). "L" and "N" respectively denote the linear and nonlinear part of the cortical LN model

tuning operation for simple cells and a max-like operation (or "soft-max") at the level of position and scale-tolerant complex cells (Riesenhuber and Poggio 1999). Interestingly, both operations can be approximated as specific instances of the more general Eq. 5:

$$y = \frac{\sum_{i,j} w_{i,j}\, x_{i,j}^{p}}{k + \left(\sum_{i,j} x_{i,j}^{q}\right)^{r}}, \tag{6}$$

where p, q and r represent static nonlinearities in the underlying neural circuit. An extra sigmoid transfer function on the output $g(y) = 1/(1 + \exp^{\alpha(y-\beta)})$ controls the sharpness of the unit response. By adjusting these nonlinearities, Eq. 6 can approximate well either a max operation or a tuning function (see Kouh and Poggio 2008 for details).

1.3 LN Cascade

More sophisticated computations in visual cortex can also be captured by cascading several LN models into a single pipeline, such that the output of one stage, described by an LN model, can be fed into the following LN model describing the next stage. In the notations used above, it means that the input $x_{i,j}$ to a model unit does not represent a direct input from the visual field anymore; instead, each position i, j in that unit's receptive field corresponds to an actual output y from a model unit from the previous stage. The corresponding linear weights $w_{i,j}$ then constitute a "generalized receptive field".

One example includes Hubel and Wiesel's model of position invariance at the level of complex cells. Such invariance is obtained by locally pooling over simple cells with the same preferred orientation. Another instance of the LN cascade was used to account for complex cells' invariance to contrast reversal. Unlike simple cells that are sensitive to the polarity of an input stimulus (e.g., white bar on a black background as opposed to a black bar on a white background), complex cells exhibit a response which is largely invariant to such change. One circuit that has been proposed to explain this type of invariance is the energy model (Adelson and Bergen 1985; Finn and Ferster 2007; Sasaki and Ohzawa 2007). In the proposed circuit, the activity of a set of simple cells (corresponding to a first LN processing stage parametrized by a Gabor function) with the same preferred selectivity for orientation and spatial frequency but different selectivity for phases (corresponding to different preferred contrast polarity) are squared, then summed (sometimes followed by a square root nonlinearity) by a second LN stage.

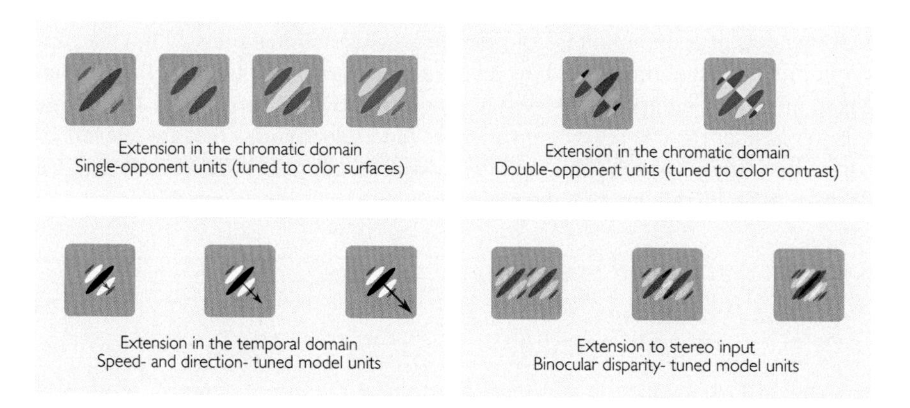

Fig. 4 Extension of the Gabor function to multiple visual modalities. The notion of spatial filtering highlighted for the processing of two-dimensional shape information can be extended to the color, disparity and space-time (i.e., motion) domains

$$y = \sqrt{\sum_{\phi} f\left(x_{\phi}\right)^{2}}, \tag{7}$$

Equation 7 guarantees the result to be invariant to the reversal of the image contrast as contrast dependence is modeled by the phase parameter in the Gabor function. In signal processing theory, this computation is called the energy function as it is equivalent to a local measure of the amplitude spectrum of the image. The energy is taken over two phases, which are said to be in quadrature, if the Gabor functions are followed by a full rectification (more realistic circuits include more afferent subunits, see Alonso and Martinez 1998; Sanada and Ohzawa 2006). Thus, the invariance to contrast reversal of complex cells is also known as an invariance to phase, by analogy with the phase parameter of the Gabor model.

2 Cortical Filter Models Across Visual Cues

As discussed above, the Gabor function given in Eq. 3 provides a good description of the response of simple cells' receptive fields that are characterized by a preferred orientation, spatial frequency, phase and bandwidth tuning. However, the conventional Gabor function only explains a limited range of the selectivity of cells observed in the primary visual cortex, namely the processing of two-dimensional shape information and local contrast. As we will show next, it is possible to generalize this simple cortical filter model to account for the processing of additional visual cues including color, motion, and binocular disparity (Fig. 4).

Beyond simple cells, there exist complex cells tuned for motion, disparity and color. However, in order to understand how complex cells should be wired across visual channels, it is important to consider their computational roles within the context of different visual functions: beyond position invariance which was the original focus in models of complex cells' visual processing (Hubel and Wiesel 1962), one needs to consider the cue-specific invariances (and selectivities) that are relevant to a particular visual channel. Although the answer is often specific to each cue, we will strive to highlight common theoretical principles whenever possible.

2.1 Color Processing

The standard Gabor function is a local function of the image contrast. Studies of color processing have shown that three types of cones in the retina, that are selective for long (L), medium (M) and short (S) wavelengths, project to the LGN via multiple processing stages. In the LGN, the visual input is reorganized into opponent color channels, which are also found in the primary visual cortex: Red (R) versus green (G) and blue (B) versus yellow (Y). The existence of additional channels such as a red versus cyan channel is also debated (Conway 2001). These channels can be traced back to inputs from individual cone types (Shapley and Hawken 2011) (either L vs. M, or S vs. L + M, respectively). The conventional Gabor function, defined by its shape parameters, can be extended to the chromatic domain in order to account for the color-sensitive receptive fields of primary visual cortex simple cells (Zhang et al. 2012).

There exist two functional classes of color-sensitive cells: single- and double- opponent receptive fields (Humanski and Wilson 1993; Mullen and Losada 1999; Shapley and Hawken 2011). Single-opponent receptive fields exhibit a center-surround configuration with excitatory center and inhibitory surround corresponding to one of the following pairs: R+G− (red ON, green OFF, meaning the cell is driven by a red light increment in the center and inhibited by an increment of green in the surround), G+R−, B+Y−, and Y+B−. Electrophysiological studies have shown that such cells can be modeled well by considering a standard Gabor function, and using the positive component of the function for the ON component of the receptive field, and its negative component for the OFF component (Johnson et al. 2008; Shapley and Hawken 2011). Note that this yields only a weakly orientation-selective receptive field. In a computer model, the Gabor function can be defined across the L, M and S channels of an LMS image (or more simply across the R, G and B channels of an RGB image for a computer vision application (Zhang et al. 2012).

Beside single-opponent cells, one can find double-opponent receptive fields, which, in addition to exhibiting chromatic opponency, also exhibit spatial opponency. Zhang et al. (2012) have shown that this type of receptive field organization can be derived by cascading the output of the single-opponency (LN + normalization) stage with an additional LN stage (Gabor function + half-wave rectification, see Fig. 5).

Note that although we described how to extend a Gabor receptive field to account for color opponent processing, color-selective complex cells require additional work.

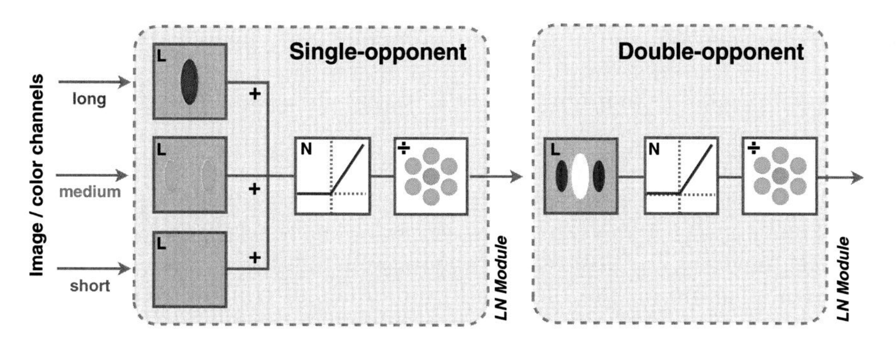

Fig. 5 The circuitry of single-opponent and double-opponent color-responsive units. Single-opponent linear receptive fields in the primary visual cortex combine chromatically-opponent subunits (here in *green* and *red* for a R+G− unit), whereas double-opponent subunits include an additional step with a spatially-opponent, orientation-selective linear receptive field. Double-opponent processing is a good example of an LN cascade, where each LN module (boxes shaded in gray) is made of a canonical sequence of computations: a linear operation (L), then a nonlinear transfer function (N), and finally divisive normalization among several units (÷)

The chromatic analogue of the invariance to contrast reversal as observed in the grayscale domain is an invariance to chromatic contrast reversal, or, equivalently, to the phase of a color-opponent equiluminant grating (Johnson et al. 2001, 2004, 2008). To achieve such an invariance, one may consider an energy model atop an LN stage based on a chromatic Gabor receptive field to combine the output of a model cell selective for a certain opponent pair with that of a model cell selective for the reverse pair. For example, a red-green double opponent complex cell can be created by combining the activities of double-opponent simple cells R+G− and G+R− with the energy model. In addition, the resulting model complex cell can also be made invariant to the position of its preferred stimulus within its receptive field, just like a conventional contrast-sensitive complex cell, by using a max-pooling operation as discussed before. In the end, the resulting color-responsive complex cell responds most to a chromatic edge or bar of the preferred orientation and opponent color pair anywhere in its receptive field.

2.2 *Binocular Disparity Processing*

The monocular Gabor function can be extended to binocular visual inputs. By allowing the Gabor function associated with either eye to have independent parameters, we introduce two new dimensions to the cortical filter described in Sect. 1. Phase disparity is defined as the phase difference between the monocular inputs from each eye, and position disparity as the offset between the locations of either monocular receptive field within the visual field. A model binocular unit then linearly combines the output of each afferent (given by a Gabor function for the left eye and another one for the right eye), followed by the energy summation described in Eq. 7; by extension, this particular model of binocular disparity processing is also called the energy model (Ohzawa et al. 1996; Ohzawa 1998). Such unit exhibits a preference for a given phase and position disparity, and a population of such cells is able to represent all disparities across the visual field (Qian 1994). Disparity is a good representation to have as any object seen from a stereoscopic sensory device creates a specific pattern of disparity signals that is related to its viewing depth and appearance; thus, by leveraging disparity, depth can be recovered (more specifically, it is inversely related to disparity).

Given certain phase and position disparities, the notion of complex cell can be directly derived from the monocular case, by taking the energy of the cells with the same tuning preferences (both disparity, orientation and spatial frequency) and with their phases in quadrature (Ohzawa et al. 1997; Ohzawa 1998; note that the quadrature is defined across afferents, which have binocular receptive fields, and not across the two eyes). Such a unit is selective for certain phase and position disparities while being phase-invariant; invariance in position can still be achieved through max-pooling. Note that some models also pool over orientations and/or spatial frequencies in order to reduce noise and make sure the population will peak at veridical disparities (Fleet et al. 1996); such cells are then robust to changes to the power spectrum of the input stimulus (if pooling over both orientation and spatial frequency). Complex cells tuned to binocular disparity, invariant to reversal of contrast, were found in the primary visual cortex (Ohzawa et al. 1997).

2.3 Motion Processing

The static Gabor function can be extended to the space-time domain by introducing a time-dependent phase term that makes the periodic part of the Gabor function drift over time within the unit receptive field (Simoncelli and Heeger 1998; Dayan and Abbott 2001; Jhuang et al. 2007b; Bradley and Goyal 2008). This generalizes the idea of sampling from the power spectrum of a static image by a population of simple cells to that of sampling from the power spectrum of an image sequence over time (Adelson and Bergen 1985; Bradley and Goyal 2008). The resulting cells are selective for spatial frequency, orientation, and temporal frequency (Adelson and Bergen 1985; Movshon et al. 1985; Bradley and Goyal 2008).

As before, phase-invariant complex cells can be obtained using the energy model by combining the output activities of a pair of simple cells with the same preferred location and tuning preferences, except for a phase difference of 90 degrees. Such a population of cells implicitly codes for local velocity (speed and direction) in a manner invariant to the textural content of the moving element. The reason for this is that a video sequence of any visual element translating at a uniform velocity has a planar power spectrum (Watson et al. 1983; Watson and Ahumada 1985; Simoncelli and Heeger 1998); changes in texture merely redistributes its power spectrum within that plane.

Therefore, model units can be designed that are truly speed and direction selective and locally invariant to texture, which help to alleviate the aperture problem,[1] by pooling together simple cells whose spectral receptive fields lie on the appropriate plane (Simoncelli and Heeger 1998; Nishimoto and Gallant 2011). By also pooling over nearby locations, such motion-sensitive complex cells gain local position invariance. Evidence for such cells has been found in MT (Simoncelli and Heeger 1998; Nishimoto and Gallant 2011), therefore this is not strictly speaking a computational model of the primary visual cortex.

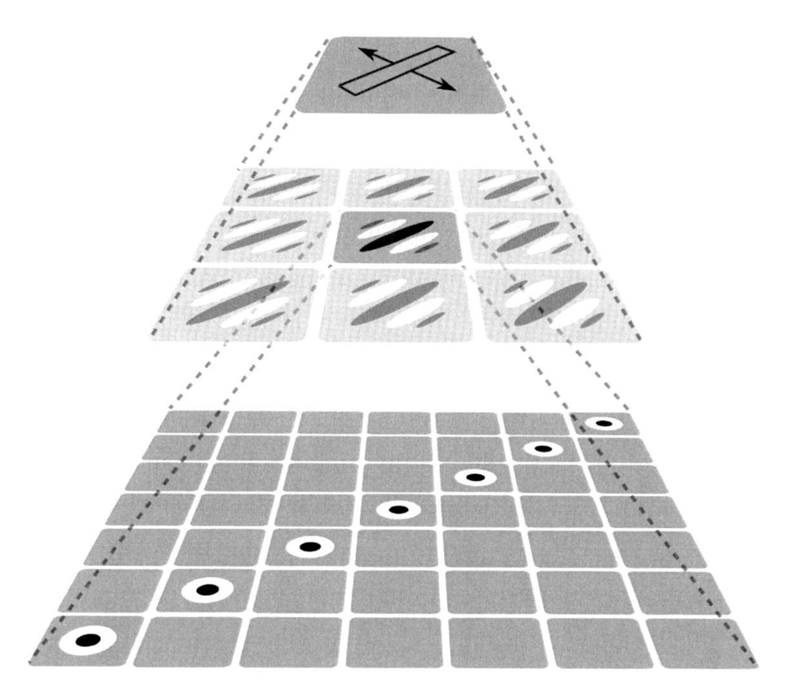

Fig. 6 Hubel and Wiesel model. A simple unit (*middle layer*) pools over afferent units with center-surround receptive fields (*bottom layer*) aligned along a preferred axis of orientation. At the next stage (*upper layer*), a complex unit pools over afferent simple units with the same preferred orientation within a small spatial neighborhood. Thus, the complex unit shown here is tolerant to local shifts of the preferred stimulus within its receptive field. A more complete model would also include pooling over simple cells tuned to slightly different spatial frequencies and phases (Rust et al. 2005; Chen et al. 2007)—consistently with the observed broadening in frequency bandwidth (DeValois et al. 1982) and tolerance to contrast reversal found in complex cells

[1] The "aperture problem" reflects the inherent ambiguity associated with the direction of motion of a moving stimulus within the receptive fields (a small aperture) of neurons in early visual areas. Because of its limited receptive field, a motion-selective neuron will often produce identical responses for stimuli that vary greatly in their shape, speed and orientation.

3 Completing the Hierarchy: Models of the Visual Cortex

3.1 *Hubel and Wiesel Model*

Hubel and Wiesel (1962) provided the first qualitative description of the receptive field (RF) organization of neurons in the primary visual cortex. As mentioned in Sect. 1, they described two functional classes of neurons: the simple and complex cells. Simple cells respond best to oriented stimuli (e.g., bars, edges, gratings) at one particular orientation, position and phase (e.g., a light bar on a black background or a dark bar on a light background) within their relatively small receptive fields (typically a fraction of a degree up to about one degree of visual angle in the monkey). Complex cells, on the other end, while also selective for orientation, tend to have larger receptive fields (about twice as large) and exhibit some tolerance with respect to the exact position of the stimulus within their receptive fields. They are also invariant to contrast reversal, i.e., the same cell responds to a white bar on a black background or the opposite.

Figure 6 illustrates a plausible neural circuit proposed in (Hubel and Wiesel 1962) to explain the receptive field organization of these two functional classes of cells. Simple cell-like receptive fields can be obtained by pooling the activity of a small set of cells tuned to spots of lights with a center-surround organization (as observed in ganglion cells in the LGN and layer IV of the striate cortex) aligned along a preferred axis of orientation (Fig. 6, bottom layer). At the next stage, position tolerance at the complex cell level, can be obtained by pooling over afferent simple cells (from the level below) with the same preferred (horizontal) orientation but slightly different positions (Fig. 6, middle layer).

Today, nearly half a century after Hubel and Wiesel's initial proposal, the coarse circuitry underlying the organization of RFs in the primary visual cortex is relatively well established (Alonso and Martinez 1998; Rust et al. 2005; Chen et al. 2007). Using this circuit as a building block, numerous hierarchical models of the visual cortex have been proposed (see Serre 2014 for a recent review) and used to demonstrate the ability of this type of architectures to be invariant to increasingly challenging transformations in the visual input (e.g., changes in the viewing angle of an object), while remaining selective for relevant aspects of it (e.g., the identity of said object), a quandary also known as the invariance-selectivity trade-off.

3.2 *Hierarchical Models: Formalism*

Hierarchical models of the visual system come in many different forms: they differ primarily in terms of their specific wiring and corresponding parametrizations as well as the mathematical operations that they use. However, all these computational models exhibit a common underlying architecture corresponding to multiple cascaded stages of processing such as the one shown on Fig. 7. Units at any stage $k + 1$

pool selectively over afferent units from the previous stage k over a local neighborhood (shown in pink). In general, pooling may occur over multiple dimensions of the afferent units (e.g., position, scale, orientation, etc.). Pooling over multiple locations (as shown in stages k or $k+1$ on Fig. 7) leads to an increase in the receptive field size of the units at the next stage (compare the receptive field size of a unit at stage k shown in red with that of a unit at a higher stage $k+1$).

For instance, a computational instantiation of the Hubel and Wiesel hierarchical model of the primary visual cortex corresponds to three processing stages. Simple units in layer $k = 1$ (highlighted in pink in Fig. 7) receive their inputs from center-surround cells in LGN sensitive to light increment (ON) or light decrement (OFF) in the previous layer $k = 0$ (in red). Complex cells in layer $k = 2$ pool over afferent simple cells at the same orientation over a local neighborhood (shown on Fig. 7 in purple is a 4×4 neighborhood). These types of circuits have yielded several models of the primary visual cortex that have focused on explaining in reasonably good neurophysiological details the tuning properties of individual cells (e.g., orientation, motion or binocular disparity, see Landy and Movshon 1991 for a review).

In recent years, because of the increasing amount of computing power available, the scale of models of visual processing has increased with models now encompassing large portions of the visual field and entire streams of visual processing (see Serre et al. 2007 for a review). Alternating between multiple layers of simple units and complex units leads to an architecture that is able to achieve a difficult trade-off between selectivity and invariance: along the hierarchy, units become tuned to features of increasing complexity (e.g., from single oriented bars, to combinations of oriented bars to form corners and features of intermediate complexities) by combining afferents (complex units) with different selectivities (e.g., units tuned to edges at different orientations). Conversely, at each "complex unit" stage, complex units become increasingly invariant to two-dimensional transformations (position and scale) by combining afferents (simple units) with the same selectivity (e.g., a vertical bar) but slightly different positions and scales.

While recent work has suggested that 'simple' and 'complex' cells may represent the ends of a continuum instead of two discrete classes of neurons (see Ringach 2004 for a discussion), this dichotomy is probably not critical for hierarchical models of the visual system. Indeed, recent models do not distinguish between simple and complex cell pooling (OReilly et al. 2013).

Units in hierarchical models of the visual cortex are typically organized in columns and/or feature maps. An hyper-column (shown in black in Fig. 7) corresponds to a population of units tuned to a basic set of features (e.g., units spanning the full range of possible orientations or directions of motion, etc.)[2] in models of the primary visual cortex (Hubel and Wiesel see 1962). These hyper-columns are then replicated at all positions in the visual field and multiple scales. An alternative perspective is to think of processing stages in terms of feature maps. Typically, maps correspond to retinotopically organized population of units tuned to the same feature (e.g., spe-

[2]A full model would also include eye dominance.

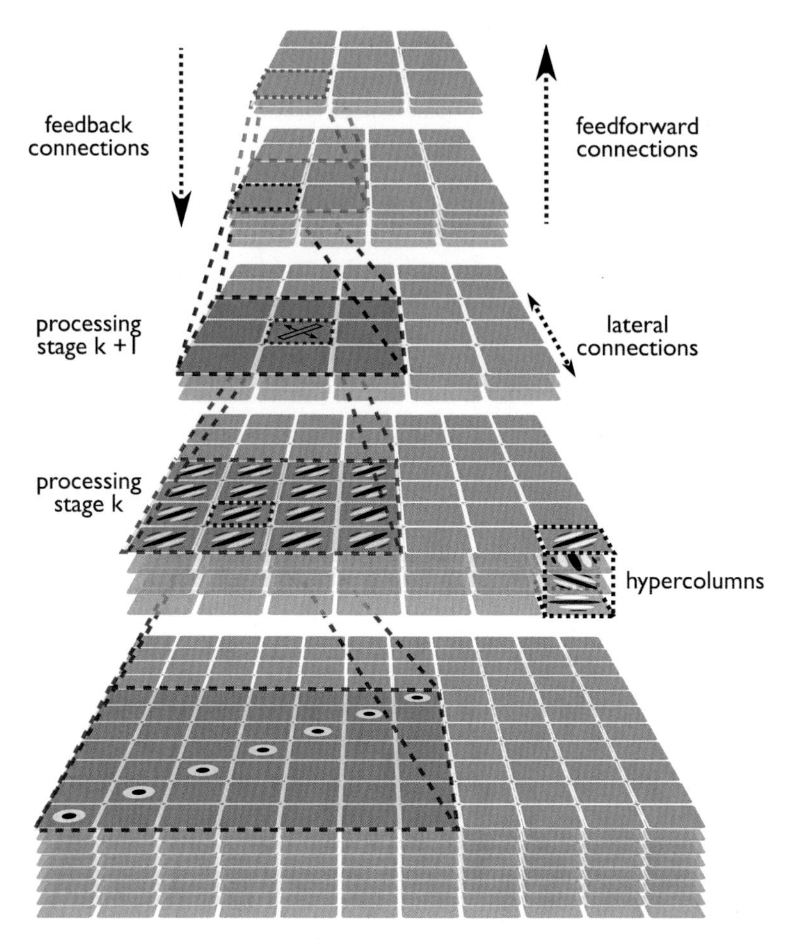

Fig. 7 Hierarchical models of the visual system. They are characterized by multiple stages of processing whereby units in one stage (shown as *squares*) pool the response of units from the previous stage (*colored projections*). Individual stages, also called layers (shown as *gray* stacks), contain multiple *feature maps* organized in terms of both space and scale. An hyper-column contains all possible features from all feature maps for that location. Hence each stage can be thought of as containing hyper-columns replicated at all positions and scales. In the most general case, hierarchical models allow for communication both ways between any two consecutive stages (feedforward and feedback connections), as well as between units part of the same stage (lateral connections). The building block of many successful hierarchical models, i.e., the circuit proposed by Hubel and Wiesel decades ago, is embedded as a particular example and shown in *red* (*center-surround stage*), *pink* (simple cell stage), and *purple* (complex cell stage)

cific motion direction, orientation, binocular disparity, etc.) but at multiple positions (tiling the visual space) and/or multiple scales.

The first instance of such a columnar model was indeed proposed by Hubel and Wiesel (1962) to explain orientation tuning and ocular dominance in the primary visual cortex, and was named the ice cube model. While more complex models of columnar organization have been proposed in recent years (e.g., to account for pinwheel centers), hierarchical models of the visual system follow the inspiration of the ice cube model for its simplicity of implementation. Thus, the set of all units within a stage typically exhibits this kind of dual organization both in terms of the visual field which they tile with their receptive fields, and in terms of their selectivities, which can be thought to span all the possible values at every location in the visual field.

In addition to the feedforward (bottom-up) connections, which correspond to projections from processing stage k to $k^* > k$, units can also be connected via lateral (horizontal) connections (both short-range connections within an hyper-column and long range between hyper-columns at different retinal locations) or feedback (top-down) connections from processing stage k to $k^* < k$.

3.3 Models of Object Recognition

Historically, most hierarchical models that have been proposed have focused on the processing of two-dimensional shape information in the ventral stream of the visual cortex, which follows a hierarchy of brain stages, starting from the retina, through the LGN in the thalamus to primary visual cortex (primary visual cortex, or striate cortex) and extra-striate visual areas, secondary visual cortex (V2), quaternary visual cortex (V4) and the inferotemporal cortex (IT). In turn, IT provides a major source of input to prefrontal cortex (PFC) involved in linking perception to memory and action (see Dicarlo et al. 2012 for a recent review).

As one progresses along the ventral stream visual hierarchy, neurons become selective for increasingly complex stimuli—from simple oriented bars and edges in early visual areas to moderately complex features in intermediate areas (such as combinations of orientations) and complex objects and faces in higher visual areas such as IT. In parallel to this increase in the complexity of the preferred stimulus, the invariance properties of neurons also increase with neurons gradually becoming more and more tolerant with respect to the exact position and scale of the stimulus within their receptive fields. As a result of this increase in invariance properties, the receptive field size of neurons increases, from about one degree or less in the primary visual cortex to several degrees in IT.

Explaining the selectivity and invariance properties of the ventral stream of the visual cortex has been one of the driving forces behind the development of hierarchical models of object recognition (see Serre 2014 for review). These models have a long history: the initial idea was proposed by Marko and Giebel with their homogeneous multi-layered architecture (Marko and Giebel 1970) and was later used in several visual architectures including Fukushima's *Neocognitron* (Fukushima 1983), convolutional networks (LeCun et al. 1998) and other models of object recognition (Wallis and Rolls 1997; Mel 1997; Riesenhuber and Poggio 1999; Wersing and Koerner 2003; Ullman 2007; Serre et al. 2007; Masquelier and Thorpe 2007). Over the years, these hierarchical models were shown to perform well for the categorization of multiple object categories (see Serre and Poggio 2010 for a review).

The HMAX model (Riesenhuber and Poggio 1999; Serre et al. 2007) shown in Fig. 8 constitutes a representative example of feedforward hierarchical models. It combines mechanisms for building up invariance and selectivity through the hierarchy, inspired by the *Neocognitron* with view-based theories of 3D object recognition (Riesenhuber and Poggio 2000). HMAX attempts to mimic the main information processing stages across the entire ventral stream visual pathway and bridges the gap between multiple levels of understanding (Serre and Poggio 2010). This system-level model seems consistent with physiological data in non-human primates in different cortical areas of the ventral visual pathway (Serre et al. 2007), as well as human behavioral data during rapid categorization tasks with natural images (Serre et al. 2007; Crouzet and Serre 2011; but see also Ghodrati et al. 2014; Cadieu et al. 2014; Yamins et al. 2014; Khaligh-Razavi and Kriegeskorte 2014).

In recent years, a number of HMAX extensions have been proposed. Most of them have focused on the learning of visual representations in intermediate stages of the model. Prominent examples include biologically-plausible learning mechanisms based on temporal continuity in video sequences (Masquelier et al. 2007), evolutionary algorithms (Ghodrati et al. 2012), as well as spike-timing dependent-based learning rules (Masquelier and Thorpe 2007; Kheradpisheh et al. 2015).

The models of object recognition described above use (Hebbian-like) *unsupervised* learning rules: they learn commonly-occurring visual features from natural images irrespective of their diagnosticity for object categorization. These learning rules seem consistent with ITC recordings that have shown that the learning of position and scale invariance, for instance, is driven by the subject's visual experience (Li and DiCarlo 2008a, 2010) and is unaffected by reward signals (Li and Dicarlo 2012).

However, a class of neural networks called deep learning architectures have recently brought about a small revolution in machine learning by becoming the new state-of-the-art on a variety of categorization tasks ranging from speech, music, text, genomes and images (see LeCun et al. 2015 for an up-to-date review). They differ in two ways from more traditional hierarchical models of the visual cortex such as the aforementioned HMAX and *Neocognitron*. First, learning across processing stages is fully supervised and uses the back-propagation algorithm (see LeCun et al. 2015 for a history), which propagates an error signal from upper-level (categorization) layers towards lower-level (perceptual) ones. Thus, only visual features that are diagnostic for the trained categorization tasks will be learned.

Second, deep learning architectures do not try to imitate biology as well as older hierarchical models of the visual cortex, which are constrained to match neuroscience data on a wide range of parameters (receptive field sizes, invariance and other tuning properties, number of layers, etc). For instance, state-of-the-art deep learning architectures incorporate many more layers (over 20 layers, see (Szegedy et al. 2014; He et al. 2015a)) in comparison to hierarchical models of the visual cortex (e.g., 7 layers for the HMAX). They possibly incorporate entire ensembles of deep networks for a given categorization task (Szegedy et al. 2014; He et al. 2015a); such

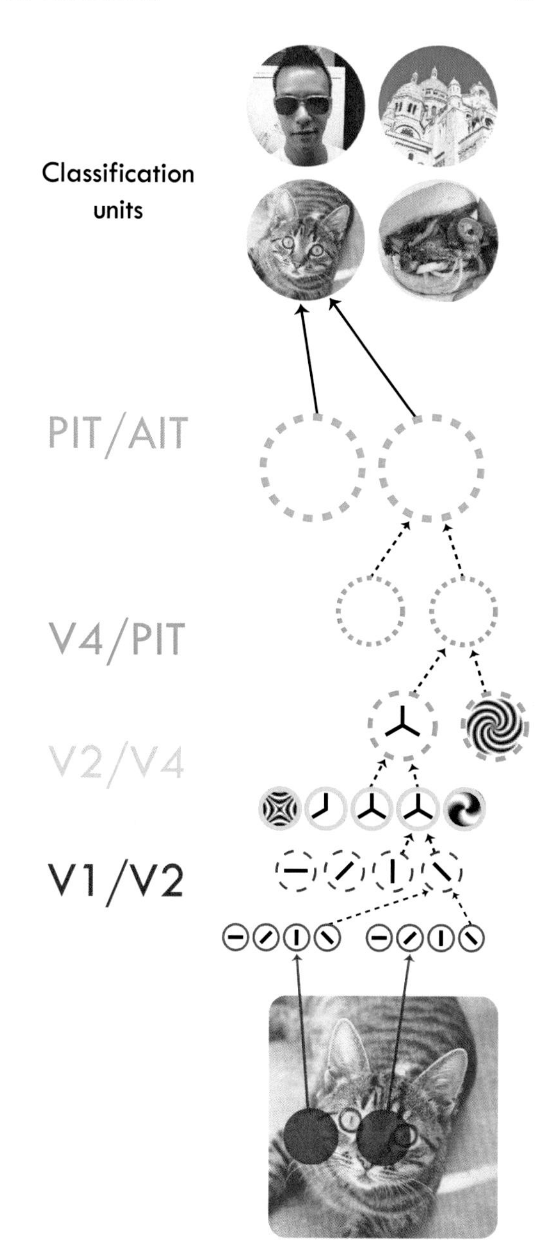

Fig. 8 Sketch of the HMAX hierarchical model of visual processing. Acronyms: V1, V2 and V4 correspond to primary, secondary and quaternary visual areas, PIT and AIT to posterior and anterior inferotemporal areas, respectively. Tentative mapping to neurophysiology is shown in color, some areas of the parietal cortex and dorsal streams are not shown. The model relies on two types of computations: a max-like operation (shown in *dash circles*) over similar features at different position and scale to gradually build tolerance to position and scale, and a bell-shape tuning operation (shown in *plain circles*) over multiple features to increase the complexity of the underlying representation, see (Serre et al. 2007) and text for details. Since it was originally developed, the model has been able to explain a number of new experimental data. This includes data that were not used to derive or fit model parameters. The model seems to be qualitatively and quantitatively consistent with (and in some cases actually predicts) several properties of subpopulations of cells in the primary visual cortex, V4, IT, and PFC as well as fMRI and psychophysical data

very deep networks can implement more complex classification functions. This, in turn, comes at the cost of sample complexity: the number of samples required for proper training increases with the number of parameters to be fitted. Not surprisingly, significant efforts have thus been recently dedicated to building ever-growing large-scale annotated image and video datasets (the ImageNet Large Scale Visual Recognition Challenge Russakovsky et al. 2014) contains >1M images and 1,000 categories), enabling the training of increasingly large networks (compare with the 2010 PASCAL VOC challenge (Everingham et al. 2010) with <20,000 images and 20 categories).

Perhaps surprisingly, despite the absence of neuroscience constraints on modern deep learning architectures, recent work has shown that these architectures are better able to explain ventral stream neural data (Yamins et al. 2014; Cadieu et al. 2014; Khaligh-Razavi and Kriegeskorte 2014; Guclu and Gerven 2015). In addition, these networks outperform all other models by a large margin (Cadieu et al. 2014) and are starting to match human level of accuracy for difficult object categorization tasks (He et al. 2015a).

3.4 Models Across Visual Cues

The effectiveness of hierarchical models of two-dimensional shape processing and object recognition has recently led to considerable interest in building hierarchical extensions to multiple visual cues beyond the early processing models described in Sect. 2. The main idea in these models is to reuse basic computational building blocks (such as the ones described in Sects. 1 and 2) across several processing stages. Moreover, the ever-increasing trove of electrophysiology data in mid-level visual areas now makes it possible to effectively constrain the space of all possible models.

Going beyond two-dimensional shape processing, several hierarchical models of motion processing have been proposed. For instance, computational models composed of the core operations described in Sects. 1 and 2 have been shown to be able to reproduce the selectivity of motion-selective neurons in the dorsal stream of the visual cortex to complex moving stimuli such as drifting plaids (Simoncelli and Heeger 1998; Rust et al. 2006) and continuous deformations (Mineault et al. 2012). Closely related models of the ventral and dorsal streams for the processing of form and motion, respectively, were used to model the brain mechanisms underlying action recognition (Giese and Poggio 2003).

Building on models of the dorsal stream of the visual cortex (Simoncelli and Heeger 1998; Giese and Poggio 2003; Rust et al. 2006; Mineault et al. 2012), a computer vision system was shown to perform well and, at the time, compete with state-of-the-art computer vision systems for the recognition of actions. The approach was later extended to the automated monitoring and analysis of rodents in their home-cage with accuracy on par with that of trained human annotators for a repertoire of about a dozen behaviors (Jhuang et al. 2010).

More recently, an extension of this approach included speed-tuned units as found in MT (Maunsell and Essen 1983; Perrone and Thiele 2001; Priebe et al. 2003) and yielded a system for the visual control of locomotory behavior that produced trajectories consistent with those produced by human participants when asked to reach a goal while avoiding obstacles in natural-looking environments (Barhomi et al. A data-driven approach to learning strategies for the visual control of navigation. Abstract presented at the Vision Science Society, 2014).

As for hierarchical models of color, the few efforts that have endeavored to go beyond one stage of processing pertain to computer vision (see Zhang et al. 2012 for an attempt to bridge this gap). These are very largely limited to solving specific tasks such as boundary detection in natural scenes and the representations they yield are *ad hoc* and cannot be compared against electrophysiology. However, work in progress from Zhang and colleagues suggests that a model consisting of the single- or double-opponent cells as described in Sect. 2.1 followed by the proper divisive normalization over an extended spatial neighborhood seems sufficient to account for psychophysics data of color constancy (Mély and Serre. A canonical circuit for visual contextual integration explains induction effects across visual modalities. Abstract presented at the Vision Science Society, 2015).

Regarding binocular disparity, our group has started to design a hierarchical model of disparity tuning (Kim et al. 2015) that builds on a population of model cells with linear receptive fields based on the binocular Gabor filters described in Sect. 2.2. Even though these units display varied selectivity to position disparity, phase disparity, orientation, spatial frequency, scale and phase, they are individually prone to incorrectly matching visually discordant inputs from either eye. To address this problem (see Read and Cumming 2007 for a formalization), we leveraged the divisive normalization circuit between units that prefer the same position disparity but opposite phase disparities in order to reduce sensitivity to false matches. We further included an energy computation as well to implement local invariance to stimulus phase. As a result, units from this additional stage of the model tend to be much more selective to the correct binocular disparity.

4 Discussion and Concluding Remarks

4.1 Why Hierarchies?

It has been postulated that the goal of the visual cortex is to achieve an optimal trade-off between selectivity and invariance via a hierarchy of processing stages whereby neurons at higher and higher levels exhibit an increasing degree of invariance to image transformations such as translations and scale changes (Riesenhuber and Poggio 1999; Serre and Poggio 2010).

Now, why hierarchies? The answer—for models in the Hubel and Wiesel spirit—is that the hierarchy may provide a solution to the invariance-selectivity trade-off problem by decomposing a complex task such as invariant object recognition in a hierarchy of simpler ones (at each stage of processing). Hierarchical organization in cortex is not limited to the visual pathways, and thus a more general explanation may be needed. Interestingly, from the point of view of classical learning theory (Poggio and Smale 2003), there is no need for architectures with more than three layers. So, why hierarchies? There may be reasons of efficiency, such as the efficient use of computational resources. For instance, the lowest levels of the hierarchy may represent a dictionary of features that can be shared across multiple classification tasks (Geman and Koloydenko 1999).

There may also be the more fundamental issue of sample complexity, the number of training examples required for good generalization (see Serre and Poggio 2010 for discussion). An obvious difference between the best classifiers derived from statistical learning theory and human learning is in fact the number of examples required in tasks such as object recognition. Statistical learning theory shows that the complexity of the hypothesis space sets the speed limit and the sample complexity for learning. If a task—like a visual recognition task—can be decomposed into low-complexity learning tasks for each layer of a hierarchical learning machine, then each layer may require only a small number of training examples. Neuroscience suggests that what humans can learn may be represented by hierarchies that are locally simple. Thus, our ability to learn from just a few examples, and its limitations, may be related to the hierarchical architecture of cortex.

4.2 Limitations

To date, most existing hierarchical models of visual processing both from the perspective of biological and machine vision are instances of feedforward models. These models have been useful to explore the power of fixed hierarchical organization as originally suggested by Hubel and Wiesel. These models assume that our core visual capabilities proceed through a cascade of hierarchically organized areas along various streams of processing in the visual cortex with computations at each successive stage being largely feedforward (Riesenhuber and Poggio 1999; Dicarlo et al. 2012). They have led, for instance, to algorithms that were at the time competitive with the best computer vision systems (Serre and Poggio 2010) and culminating with deep learning architectures that are bringing about a small revolution in artificial intelligence (LeCun et al. 2015).

The limitations of these visual architectures, however, are becoming increasingly obvious. Not only top-down mechanisms are key to normal, everyday vision, but back-projections are also likely to be a key part of what cortex is computing and how. Thus, a major question for modeling visual cortex revolves around the role of back-projections and the related fact that vision is more than categorization and requires interpreting and parsing visual scenes (as opposed to simply finding out whether a specific object is present in the visual scene or not). A human observer can essentially answer an infinite number of questions about an image. Such image interpretation tasks have proven challenging for modern computer vision architectures (Fleuret et al. 2011; Gülçehre and Bengio 2013).

In addition, while the overall hierarchical organization of the visual cortex is now well established (Felleman and Essen 1991), the parallel between the anatomical and functional hierarchy is, however, looser than one might expect. While the trend is, from lower to higher visual areas, for neurons' receptive fields to become increasingly large and tuned to increasingly complex preferred stimuli, there remains a very broad distribution of tuning and receptive field sizes in all areas of the visual hierarchy. For instance, IT, which is commonly assumed to have solved the problem of invariant recognition (Dicarlo et al. 2012), also contains neurons with relatively small receptive fields and tuned to relatively simple visual features such as simple orientations (Desimone et al. 1984). A close comparison of shape representation between primary visual cortex, V2 and V4 also demonstrated a complex pattern of shape selectivity with significant deviation from strict hierarchical organization with some cells in the primary visual cortex exhibiting more complex tuning than some cells in V4 (Hegdé and Essen 2007b). Furthermore, beside the visual cortical hierarchy, there exist additional subcortical pathways (including cortico-thalamo-cortical loops). Hence, the anatomical hierarchy should be taken as an idealization and cannot be taken as a strict flowchart of visual information (Hegdé and Felleman 2007).

Another weakness shared by both larger-scale models of biological and machine vision are their reliance on a surprisingly limited number of computations, viz., the linear-nonlinear (LN) modules we mentioned in Sect. 1 and divisive normalization under various forms. As a result, a potentially fruitful way to improve such hierarchical models would be to extend their repertoire to include new computations inspired by cutting-edge neurophysiology research on cortical microcircuits. Among the cortical operations yet untapped on a large scale by modeling efforts are dynamic or stochastic synapses (current models assume synaptic weights to be fixed and static after learning), heavily nonlinear computations in dendrites (current models only assume "weak" nonlinearity in their units, viz., linearity followed by a rectification, as opposed to more complex transformation), or synchrony between model neurons (though many researchers have discussed its potential use to tackle the "binding problem' between overlapping, noisy visual representations (Crick 1984; von der Malsburg 1994; Singer and Gray 1995; Fries 2005; Uhlhaas et al. 2009; Stanley 2013; Reichert and Serre 2014)). The availability of large-scale architectures such as deep learning nets, combined with extensive human-annotated datasets, should make for an ideal testbed for any potential, cortically-inspired computational mechanism.

References

Adelson EH, Bergen JR (1985) Spatiotemporal energy models for the perception of motion. J Opt Soc Am A 2(2):284–299

Alonso JM, Martinez LM (1998) Functional connectivity between simple cells and complex cells in cat striate cortex. Nat Neurosci 1(5):395–403

Bradley D, Goyal M (2008) Velocity computation in the primate visual system. Nat Rev Neurosci 9(9):686–695

Cadieu CF, Hong H, Yamins DLK, Pinto N, Ardila D, Solomon EA, Majaj NJ, DiCarlo JJ (2014) Deep neural networks rival the representation of primate IT cortex for core visual object recognition. PLoS Comput Biol 10(12):e1003963

Carandini M (2012) From circuits to behavior: a bridge too far? Nat Neurosci 15(4):507–509

Carandini M, Heeger D (2012) Normalization as a canonical neural computation. Nat Rev Neurosci 13:51–62

Chen X, Han F, Poo M-MM, Dan Y (2007) Excitatory and suppressive receptive field subunits in awake monkey primary visual cortex (V1). Proc Natl Acad Sci USA 104(48):19120–19125

Conway BR (2001) Spatial structure of cone inputs to color cells in alert macaque primary visual cortex (V-1). J Neurosci 21(8):2768–2783

Crick F (1984) Function of the thalamic reticular complex: the searchlight hypothesis. Proc Natl Acad Sci USA 81:4586–4590

Crouzet SM, Serre T (2011) What are the visual features underlying rapid object recognition? Front Psychol 2:326

Daugman JG (1980) Two-dimensional spectral analysis of cortical receptive field profile. Vis Res 20:847–856

Daugman JG (1985) Uncertainty relation for resolution in space, spatial frequency, and orientation optimization by two-dimensional visual cortical filters. J Opt Soc Am A 2(7):1160–1169

Dayan P, Abbott LF (2001) Theoretical neuroscience: computational and mathematical modeling of neural systems. MIT Press

Desimone R, Albright TD, Gross CG, Bruce C (1984) Stimulus-selective properties of inferior temporal neurons in the macaque. J Neurosci 4(8):2051–2062

DeValois RL, Albrecht DG, Thorell LG (1982) Spatial-frequency selectivity of cells in macaque visual cortex. Vis Res 22:545–559

Dicarlo JJ, Zoccolan D, Rust NC (2012) How does the brain solve visual object recognition ? Neuron 73(3):415–434

Douglas RJ, Martin KAC (2007) Mapping the matrix: the ways of neocortex. Neuron 56(2):226–238

Everingham M, Van Gool L, Williams C, Winn J, Zisserman A (2010) The PASCAL visual object classes (VOC) challenge. Int J Comput Vis 88(2):303–338

Felleman DJ, Van Essen DC (1991) Distributed hierarchical processing in the primate cerebral cortex. Cereb Cortex 1:1–47

Fenno L, Yizhar O, Deisseroth K (2011) The development and application of optogenetics. Annu Rev Neurosci 34:389–412

Finn I, Ferster D (2007) Computational diversity in complex cells of cat primary visual cortex. J Neurosci 27(36):9638–9648

Fleet DJ, Wagner H, Heeger DJ (1996) Neural encoding of binocular disparity: energy models, positionshifts and phase shifts. Vis Res 36(12):1839–1857

Fleuret F, Li T, Dubout C, Wampler EK, Yantis S, Geman D (2011) Comparing machines and humans on a visual categorization test. Proc Natl Acad Sci USA 108(43):17621–17625

Fries P (2005) A mechanism for cognitive dynamics: neuronal communication through neuronal coherence. Trends Cogn Sci 9(10):474–480

Fukushima K (1983) Neocognitron: a neural network model for a mechanism of visual pattern recognition. IEEE Trans Syst Man Cybern B Cybern 13:826–834

Geman D, Koloydenko A (1999) Invariant statistics and coding of natural microimages. Proc IEEE Work Stat Comput Theor Vis

Ghodrati M, Farzmahdi A, Rajaei K, Ebrahimpour R, Khaligh-Razavi S-M (2014) Feedforward object-vision models only tolerate small image variations compared to human. Front Comput Neurosci 8:74

Ghodrati M, Khaligh-Razavi S-M, Ebrahimpour R, Rajaei K, Pooyan M (2012) How can selection of biologically inspired features improve the performance of a robust object recognition model? PLoS One 7(2):e32357
Giese MA, Poggio T (2003) Neural mechanisms for the recognition of biological movements. Nat Rev Neurosci 4(3):179–192
Guclu U, van Gerven MAJ (2015) Deep neural networks reveal a gradient in the complexity of neural representations across the ventral stream. J Neurosci 35(27):10005–10014
Gülçehre C, Bengio Y (2013) Knowledge matters: importance of prior information for optimization. arXiv:1301.4083v6
He K, Zhang X, Ren S, Sun J (2015) Delving deep into rectifiers: surpassing human-level performance on imagenet classification
Heeger DJ (1992) Normalization of cell responses in cat striate cortex. Vis Neurosci 9(2):181–197
Hegdé J, Essen DV (2007) A comparative study of shape representation in macaque visual areas V2 and V4. Cereb Cortex 2(May)
Hegdé J, Felleman DJ (2007) Reappraising the functional implications of the primate visual anatomical hierarchy. Neuroscience 13(5):416–421
Hubel D, Wiesel T (1962) Receptive fields, binocular interaction and functional architecture in the cat's visual cortex. J Physiol 160:106–154
Humanski RA, Wilson HR (1993) Spatial-frequency adaptation: evidence for a multiple-channel model of short-wavelength-sensitive-cone spatial vision. Vis Res 33(5–6):665–675
Jhuang H, Garrote E, Yu X, Khilnani V, Poggio T, Steele AD, Serre T (2010) Automated home-cage behavioural phenotyping of mice. Nat Commun 1(6):1–9
Jhuang H, Serre T, Wolf L, Poggio T (2007) A biologically inspired system for action recognition. In: 2007 IEEE 11th International Conference Computer Vision, pp 1–8
Johnson EN, Hawken MJ, Shapley R (2001) The spatial transformation of color in the primary visual cortex of the macaque monkey. Nat Neurosci 4(4):409–416
Johnson EN, Hawken MJ, Shapley R (2004) Cone inputs in macaque primary visual cortex. J Neurophysiol 91(6):2501–2514
Johnson EN, Hawken MJ, Shapley R (2008) The orientation selectivity of color-responsive neurons in macaque V1. J Neurosci 28(32):8096–8106
Jones JP, Palmer LA (1987) An evaluation of the two-dimensional Gabor filter model of simple receptive fields in cat striate cortex. J Neurophysiol 58(6):1233–1258
Khaligh-Razavi S-M, Kriegeskorte N (2014) Deep supervised, but not unsupervised, models may explain IT cortical representation. PLoS Comput Biol 10(11):e1003915
Kheradpisheh SR, Ganjtabesh M, Masquelier T (2015) Bio-inspired unsupervised learning of visual features leads to robust invariant object recognition. arXiv:1504.03871v3
Kim J, Mely DA, Serre T (2015) A critical evaluation of computational mechanisms of binocular disparity
Kouh M, Poggio T (2008) A canonical neural circuit for cortical nonlinear operations. Neural Comput 20(6):1427–1451
Landy MS, Movshon JA (1991) Computational models of visual processing. MIT Press
LeCun Y, Bengio Y, Hinton G (2015) Deep learning. Nature 521(7553):436–444
LeCun Y, Bottou L, Bengio Y, Haffner P (1998) Gradient-based learning applied to document recognition. Proc IEEE 86(11):2278–2324
Li N, DiCarlo JJ (2008) Unsupervised natural experience rapidly alters invariant object representation in visual cortex. Science (80) 321(5895):1502–1507
Li N, DiCarlo JJ (2010) Unsupervised natural visual experience rapidly reshapes size-invariant object representation in inferior temporal cortex. Neuron 67(6):1062–1075
Li N, Dicarlo JJ (2012) Neuronal learning of invariant object representation in the ventral visual stream is not dependent on reward. J Neurosci 32(19):6611–6620
Malik J, Perona P (1990) Preattentive texture discrimination with early vision mechanisms. J Opt Soc Am A 7(5):923–932

Marcelja S (1980) Mathematical description of the responses of simple cortical cells. J Opt Soc Am 70:1297–1300
Marko H, Giebel H (1970) Recognition of handwritten characters with a system of homogeneous layers. Nachrichtentechnische Z 23:455–459
Marr D (1982) Vision: a computational investigation into the human representation and processing of visual information. W.H.Freeman & Co Ltd, San Francisco
Masquelier T, Serre T, Poggio T (2007) Learning complex cell invariance from natural videos: a plausibility proof. Technical report, Massachusetts Institute of Technology, Cambridge MA
Masquelier T, Thorpe SJ (2007) Unsupervised learning of visual features through spike timing dependent plasticity. PLoS Comput Biol 3(2):e31
Maunsell JH, Essen DCV (1983) Functional properties of neurons in middle temporal visual area ofthe macaque monkey. II. Binocular interactions and sensitivity tobinocular disparity
Mel BW (1997) SEEMORE: combining color, shape, and texture histogramming in a neurally inspired approach to visual object recognition. Neural Comput. 9(4):777–804
Mineault P, Khawaja F, Butts D, Pack C (2012) Hierarchical processing of complex motion along the primate dorsal visual pathway. Proc Natl Acad Sci 109(16):E972–E980
Movshon JA, Adelson EH, Gizzi MS, Newsome WT (1985) The analysis of moving visual patterns. Pattern Recogn Mech
Mullen KT, Losada MA (1999) The spatial tuning of color and luminance peripheral vision measured with notch filtered noise masking. Vis Res 39(4):721–731
Nishimoto S, Gallant JL (2011) A three-dimensional spatiotemporal receptive field model explains responses of area MT neurons to naturalistic movies. J Neurosci 31(41):14551–14564
Ohzawa I (1998) Mechanisms of stereoscopic vision: the disparity energy model. Curr Opin Neurobiol 8(4):509–515
Ohzawa I, DeAngelis G, Feeman R (1997) Encoding of binocular disparity by complex cells in the cat's visual cortex. J Neurophysiol 77(6):2879–2909
Ohzawa I, DeAngelis GC, Freeman RD (1996) Encoding of binocular disparity by simple cells in the cat's visual cortex. J Neurophysiol 75(5):1779–1805
OReilly RC, Wyatte D, Herd S, Mingus B, Jilk DJ (2013) Recurrent processing during object recognition. Front Psychol 4(April):1–14
Ostojic S, Brunel N (2011) From spiking neuron models to linear-nonlinear models. PLoS Comput Biol 7(1):e1001056
Perrone JA, Thiele A (2001) Speed skills: measuring the visual speed analyzing properties of primate MT neurons. Nat Neurosci 4(5):526–532
Plaza SM, Scheffer LK, Chklovskii DB (2014) Toward large-scale connectome reconstructions. Curr Opin Neurobiol 25:201–210
Poggio T, Serre T (2013) Models of the visual cortex. Scholarpedia 8(4):3516
Poggio T, Smale S (2003) The mathematics of learning: dealing with data. Not Am Math Soc 50(5)
Priebe NJ, Cassanello CR, Lisberger SG (2003) The neural representation of speed in macaque area MT/V5. J Neurosci 23(13):5650–5661
Qian N (1994) Computing stereo disparity and motion with known binocular cell properties. Neural Comput 6(3):390–404
Read JC, Cumming BG (2007) Sensors for impossible stimuli may solve the stereo correspondence problem. Nat Neurosci 10(10):1322–1328
Reichert DP, Serre T (2014) Neuronal synchrony in complex-valued deep networks. In: International Conference on Learning Vision Representations
Rieke F, Warland D, van Steveninck R, Bialek W, van Steveninck R (1997) Spikes. The MIT Press, Cambridge, Massachusetts
Riesenhuber M, Poggio T (1999) Hierarchical models of object recognition in cortex. Nat Neurosci 2(11):1019–1025
Riesenhuber M, Poggio T (2000) Models of object recognition. Nat Neurosci 3:1199–1204
Ringach DL (2004) Haphazard wiring of simple receptive fields and orientation columns in visual cortex. J Neurophysiol 92:468–476

Ringach DL (2004) Mapping receptive fields in primary visual cortex. J Physiol 558(3):717–728
Russakovsky O, Deng J, Su H, Krause J, Satheesh S, Ma S, Huang Z, Karpathy A, Khosla A, Bernstein M, Berg AC, Fei-Fei L (2014) Imagenet large scale visual recognition challenge. arXiv:1409.0575v3
Rust NC, Mante V, Simoncelli EP, Movshon JA (2006) How MT cells analyze the motion of visual patterns. Nat Neurosci 9(11):1421–1431
Rust NC, Schwartz O, Movshon JA, Simoncelli EP (2005) Spatiotemporal elements of macaque V1 receptive fields. Neuron 46(6):945–956
Sanada TM, Ohzawa I (2006) Encoding of three-dimensional surface slant in cat visual areas 17 and 18. J Neurophysiol 95(5):2768–2786
Sasaki K, Ohzawa I (2007) Internal spatial organization of receptive fields of complex cells in the early visual cortex. J Neurophysiol 98(3):1194–1212
Series P, Lorenceau J, Frégnac Y (2003) The silent surround of V1 receptive fields: theory and experiments. J Physiol 97:453–474
Serre, T (2014) Hierarchical models of the visual system
Serre T, Kreiman G, Kouh M, Cadieu C, Knoblich U, Poggio T (2007) A quantitative theory of immediate visual recognition. Prog Brain Res 165:33
Serre T, Kreiman G, Kouh M, Cadieu C, Knoblich U, Poggio T (2007) A quantitative theory of immediate visual recognition. Prog Brain Res 165(06):33–56
Serre T, Oliva A, Poggio T (2007) A feedforward architecture accounts for rapid categorization. Proc Natl Acad Sci USA 104(15):6424–6429
Serre T, Poggio T (2010) A neuromorphic approach to computer vision. Commun ACM 53(10):54
Shapley R, Hawken MJ (2011) Color in the cortex: single- and double-opponent cells. Vis Res 51:701–717
Simoncelli EP, Heeger DJ (1998) A model of neuronal responses in visual area MT. Vision Res 38(5):743–761
Singer W, Gray CM (1995) Visual feature integration and the temporal correlation hypothesis. Ann Rev Neurosci 18:555–586
Stanley GB (2013) Reading and writing the neural code. Nat Neurosci 16(3):259–263
Stevenson IH, Kording KP (2011) How advances in neural recording affect data analysis. Nat Neurosci 14(2):139–142
Szegedy C, Liu W, Jia Y, Sermanet P, Reed S, Anguelov D, Erhan D, Vanhoucke V, Rabinovich A (2014) Going deeper with convolutions. arXiv:1409.4842v1
Uhlhaas PJ, Pipa G, Lima B, Melloni L, Neuenschwander S, Nikolić D, Singer W (2009) Neural synchrony in cortical networks: history, concept and current status. Front Integr Neurosci 3:17
Ullman S (2007) Object recognition and segmentation by a fragment-based hierarchy. Trends Cogn Sci 11(2):58–64
von der Malsburg C (1994) The correlation theory of brain function. In: Domany E (ed) Models of neural networks II, pp 94–119. Springer
Wallis G, Rolls ET (1997) A model of invariant recognition in the visual system. Prog Neurobiol 51:167–194
Watson AB (1987) Efficiency of a model human image code. J Opt Soc Am A. 4(12):2401–2417
Watson AB, Ahumada AJ (1985) Model of human visual-motion sensing. J Opt Soc Am A 2(2):322–341
Watson AB, Barlow HB, Robson JG (1983) What does the eye see best? Nature 302(5907):419–422
Wersing H, Koerner E (2003) Learning optimized features for hierarchical models of invariant recognition. Neural Comput 15(7):1559–1588
Yamins DLK, Hong H, Cadieu CF, Solomon EA, Seibert D, DiCarlo JJ (2014) Performance-optimized hierarchical models predict neural responses in higher visual cortex. Proc Natl Acad Sci USA 111(23):8619–8624
Young RA (1987) The Gaussian derivative model for spatial vision: I. Retinal mechanisms. Spat Vis 2(4):273–293

Young RA, Lesperance RM (2001) The Gaussian derivative model for spatial-temporal vision: II. Cortical data. Spat Vis 14(3):321–389

Zhang J, Barhomi Y, Serre T (2012) A new biologically inspired color image descriptor. In: European Conference on Computer Vision. LNCS, vol 7576, pp 312–324

Invariant Recognition Predicts Tuning of Neurons in Sensory Cortex

Jim Mutch, Fabio Anselmi, Andrea Tacchetti, Lorenzo Rosasco, Joel Z. Leibo and Tomaso Poggio

Abstract Tuning properties of simple cells in cortical V1 can be described in terms of a "universal shape" characterized quantitatively by parameter values which hold across different species (Jones and Palmer 1987; Ringach 2002; Niell and Stryker 2008). This puzzling set of findings begs for a general explanation grounded on an evolutionarily important computational function of the visual cortex. We show here that these properties are quantitatively predicted by the hypothesis that the goal of the ventral stream is to compute for each image a "signature" vector which is invariant to geometric transformations (Anselmi et al. 2013b). The mechanism for continuously learning and maintaining invariance may be the memory storage of a sequence of neural images of a few (arbitrary) objects via Hebbian synapses, while undergoing transformations such as translation, scale changes and rotation. For V1 simple cells this hypothesis implies that the tuning of neurons converges to the eigenvectors of the covariance of their input. Starting with a set of dendritic fields spanning a range of sizes, we show with simulations suggested by a direct analysis, that the solution of the associated "cortical equation" effectively provides a set of Gabor-like shapes with parameter values that quantitatively agree with the physiology data. The same theory provides predictions about the tuning of cells in V4 and in the face patch AL (Leibo et al. 2013a) which are in qualitative agreement with physiology data.

The original work of Hubel and Wiesel, as well as subsequent research, left open the questions of what is the function of the ventral stream in visual cortex and of how the properties of its neurons are related to it. A recent theory (Anselmi et al. 2013a; Poggio et al. 2013) proposes that the main computational goal of the ventral stream is to provide, at each level in the hierarchy of visual areas, a signature that is unique for the given image, invariant under geometric transformations and robust to small perturbations. The theory suggests a mechanism for learning the relevant invariances during unsupervised visual experience: storing sequences of images (called "templates") of a few objects transforming, for instance translating, rotating and looming. It has been shown that in this way invariant hierarchical architectures similar to models of the ventral stream such as Fukushima's Neocognitron (Fukushima 1980) and

J. Mutch · F. Anselmi (✉) · A. Tacchetti · L. Rosasco · J. Z. Leibo · T. Poggio
Massachusetts Institute of Technology (MIT), 77 Massachusetts Ave. MIT Bldg 46, Cambridge, MA 02139, USA
e-mail: anselmi@mit.edu

Q. Zhao (ed.), *Computational and Cognitive Neuroscience of Vision*, Cognitive Science and Technology, DOI 10.1007/978-981-10-0213-7_5

HMAX (Riesenhuber and Poggio 1999; Serre et al. 2007)—as well as deep neural network architectures called convolutional networks (LeCun et al. 1989; LeCun and Bengio 1995) and related models—e.g. Poggio and Edelman (1990), Perrett and Oram (1993), Mel (1997), Stringer and Rolls (2002), Pinto et al. (2009), Saxe et al. (2011), Le et al. (2011), Abdel-Hamid et al. (2012)—can be learned from unsupervised visual experience. We focus on V1 and assume that the development of an array of cells with spatially localized dendritic trees of different sizes is genetically determined, reflects the organization of the retinal arrays of photoreceptors and cells and is connected to the outputs of the Lateral Geniculate Body (see also Poggio et al. 2014).

As discussed in Anselmi et al. (2013a), Poggio et al. (2013), Anselmi et al. (2013b) the templates and their transformations—corresponding to a set of "simple" cells—may be memorized from unsupervised visual experience. In a second learning step a complex cell is wired to "simple" cells that are activated in close temporal contiguity and thus are likely to correspond to the same patch of image undergoing a transformation in time (Földiák 1991). However, the idea of a direct storage of sequences of images patches—seen through a Gaussian window—in a set of V1 cells is biologically implausible. Here we propose that the neural memorization of frames (of transforming objects) is performed online via Hebbian synapses that change as an effect of visual experience. Specifically, we assume that the distribution of signals "seen" by a maturing simple cell is Gaussian in the spatial directions x, y reflecting the distribution on the dendritic tree of synapses from the lateral geniculate nucleus. We also assume that *there is a range of Gaussian distributions with different σ which increase with retinal eccentricity* (see also Poggio et al. 2014). As an effect of visual experience the weights of the synapses are modified by a Hebb rule (Hebb 1949). Hebb's original rule, which states in conceptual terms that "neurons that fire together, wire together", can be written as $\dot{\mathbf{w}} = u(\mathbf{v})\mathbf{v}$, where $\mathbf{v}$ is the input vector $\mathbf{w}$ is the presynaptic weights vector, u is the postsynaptic response and $\dot{\mathbf{w}} = d\mathbf{w}/dt$. In order for this dynamical system to actually converge, the weights have to be normalized. In fact, there is considerable experimental evidence that cortex employs normalization (cf. Turrigiano and Nelson 2004 and references therein).

Mathematically, this requires a modified Hebbian rule. We consider here the simplest such rule, proposed by Oja (1982) but others (such as Independent Component Analysis, see later and Bell and Sejnowski 1997; Hyvrinen and Oja 1998) would also be appropriate for our argument. Oja's equation $\dot{\mathbf{w}} = \gamma u(\mathbf{v})[\mathbf{v} - u(\mathbf{v})\mathbf{w}]$ defines the change in presynaptic weights $\mathbf{w}$ where γ is the "learning rate" and the "output" u is assumed to depend on the "input" $\mathbf{v}$ as $u(\mathbf{v}) = \mathbf{w}^{\mathbf{T}}\mathbf{v}$. The equation follows from expanding to the first order the Hebb rule normalized to avoid divergence of the weights. The important point is that Oja's version of Hebb's rule has been proven to converge to the top principal component of its input (technically to the eigenvector of the covariance of its inputs with the largest eigenvalue). Lateral inhibitory connections can enforce convergence of different neighboring neurons to several of the top eigenvectors (Oja 1982, 1992 and see Anselmi and Poggio 2010; Poggio et al. 2013). Our simulations with parameter values in the physiological range suggest that eigenvectors above the first three are almost always in the range of the noise. Because of this we can study directly the result of Hebbian learning by studying the properties of the top eigenvectors of the covariance of the visual inputs to a cortical cell, seen through a Gaussian window (Fig. 1).

In particular we consider the continuous version of the problem where images are transformed by the group of $2D$-translations. Notice that for small Gaussian apertures all motions are effectively translations (as confirmed by our simulations for natural images seen through physiologically sized Gaussian). Thus, in this case, the tuning of each simple cell in V1—given by the vector of its synaptic weights $\mathbf{w}$—is predicted

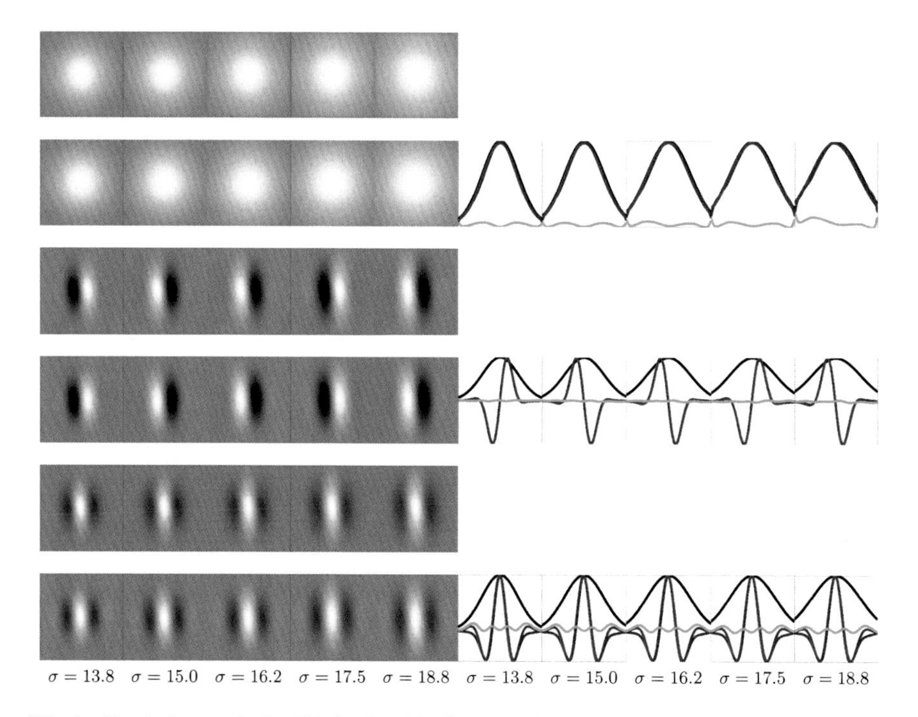

Fig. 1 Simulation results for V1 simple cells "learned" via PCA. Each "cell" receives as input all frames from 40 movies, each generated by a patch from a natural image undergoing a translation along the horizontal axis. A Gaussian filter with small sigma simulates the optics, a difference of Gaussians filter and a spatial lowpass filter are applied to every frame to simulate retinal processing. Each frame is multiplied by a Gaussian mask to model a cell's initial distribution of input synapses on its dendritic tree. The weighted difference between subsequent frames is fed to the learning stage, to simulate an imperfect temporal derivative (the weights we used are (−0.95, 1.00)) (see Appendix 1.2). Each cell "learns" its weight vector extracting the principal components of its input. On the *left*, for each *row pair*: the *top* row shows the best Gabor fit (*least squares*) and the *bottom row* shows the actual principal component vector; different *columns* represent different σ values for the Gaussian mask aperture. On the *right* we show 1D sections of the 2D tuning functions just described. The *blue line* is the learned function, *red* indicates the best least-squares fit to a Gabor wavelet, and *green* shows the difference (fitting error). The processing pipeline is described in the text and in the Appendix 1.1. An orientation orthogonal to the direction of motion emerges

to converge to one of the top few eigenfunctions $\psi_n(x, y)$ of the following equation (see Appendix 1.3 for solutions in the $1D$ case):

$$\int d\xi d\eta g(x, y) g(\xi, \eta) t^{\circledast}(\xi - x, \eta - y) \psi_n(\xi, \eta) = v_n \psi_n(x, y). \tag{1}$$

where the functions g are Gaussian distributions with the same, fixed width σ and $t^{\circledast}$ is the autocorrelation function of the input from the LGN. ψ_n is the eigenfunction and v_n the associated eigenvalue. Equation (1), which depends on $t^{\circledast}$, defines a set of eigenfunctions parametrized by σ. We assume that the images generating the LGN signal $t(x, y)$ are natural images, with a power spectrum $\mathcal{F}t^{\circledast}(x) = 1/\omega^2$, where $\mathcal{F}$ is the Fourier transform (Torralba and Oliva 2003). In 1-D the solutions of Eq. 1 with this $t^{\circledast}$ are windowed Fourier transforms but for different σ they provide a very good approximation of Gabor wavelets for $n = 0, 1, 2$, since λ increase roughly proportionally to σ. An analytic solution for a $\frac{1}{\omega^2}$ input spectrum is given in Appendix 1.3.2. An analytical insight in how the different eigenvectors emerge is sketched in Appendix 1.3.1. In 2D we need to take into account that the known temporal high-pass properties of retinal processing (modeled as an imperfect high-pass, derivative-like operation in time) are compensated in the direction of motion by a local spatial average followed by a Difference of Gaussian (DOG) filter (see for instance Dan 1996). Simulations, suggested by a direct analysis of the equations (see Appendix 1.3) show that, independently of the parameter values of the filtering, Gabor functions with modulation in the direction of motion (e.g. x), $G_n(x, y) \propto exp(-y^2/\sigma_{ny}^2 - x^2/\sigma_{nx}^2) \sin[(2\pi/\lambda_n)x]$ are approximate solutions of the equation. If for each aperture σ only the first three eigenvectors are significant, then the set of solutions is well described well by a set of Gabor wavelets, that is a set of Gabor functions in which lambda is proportional to σ_x, which in turn is proportional to σ_y. These relations are captured in the ratio n_x/n_y (where $n_x = \sigma_x/\lambda$ and $n_y = \sigma_y/\lambda$) which was introduced by Dario Ringach to characterize tuning properties of simple cells in V1 in the macaque (Ringach 2002). It turns out that simple cells show Gabor-like tuning curves which are wavelet-like. Remarkably, the key parameter values are similar across three different species as shown by Fig. 2 which includes, in addition to Ringach', also Niell and Stryker's data on mouse V1 (Niell and Stryker 2008) and the original Palmer et al. experiments in cat cortex (Jones and Palmer 1987). The theory described here predicts the data well. Equation (1) gives Gaussian eigenfunctions with no modulation as well as with odd and even modulations, similar to data from simple cells. In addition to the key plot in Fig. 2, the equations directly predict that the sigma in the unmodulated direction (σ_y) is always somewhat larger than σ_x (see inset). As shown in Appendix 1.4, some of the subtle properties of the tuning depend crucially on the exposure to continuous motion of objects. However, the general tuning is rather robust (see Sect. 1.3.2 in Poggio et al. 2012). In particular, we expect to find very similar tuning if instead of natural images with a $\frac{1}{\omega^2}$ spectrum, the input from the retina is determined during the early stages of development by retinal waves (Meister et al. 1991; Wong et al. 1993, Fig. 3).

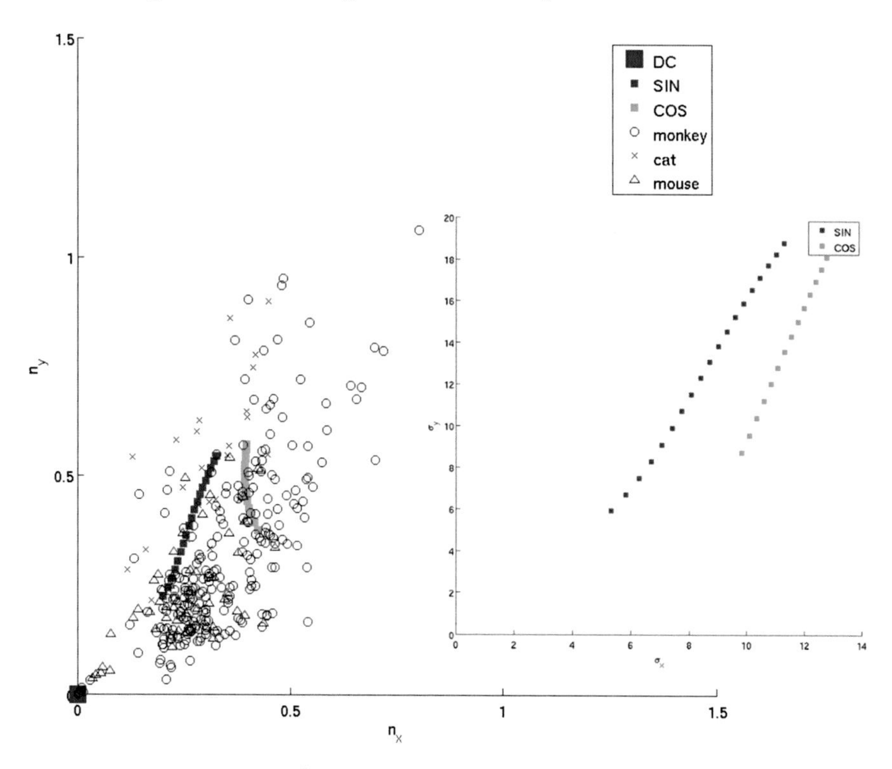

Fig. 2 This figure shows $n_y = \frac{\sigma_y}{\lambda}$ vs. $n_x = \frac{\sigma_x}{\lambda}$ for the modulated (x) and unmodulated (y) direction of the Gabor shape. Notice that the slope $\frac{n_y}{n_x} = \frac{\sigma_y}{\sigma_x}$ seems to be the same across species and is close to 1 (but $\frac{\sigma_y}{\sigma_x} > 1$)—a robust finding in the theory. Neurophysiology data from monkeys Ringach (2002), cats Jones and Palmer (1987), and mice Niell and Stryker (2008) are reported together with our simulations (for $n = 0, 1, 2$ for which the eigenvectors had significant power). Simulated cells learn their weight vector according to the algorithm described in Fig. 1. The *inset* shows values for σ_x and σ_y from our simulations. These quantities vary significantly across species and are not easy to obtain; Jones and Palmer (1987) present two different methods to estimate them and report inconsistent results. Conversely n_x and n_y, as defined above, are dimensionless and consistent across different fitting methods

Notice that the theory does not necessarily require visual experience for the initial tuning to emerge during development: it is quite possible that a tuning originally discovered by evolution because of visual experience was eventually compiled into the genes. The theory however predicts that the *tuning is maintained and updated* by continuous visual experience (under the assumption that Hebbian plasticity is present). In particular, it predicts that tuning can be modified by disrupting normal visual experience. At the level of Inferior Temporal (IT) cortex, such a prediction is consistent with the rapid disruption of position and scale invariance induced by exposure to altered visual experience Li and DiCarlo (2008).

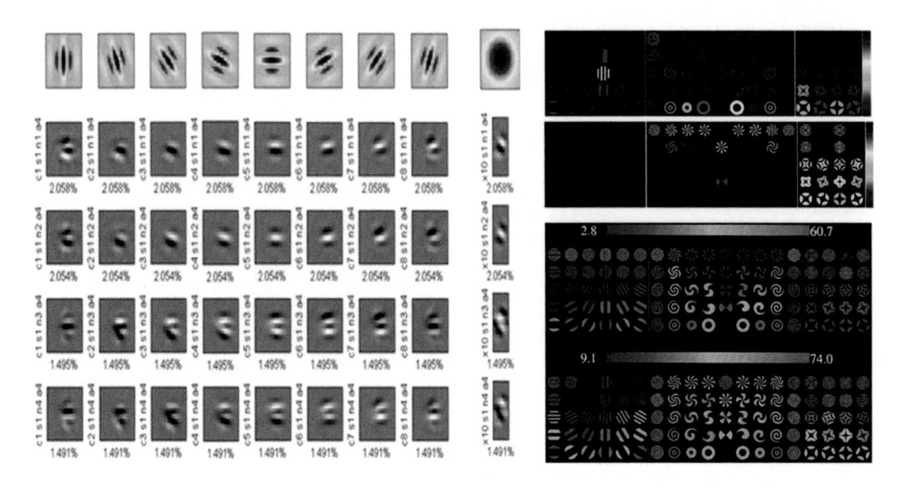

Fig. 3 At the next higher cortical level, similar Hebbian learning on the V1 output representation generates 4-dimensional wavelets (in x, y, θ, s, where s is scale). *Left* the top 4 eigenvectors obtained for a level-2 cell observing off-center rotation (the center of rotation is outside the receptive field, to the *left*). Each *row* is an eigenvector in 3D (x, y, θ: for ease of visualization scale was omitted). The first *8 columns* show x, y slices for the 8 different V1 orientations. The *last column* is a θ, y slice at $x =$ center, showing that the sinusoidal component of the wavelets is oriented along both y and θ. *Right* the responses of two model complex cells tuned to such 3D wavelets (*top*) and of two real V4 cells (Gallant et al. 1996) (*bottom*) to various stimuli used by Gallant et al. (1996). *Red/orange* indicates a high response and *blue/green* indicates a low response. Note that we have not attempted to match particular model cells to real cells. We note that by varying only the orientation of a 3D higher-order wavelet, we are able to obtain a wide variety of selectivity patterns

The original theory (Poggio et al. 2013) posits that local invariance is obtained in "complex" cells by pooling the outputs of several simple cells in a way similar to "energy models". The wiring between a group of simple cells with the same orientation and a complex cell may develop according to a Hebbian trace rule (Földiák 1991). Complex cells would thus inherit several of the properties of simple cells. Notice that a complex cell is invariant to translations in every direction even if its set of simple cells was "learned" while being exposed to motion in a specific direction (see Appendix 1.4). Thus the theory predicts the emergency of multiresolution analysis during development of V1 spanning a range of frequencies determined by a set of Gaussian distributions of synapses on dendritic trees with a range of σ *which is assumed to be present at the beginning of visual development*. More complex activity-dependent mechanisms than the Oja rule may automatically determine different sizes of receptive fields during development (Zylberberg et al. 2011; Rehn 2007): the details of the rules operating during development are of course less important than the experimental confirmation of a key role of Hebbian rules in determining *and/or maintaining* the tuning of V1 cells.

A similar set of assumptions about invariance and Hebbian synapses leads to wavelets-of-wavelets like shapes at higher layers, representing local shifts in the 4-cube of x, y, scale, orientation learned at the level of the simple cells in V1. Simulations show tuning that is qualitatively similar to physiology data in V2 and V4. A prediction that should be verifiable experimentally is that the tuning of cells in V2 corresponds to Gabor wavelets with a fixed relation between λ and σ in the four-dimensional cube of x, y, θ, s. Similar mechanisms, based on "simple" and "complex" cells modules can provide invariance to pose in face recognition; together with the Hebbian assumption, they may explain puzzling properties of neurons in one of the face patches recently found (Freiwald and Tsao 2010) in macaque IT (Leibo et al. 2013a, b).

In summary, we show that "universal" properties of simple cells in cortical V1 can be predicted in an almost parameter-free way by assuming that the computational goal of the ventral stream is to learn via Hebbian synapses how objects transform—during and after development—in order to later compute for each image a "signature" vector which is invariant to geometric transformations. Taking into account the statistics of natural images, we derive that the solutions of an associated "cortical equation" are Gabor-like wavelets with parameter values that closely agree with the physiology data across different species. Hebbian plasticity predicts the tuning of cells in V2, V4 and in the face patch AL, qualitatively in agreement with physiology data (Freiwald and Tsao 2010; Leibo et al. 2013b). The theory gives computational and biological foundations for previous theoretical and modeling work. It is important to notice that the emergence and maintenance of the tuning of simple cells is one of several predictions of the theory. The main result is the characterization of a class of systems that performs visual recognition at the human level, while accounting for the architecture of the ventral stream and for several tuning and invariance properties of the neurons in different areas. In fact, related architectures have been shown to perform well in computer vision recognition tasks and to mimic human performance in rapid categorization (Serre et al. 2007; Mutch and Lowe 2008; Krizhevsky et al. 2012). The results here are indirectly supported by Stevens' (2004) symmetry argument showing that preserving shape invariance to rotation, translation and scale changes requires simple cells in V1 perform a wavelet transform (Stevens also realized the significance of the Palmer and Ringach data and their "universality"). Similar indirect support can be found in Mallat's elegant mathematical theory of a scattering transform (Mallat 2012). Independent Component Analysis (ICA) (Bell and Sejnowski 1997; Hyvrinen and Oja 1998), Sparse Coding (SC) (Olshausen 1996) and similar unsupervised mechanisms (Saxe et al. 2011; Zylberberg et al. 2011; Rehn 2007; Olshausen et al. 2009) may result from plasticity rules similar to the simple Hebbian rule used in the theory described here. They can generate Gabor-like receptive fields (some of them fit the data and some other less so Ringach 2002) and they do not need the assumption of different sizes of Gaussian distributions of LGN synapses; however, the required biophysical mechanisms and circuitry are unknown or rather complex and, more importantly their motivation depends on sparsity whose computational and evolutionary significance is unclear—unlike our assumption of invariant recognition. It is interesting that in our theory a high level computational goal—invariant recognition—determines rather directly low-level properties of sensory cortical neurons.

Acknowledgments This work was supported by the Center for Brains, Minds and Machines (CBMM), funded by NSF STC award CCF 1231216.

1 Appendix

1.1 Retinal Processing

Our simulation pipeline consists of several filtering stages steps that mimic retinal processing, followed by a Gaussian mask, as shown in Fig. 4. Values for the DoG filter were those suggested by Croner and Kaplan (1995); the spatial lowpass filter has frequency response: $1/\sqrt{\omega_x^2 + \omega_y^2}$. The temporal derivative is performed using imbalanced weights (−0.95,1) so that the DC components is not zero. Each cells learns by extracting the principal components of a movie generated by a natural image patch undergoing a rigid translation. Each frame goes through the pipeline described here and is then fed to the unsupervised learning module (computing eigenvectors of the covariance). We used 40 natural images and 19 different Gaussian apertures for the simulations presented in this book chapter (Fig. 5).

1.2 Additional Evidence for Gabor Shapes as Templates in V1

In addition to Jones and Palmer (1987), Niell and Stryker (2008), Ringach (2002), a recent paper (Kay et al. 2008) shows that the assumption of a system of Gabor wavelets in V1 provides a very good fitting of fMRI data. Note that the templates of the theory described in Anselmi et al. (2013a) become during unsupervised learning (because of Hebbian synapses) Gabor-like eigenfunctions, as described here.

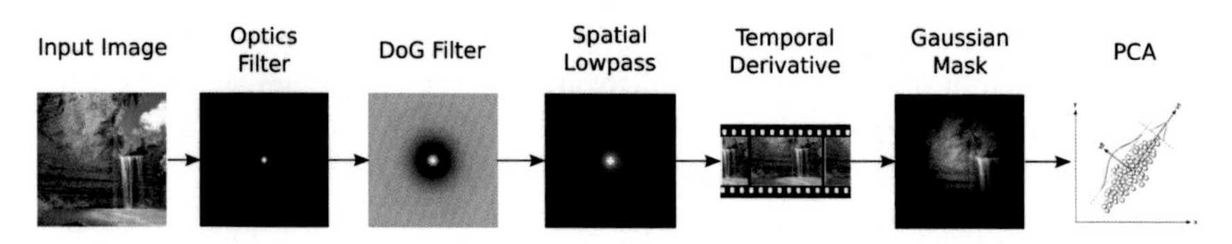

Fig. 4 Retinal processing pipeline used for V1 simulations. Though Gabor-like filters are obtained irrespectively of the presence or absence of any element of the pipeline the DoG filter is important in 1D and 2D for the emergence of actual Gabor wavelets with the correct dependence of λ on σ; the spatial low-pass filter together with the temporal derivative are necessary in our simulation to constrain λ to be proportional to σ

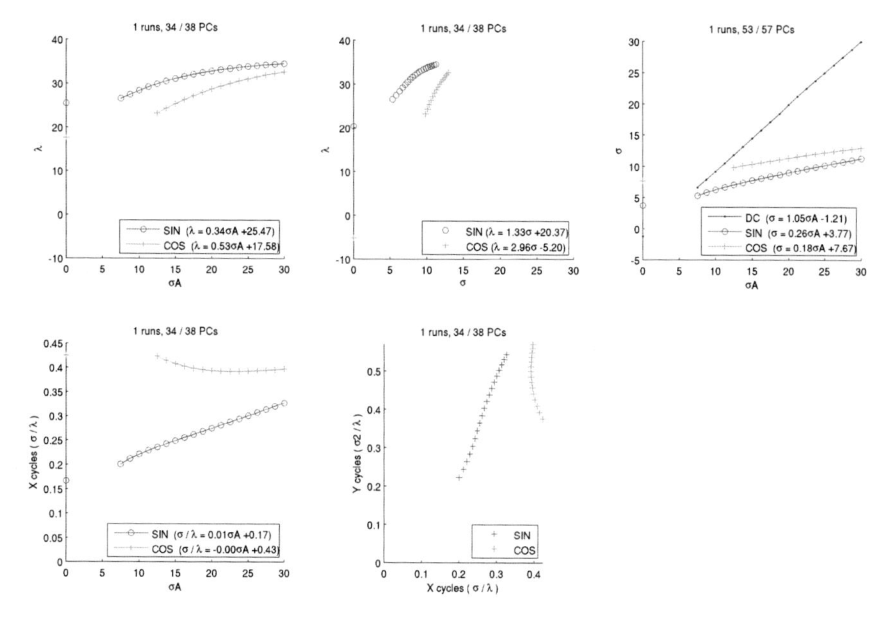

Fig. 5 Summary plots for $2D$ simulations of V1 cells trained according to the pipeline described in Fig. 1. Figures from *top left* to *bottom right*: sinusoid wavelength (λ) versus Gaussian aperture width (σ_α); sinusoid wavelength (λ) versus Gaussian envelope width on the modulated direction (σ); Gaussian envelope width for the modulated direction (σ) versus Gaussian aperture width (σ_α); ratio between sinusoid wavelength and Gaussian envelope width for the modulated direction (n_x) vs. Gaussian aperture width (σ_α); ratio between sinusoid wavelength and Gaussian envelope width on the unmodulated direction (n_y) versus ratio between sinusoid wavelength and Gaussian envelope width for the modulated direction (n_x). The pipeline consists of a Gaussian blur, a DOG filter, a spatial low-pass filter $1/\sqrt{\omega_x^2 + \omega_y^2}$ and an imperfect temporal derivative. Parameters for all filters were set to values measured in macaque monkeys by neurophysiologists

1.3 Hebbian Rule and Gabor-Like Functions

In this section we show how, from the hypothesis that the synaptic weights of a simple cell change according to a Hebbian rule, the tuning properties of the simple cells in V1 converge to Gabor-like functions.

We consider, for simplicity, the $1D$ case (see also Poggio et al. 2013 for a derivation and properties). The associated eigenproblem is

$$\int dx g(y) g(x) \psi_n(x) t^{\circledast}(y-x) = \nu_n \psi_n(y) \tag{2}$$

where $t^{\circledast}$ is the autocorrelation function of the template t, g is a gaussian function with fixed σ, ν_n are the eigenvalues and ψ_n are the eigenfunctions (see Poggio et al. 2012; Perona 1991 for solutions in the case where there is no gaussian).

1.3.1 Approximate Anzatz Solution for Piecewise Constant Spectrum

We start representing the template autocorrelation function as the inverse of its Fourier transform:

$$t^{\circledast}(x) = \frac{1}{\sqrt{2\pi}} \int d\omega\, t^{\circledast}(\omega) e^{i\omega x}. \tag{3}$$

Let $\alpha = 1/\sigma_x^2$, $\beta = 1/\sigma_\psi^2$ and assume that the eigenfunctions have the form $\psi_n(x) = e^{-\frac{\beta}{2}x^2} e^{i\omega_n x}$, where β and ω_n are parameters to be found. Assume also that $g(x) = \exp(-(\alpha/2)x^2)$. With these assumptions Eq. (2) reads:

$$\frac{1}{\sqrt{2\pi}} e^{-\frac{\alpha}{2}y^2} \int dx\, e^{-\frac{x^2(\alpha+\beta)}{2}} \int d\omega\, t^{\circledast}(\omega) e^{i\omega(y-x)} e^{iw_n x} = \nu(\omega_n) e^{-\frac{\beta y^2}{2}} e^{i\omega_n y}. \tag{4}$$

Collecting the terms in x and integrating in x we have that the l.h.s becomes:

$$\sqrt{\frac{1}{\alpha+\beta}} e^{-\frac{\alpha}{2}y^2} \int d\omega\, t^{\circledast}(\omega) e^{i\omega y} e^{-\frac{(\omega-\omega_n)^2}{2(\alpha+\beta)}}. \tag{5}$$

With the variable change $\bar{\omega} = \omega - \omega_n$ and in the hypothesis that $t^{\circledast}(\bar{\omega} + \omega_n) \approx const$ over the significant support of the Gaussian centered in 0, integrating in $\bar{\omega}$ we obtain:

$$\sqrt{2\pi} e^{-\frac{y^2\alpha}{2}} e^{i\omega_n y} e^{-\frac{y^2(\alpha+\beta)}{2}} \sim \nu(\omega_n) e^{-\frac{y^2\beta}{2}} e^{i\omega_n y}. \tag{6}$$

Notice that this implies an upper bound on β since otherwise t would be white noise which is inconsistent with the diffraction-limited optics of the eye. Thus the condition in Eq. (6) holds approximately over the relevant y interval which is between $-\sigma_\psi$ and $+\sigma_\psi$ and therefore Gabor functions are an approximate solution of Eq. (2).

We prove now that the orthogonality conditions of the eigenfunctions lead to Gabor wavelets. Consider, e.g., the approximate eigenfunction ψ_1 with frequency ω_0. The minimum value of ω_0 is set by the condition that ψ_1 has to be roughly orthogonal to the constant (this assumes that the visual input does have a DC component, which implies that there is no exact derivative stage in the input filtering by the retina).

$$\langle \psi_0, \psi_1 \rangle = C_{(0,1)} \int dx\, e^{-\beta x^2} e^{-i\omega_0 x} = 0 \Rightarrow e^{-\frac{\omega_0^2}{4\beta}} \approx 0 \tag{7}$$

where $C_{(0,1)}$ is the multiplication of the normalizing factors of the eigenfunctions.

Using $2\pi f_0 = \frac{2\pi}{\lambda_0} = \omega_0$ the condition above implies $e^{-(\frac{\pi\sigma_\psi}{\lambda_0})^2} \approx 0$ which can be satisfied with $\sigma_\psi \geq \lambda_0$; the condition $\sigma_\psi \sim \lambda_0$ is enough since it implies $e^{-(\frac{\pi\sigma_\psi}{\lambda_0})^2} \approx e^{-\pi^2}$.

Imposing orthogonality of any pair of eigenfunctions:

$$\int dx\, \psi_n^*(x)\psi_m(x) = const(m,n) \int dx e^{-\beta x^2} e^{in\omega_0 x} e^{-im\omega_0 x} \propto e^{-\frac{((m-n)\omega_0)^2 \sigma_\psi^2}{4}},$$

we have a similar condition to the above. This implies that λ_n should increase with σ_ψ of the Gaussian aperture, *which is a property of gabor wavelets!*, even if this is valid here only for $n = 0, 1, 2$.

1.3.2 Differential Equation Approach

In this section we describe another approach to the analysis of the cortical equation which is somewhat restricted but interesting for the potential connections with classical problems in mathematical physics.

Suppose as in the previous paragraph $g(x) = e^{-\frac{\alpha}{2}x^2}$. The eigenproblem (2) can be written as:

$$\nu_n \psi_n(y) - e^{-\frac{\alpha}{2}y^2} \int dx\, e^{-\frac{\alpha}{2}x^2} t^{\circledast}(y-x)\psi_n(x) = 0, \tag{8}$$

or equivalently, multiplying both sides by $e^{+\frac{\alpha}{2}y^2}$ and defining the function $\xi_n(x) = e^{+\frac{\alpha}{2}x^2}\psi_n(x)$, as

$$\nu_n \xi_n(y) - \int dx\, e^{-\alpha x^2} t^{\circledast}(y-x)\xi_n(x) = 0. \tag{9}$$

Decomposing $t^{\circledast}(x)$ as in Eq. (3) in Eq. (9):

$$\nu_n \xi_n(y) - \frac{1}{\sqrt{2\pi}} \int dx\, e^{-\alpha x^2} \int d\omega\, t^{\circledast}(\omega) e^{i\omega(y-x)} \xi_n(x) = 0.$$

Deriving twice in the y variable and rearranging the order of the integrals:

$$\nu_n \xi_n''(y) + \frac{1}{\sqrt{2\pi}} \int d\omega\, \omega^2 t^{\circledast}(\omega) e^{i\omega y} \int dx\, e^{-\alpha x^2} \psi_n(x) e^{-i\omega x} = 0. \tag{10}$$

The expression above is equivalent to the original eigenproblem in Eq. (2) and will provide the same ψ modulo a first order polynomial in x (we will show the equivalence in the next paragraph where we specialize the template to natural images).

Indicating with $\mathfrak{F}$ the Fourier transform we can rewrite (10) as:

$$\nu_n \xi_n''(y) + \sqrt{2\pi}\mathfrak{F}^{-1}\Big(\omega^2 t^{\circledast}(\omega)\mathfrak{F}\big(e^{-\alpha x^2}\psi_n(x)\big)\Big) = 0.$$

Indicating with $*$ the convolution operator by the convolution theorem

$$f * g = \mathfrak{F}^{-1}(\mathfrak{F}(f)\mathfrak{F}(g)), \quad \forall f, g \in L^2(\mathbb{R})$$

we have

$$v_n \xi_n^{''}(y) + \sqrt{2\pi}\mathfrak{F}^{-1}\left(\omega^2 t^{\circledast}(\omega)\right) * \left(e^{-\alpha x^2}\xi_n(x)\right) = 0.$$

Expanding $\omega^2 t^{\circledast}(\omega)$ in Taylor series, $\omega^2 t^{\circledast}(\omega) = \sum_i c_i \omega^i$ and remembering that $\mathfrak{F}^{-1}(\omega^m) = i^m \sqrt{2\pi}\delta^m(x)$ we are finally lead to

$$v_n \xi_n^{''}(y) + 2\pi(c_0\delta + ic_1\delta^{'} + \ldots) * \left(e^{-\alpha x^2}\xi_n(x)\right) = 0. \tag{11}$$

The differential equation so obtained is difficult to solve for a generic power spectrum. In the next paragraph we study the case where we can obtain explicit solutions.

Case: $1/\omega^2$ **Power Spectrum**

In the case of average natural images power spectrum

$$t^{\circledast}(\omega) = \frac{1}{\omega^2}$$

the differential equation (11) assumes the particularly simple form

$$v_n \xi_n^{''}(y) + 2\pi e^{-\alpha y^2}\xi_n(y) = 0. \tag{12}$$

In the harmonic approximation, $e^{-\alpha y^2} \approx 1 - \alpha y^2$ (valid for $\sqrt{\alpha}y \ll 1$) we have

$$\xi_n^{''}(y) + \frac{2\pi}{v_n}(1 - \alpha y^2)\xi_n(y) = 0. \tag{13}$$

The equation above is of the form of a so called Weber differential equation:

$$\xi^{''}(y) + (ay^2 + by + c)\xi(y) = 0, \quad a, b, c \in \mathbb{R}$$

The general solutions of Eq. (12) are:

$$\xi_n(y) = C_1 D\left(-\frac{1}{2} + \frac{\pi^{\frac{1}{2}}}{\sqrt{2\alpha v_n}}, \frac{2^{\frac{3}{4}}\alpha^{\frac{1}{4}}\pi^{\frac{1}{4}}}{v_n^{\frac{1}{4}}}\right) + C_2 D\left(-\frac{1}{2} - \frac{\pi^{\frac{1}{2}}}{\sqrt{2\alpha v_n}}, i\frac{2^{\frac{3}{4}}\alpha^{\frac{1}{4}}\pi^{\frac{1}{4}}}{v_n^{\frac{1}{4}}}\right)$$

where $D(\eta, y)$ are parabolic cylinder functions and C_1, C_2 are constants. It can be proved (Müller-Kirsten (2012), p. 139) that the solutions have two different behaviors, exponentially increasing or exponentially decreasing, and that we have exponentially decreasing real solutions if $C_2 = 0$ and the following quantization condition holds:

$$-\frac{1}{2} + \frac{\pi^{\frac{1}{2}}}{\sqrt{2\alpha v_n}} = n, \; n = 0, 1, \ldots$$

Therefore, remembering that $\alpha = 1/\sigma_x^2$, we obtain the spectrum quantization condition

$$\nu_n = 2\pi \frac{\sigma_x^2}{(2n+1)^2}, \; n = 0, 1, \ldots \tag{14}$$

Further, using the identity (true if $n \in \mathbb{N}$):

$$D(n, y) = 2^{-\frac{n}{2}} e^{-\frac{y^2}{4}} H_n\Big(\frac{y}{\sqrt{2}}\Big)$$

where $H_n(y)$ are Hermite polynomials, we have :

$$\xi_n(y) = 2^{-\frac{n}{2}} e^{-\frac{2n+1}{2\sigma_x^2} y^2} H_n\Big(\frac{\sqrt{2n+1}}{\sigma_x} y\Big)$$

i.e.

$$\psi_n(y) = 2^{-\frac{n}{2}} e^{-\frac{n+1}{\sigma_x^2} y^2} H_n\Big(\frac{\sqrt{2n+1}}{\sigma_x} y\Big) \tag{15}$$

Solutions plotted in Fig. 6 *very well approximate Gabor functions.*

Remark: The solution in Eq. (15) is also an approximate solution for any template spectrum such that

$$\omega^2 t^{\circledast}(\omega) = const + O(\omega)$$

This is important since it show how the solutions are robust to small changes of the power spectrum of the natural images.

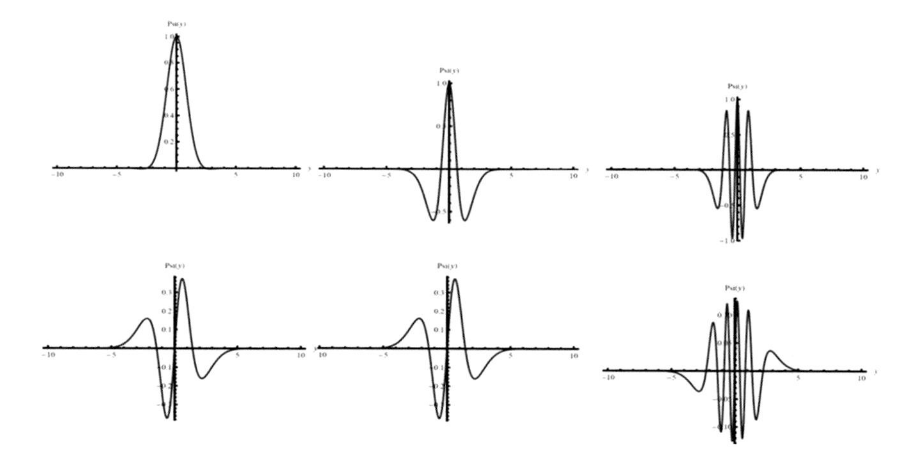

Fig. 6 Examples of odd and even solutions of the differential equation (12)

Aperture Ratio

Using Eq. (15) the ratio between the width of the Gaussian aperture and that of the eigenfunctions can be calculated as

$$\frac{\sigma_x}{\sigma_{\psi_n}} = \sqrt{n+1} \tag{16}$$

If we consider the first eigenfunction the ratio is $\sqrt{2}$.

Oscillating Behavior of Solutions

Although there isn't an explicit expression for the frequency of the oscillation part of ψ_n in Eq. (15) we can use the following approximation which calculates the frequency from the first crossing points of $H_n(x)$, i.e. $\pm\sqrt{2n+1}$, Boyd (1984). The oscillating part of (15) can be therefore written, in this approximation as

$$\cos\left(\frac{2n+1}{\sigma_x}y - n\frac{\pi}{2}\right)$$

which gives $\omega_n = (2n+1)/\sigma_x$. Using $\omega_n = 2\pi/\lambda_n$ and (16) we finally have

$$\frac{\lambda_n}{\sigma_{\psi_n}} = \frac{2\pi\sqrt{n+1}}{2n+1}.$$

The above equation gives an important information: for any fixed eigenfunction, $n = \bar{n}$ the number of oscillations under the gaussian envelope is constant.

Equivalence of the Integral and Differential Equations

To prove that the solutions of the eigenproblem (9) are equivalent to those of (13) we start with the observation that in the case of natural images power spectrum we can write explicitly (3):

$$t^{\circledast}(y-x) = \frac{1}{\sqrt{2\pi}}\int d\omega\; t^{\circledast}(\omega)e^{i\omega(y-x)} = \frac{1}{\sqrt{2\pi}}\int d\omega\;\frac{e^{i\omega(y-x)}}{\omega^2} = -\sqrt{\frac{\pi}{2}}|y-x|. \tag{17}$$

The integral equation (9) can be written for for $a > 0,\ a \to \infty$

$$\xi(y) + c\int_{-a}^{a} e^{-\alpha x^2}|y-x|\xi(x)dx$$

where for simplicity we dropped the index n and $c = \sqrt{\pi}/(\sqrt{2}\nu)$. Removing the modulus

$$\xi(y) + c\int_{-a}^{y} e^{-\alpha x^2}(y-x)\xi(x)dx + c\int_{y}^{a} e^{\alpha x^2}(x-y)\xi(x)dx.$$

Putting $y = a,\ y = -a$ we can derive the boundary conditions

$$\xi(a) + c\int_{-a}^{a} e^{-\alpha x^2}(a - x)\xi(x)dx$$
$$\xi(-a) + c\int_{-a}^{a} e^{-\alpha x^2}(a + x)\xi(x)dx$$

Substituting, using the differential equation, $e^{-\alpha x^2}\xi(x)$ with $-\xi''(x)\nu/2\pi$ and integrating by parts we have

$$\xi(a) + \xi(-a) + 2ac'\xi'(-a) = 0$$
$$\xi(-a) + \xi(-a) - 2ac'\xi'(a) = 0$$

where $c' = 1/2\sqrt{2}$. The above two boundary conditions together with the differential equation are equivalent to the initial integral eigenproblem. If we want bounded solutions at infinity: $\xi(\infty) = \xi(-\infty) = \xi'(\infty) = \xi'(-\infty) = 0$.

1.4 Motion Determines a Consistent Orientation of the Gabor-Like Eigenfunctions

Consider a $2D$ image moving through time t, $I(x(t), y(t)) = I(\mathbf{x}(t))$ filtered, as in pipeline of Fig. 4, by a spatial low-pass filter and a band-pass filter and call the output $f(\mathbf{x}(t))$.

Suppose now a temporal filter is done by a high-pass impulse response $h(t)$. For example, let $h(t) \sim \frac{d}{dt}$. We consider the effect of the time derivative over the translated signal, $\mathbf{x}(t) = \mathbf{x} - \mathbf{v}t$ where $\mathbf{v} \in \mathbb{R}^2$ is the velocity vector

$$\frac{\mathrm{d}f(\mathbf{x}(t))}{\mathrm{d}t} = \nabla f(\mathbf{x}(t)) \cdot \mathbf{v}. \tag{18}$$

If, for instance, the direction of motion is along the x axis with constant velocity, $\mathbf{v} = (\mathbf{v_x}, \mathbf{0})$, then Eq. (18) become

$$\frac{\mathrm{d}f(\mathbf{x}(t))}{\mathrm{d}t} = \frac{\partial f(\mathbf{x}(t))}{\partial x}v_x,$$

or, in Fourier domain of spatial and temporal frequencies:

$$\hat{f}(i\omega_t) = iv_x\omega_x\hat{f}. \tag{19}$$

Consider now an image I with a symmetric spectrum $1/(\sqrt{\omega_x^2 + \omega_y^2})$. Equation (19) shows that the effect of the time derivative is to break the radial symmetry of the spectrum in the direction of motion (depending on the value of v_x). Intuitively, spatial frequencies in the x direction are enhanced. Thus motion effectively selects a specific orientation since it enhances the frequencies orthogonal to the direction of motion in Eq. (1).

Thus the theory suggests that motion effectively "selects" the direction of the Gabor-like function (see previous section) during the emergence and maintenance of a simple cell tuning. It turns out that in addition to orientation other features of the eigenvectors are shaped by motion during learning. This is shown by an equivalent simulation to that presented in Fig. 1 but in which the order of frames was scrambled before the time derivative stage. The receptive fields are still Gabor-like functions but lack the important property of having $\sigma_x \propto \lambda$. This is summarized in Fig. 7.

The theory also predicts—assuming that the cortical equation provides a perfect description of Hebbian synapses—that the even eigenfunctions have slightly different n_x, n_y relations than odd eigenfunctions. It is unlikely that experiments data may allow to distinguish this small difference.

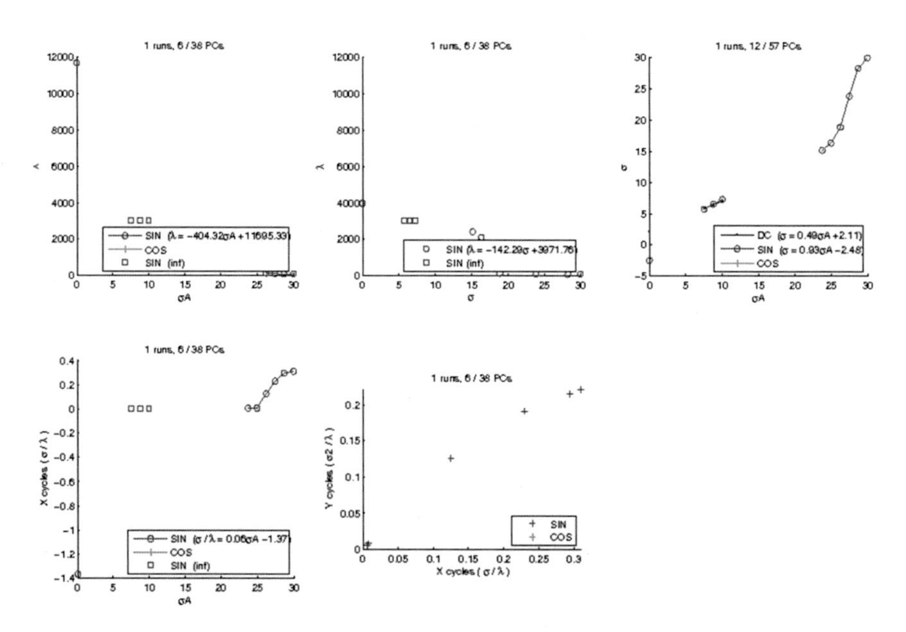

Fig. 7 Summary plots for 2*D* simulations of V1 cells trained according to the pipeline described in Figs. 1 and 4 but scrambling the order of frames before the temporal derivative. Figures from *top left* to *bottom right*: sinusoid wavelength (λ) versus Gaussian aperture width (σ_α); sinusoid wavelength (λ) versus Gaussian envelope width on the modulated direction (σ); Gaussian envelope width for the modulated direction (σ) versus Gaussian aperture width (σ_α); ratio between sinusoid wavelength and Gaussian envelope width for the modulated direction (n_x) versus Gaussian aperture width (σ_α); ratio between sinusoid wavelength and Gaussian envelope width on the unmodulated direction (n_y) versus ratio between sinusoid wavelength and Gaussian envelope width for the modulated direction (n_x). For an explanation of the details of this figure see Poggio et al. (2013)

1.5 Phase of Gabor RFs

We do not analyze here the phase for a variety of reasons. The main reason is that phase measurements are rather variable in each species and across species. Phase is also difficult to measure. The general shape shows a peak in 0 and a higher peak at 90. These peaks are consistent with the $n = 2$ eigenfunctions (even) and the $n = 1$ eigenfunctions (odd) of Eq. 2 (the zero-th order eigenfunction is not included in the graph). The relative frequency of each peak would depend, according to our theory, on the dynamics (Oja equation) of learning and on the properties of the lateral inhibition between simple cells (to converge to eigenfunctions other than the first one). It is in any case interesting that the experimental data fit *qualitatively* our predictions: the ψ_1 odd eigenfunction of the cortical equation should appear more often (because of its larger power) than the even ψ_2 eigenfunction and no other ones with intermediate phases should exist—at least in the noise-less case (Fig. 8).

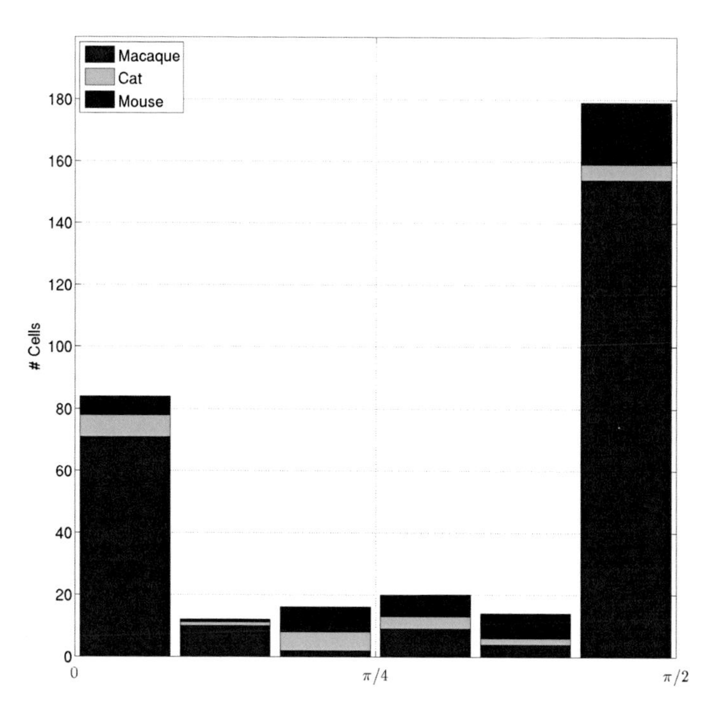

Fig. 8 Data from Jones and Palmer (1987) (Cat), Ringach (2002) (Macaque) and Niell and Stryker (2008) (Mouse). Here zero phase indicates even symmetry for the Gabor like wavelet

References

Abdel-Hamid O, Mohamed A, Jiang H, Penn G (2012) Applying convolutional neural networks concepts to hybrid nn-hmm model for speech recognition. In: 2012 IEEE International conference on acoustics, speech and signal processing (ICASSP), pp 4277–4280. IEEE

Anselmi F, Leibo JZ, Mutch J, Rosasco L, Tacchetti A, Poggio T (2013a) Part I: computation of invariant representations in visual cortex and in deep convolutional architectures. In preparation

Anselmi F, Leibo JZ, Rosasco L, Mutch J, Tacchetti A, Poggio T (2013b) Unsupervised learning of invariant representations in hierarchical architectures. Theoret Comput Sci. CBMM Memo n 1, in press. arXiv:1311.4158

Anselmi F, Poggio T (2010) Representation learning in sensory cortex: a theory. CBMM memo n 26

Bell A, Sejnowski T (1997) The independent components of natural scenes are edge filters. Vis Res 3327–3338

Boyd J (1984) Asymptotic coefficients of hermite function series. J Comput Phys 54:382–410

Croner L, Kaplan E (1995) Receptive fields of p and m ganglion cells across the primate retina. Vis Res 35(1):7–24

Dan Y, Atick JJ, Reid RC (1996) Effcient coding of natural scenes in the lateral geniculate nucleus: experimental test of a computational theory. J Neurosci 16:3351–3362

Földiák P (1991) Learning invariance from transformation sequences. Neural Comput 3(2):194–200

Freiwald W, Tsao D (2010) Functional compartmentalization and viewpoint generalization within the macaque face-processing system. Science 330(6005):845

Fukushima K (1980) Neocognitron: a self-organizing neural network model for a mechanism of pattern recognition unaffected by shift in position. Biol Cybern 36(4):193–202

Gallant J, Connor C, Rakshit S, Lewis J, Van Essen D (1996) Neural responses to polar, hyperbolic, and Cartesian gratings in area V4 of the macaque monkey. J Neurophysiol 76:2718–2739

Hebb DO (1949) The organization of behaviour: a neuropsychological theory. Wiley

Hyvrinen A, Oja E (1998) Independent component analysis by general non-linear hebbian-like learning rules. Signal Proces 64:301–313

Jones JP, Palmer LA (1987) An evaluation of the two-dimensional Gabor filter model of simple receptive fields in cat striate cortex. J Neurophysiol 58(6):1233–1258

Kay K, Naselaris T, Prenger R, Gallant J (2008) Identifying natural images from human brain activity. Nature 452(7185):352–355

Krizhevsky A, Sutskever I, Hinton G (2012) Imagenet classification with deep convolutional neural networks. Adv Neural Inf Proces Syst 25

Le QV, Monga R, Devin M, Corrado G, Chen K, Ranzato M, Dean J, Ng AY (2011) Building high-level features using large scale unsupervised learning. CoRR. arXiv:1112.6209

LeCun Y, Boser B, Denker J, Henderson D, Howard R, Hubbard W, Jackel L (1989) Backpropagation applied to handwritten zip code recognition. Neural Comput 1(4):541–551

LeCun Y, Bengio Y (1995) Convolutional networks for images, speech, and time series. The handbook of brain theory and neural networks, pp 255–258

Leibo JZ, Anselmi F, Mutch J, Ebihara AF, Freiwald WA, Poggio T (2013a) View-invariance and mirror-symmetric tuning in a model of the macaque face-processing system. Comput Syst Neurosci I-54. Salt Lake City, USA

Leibo JZ, Anselmi F, Mutch J, Ebihara AF, Freiwald WA, Poggio T (2013b) View-invariance and mirror-symmetric tuning in a model of the macaque face-processing system. Comput Syst Neurosci (COSYNE)

Li N, DiCarlo JJ (2008) Unsupervised natural experience rapidly alters invariant object representation in visual cortex. Science 321(5895):1502–1507

Mallat S (2012) Group invariant scattering. Commun Pure Appl Math 65(10):1331–1398

Meister M, Wong R, Baylor DA, Shatz CJ et al (1991) Synchronous bursts of action potentials in ganglion cells of the developing mammalian retina. Science 252(5008):939–943

Mel BW (1997) SEEMORE: combining color, shape, and texture histogramming in a neurally inspired approach to visual object recognition. Neural Comput 9(4):777–804
Müller-Kirsten HJW (2012) Introduction to quantum mechanics: Schrödinger equation and path integral, 2nd edn. World Scientific, Singapore
Mutch J, Lowe D (2008) Object class recognition and localization using sparse features with limited receptive fields. Int J Comput Vis 80(1):45–57
Niell C, Stryker M (2008) Highly selective receptive fields in mouse visual cortex. J Neurosci 28(30):7520–7536
Oja E (1982) Simplified neuron model as a principal component analyzer. J Math Biol 15(3):267–273
Oja E (1992) Principal components, minor components, and linear neural networks. Neural Netw 5(6):927–935
Olshausen BA, Cadieu CF, Warland D (2009) Learning real and complex overcomplete representations from the statistics of natural images. In: Goyal VK, Papadakis M, van de Ville D (eds) SPIE Proceedings, vol. 7446: Wavelets XIII
Olshausen B et al (1996) Emergence of simple-cell receptive field properties by learning a sparse code for natural images. Nature 381(6583):607–609
Perona P (1991) Deformable kernels for early vision. IEEE Trans Pattern Anal Mach Intell 17:488–499
Perrett D, Oram M (1993) Neurophysiology of shape processing. Image Vis Comput 11(6):317–333
Pinto N, DiCarlo JJ, Cox D (2009) How far can you get with a modern face recognition test set using only simple features? In: CVPR 2009. IEEE Conference on computer vision and pattern recognition, 2009. IEEE, pp 2591–2598
Poggio T, Edelman S (1990) A network that learns to recognize three-dimensional objects. Nature 343(6255):263–266
Poggio T, Mutch J, Anselmi F, Leibo JZ, Rosasco L, Tacchetti A (2011) Invariances determine the hierarchical architecture and the tuning properties of the ventral stream. Technical report available online, MIT CBCL, 2013. Previously released as MIT-CSAIL-TR-2012-035, 2012 and in Nature Precedings, 2011
Poggio T, Mutch J, Anselmi F, Leibo JZ, Rosasco L, Tacchetti A (2012) The computational magic of the ventral stream: sketch of a theory (and why some deep architectures work). Technical report MIT-CSAIL-TR-2012-035, MIT Computer Science and Artificial Intelligence Laboratory, 2012. Previously released in Nature Precedings, 2011
Poggio T, Mutch J, Isik L (2014) Computational role of eccentricity dependent cortical magnification. CBMM Memo No. 017. CBMM Funded. arXiv:1406.1770v1
Rehn M, Sommer FT (2007) A network that uses few active neurones to code visual input predicts the diverse shapes of cortical receptive fields. J Comput Neurosci 22(2):135–146
Riesenhuber M, Poggio T (1999) Hierarchical models of object recognition in cortex. Nature Neurosci. 2(11):1019–1025
Ringach D (2002) Spatial structure and symmetry of simple-cell receptive fields in macaque primary visual cortex. J Neurophysiol 88(1):455–463
Saxe AM, Bhand M, Mudur R, Suresh B, Ng AY (2011) Unsupervised learning models of primary cortical receptive fields and receptive field plasticity. In: Shawe-Taylor J, Zemel R, Bartlett P, Pereira F, Weinberger K (eds) Advances in neural information processing systems, vol 24, pp 1971–1979
Serre T, Wolf L, Bileschi S, Riesenhuber M, Poggio T (2007) Robust object recognition with cortex-like mechanisms. IEEE Trans Pattern Anal Mach Intell 29(3):411–426
Stevens CF (2004) Preserving properties of object shape by computations in primary visual cortex. PNAS 101(11):15524–15529
Stringer S, Rolls E (2002) Invariant object recognition in the visual system with novel views of 3D objects. Neural Comput 14(11):2585–2596
Torralba A, Oliva A (2003) Statistics of natural image categories. In: Network: computation in neural systems, pp 391–412

Turrigiano GG, Nelson SB (2004) Homeostatic plasticity in the developing nervous system. Nature Rev Neurosci 5(2):97–107

Wong R, Meister M, Shatz C (1993) Transient period of correlated bursting activity during development of the mammalian retina. Neuron 11(5):923–938

Zylberberg J, Murphy JT, DeWeese MR (2011) A sparse coding model with synaptically local plasticity and spiking neurons can account for the diverse shapes of v1 simple cell receptive fields. PLoS Comput Biol, 7(10):135–146

Speed Versus Accuracy in Visual Search: Optimal Performance and Neural Implementations

Bo Chen and Pietro Perona

This chapter is about the psychophysics of visual search.

What is **psychophysics**? In its traditional definition, psychophysics studies how mental states (psycho) are affected by physical stimuli (physics). For instance, a cute dog (physical stimulus) may elicit a feeling of warmth and affection (psycho state), and psychophysics studies how specific attributes (breed, size, smell, etc.) of the dog relate to our feelings.

Additionally, in this chapter we would also consider psychophysics as the "physics of psychology" and study how a psychological process may be implemented by a physical system. For instance, upon seeing the puppy we may wonder whether it is safe to pet it. This process of reasoning has a physical substrate—it is the outcome of the computations performed by an enormous network of neurons in our brain. Just like physicists postulate fundamental particles and their interactions that give rise to matters, we will postulate fundamental units of computation and their architecture that give rise to decisions.

What is **visual search**? Visual search is the problem of looking for a target object amongst clutters or distractors. It is a common task for our everyday life (looking for keys on a desk, friends in a crowd or signs on a map) and a vital function for animals in the wild (searching for food, mate, threats).

The focus of this chapter is to explain the **speed versus accuracy** trade-off (SAT) of visual search. First, visual search is difficult and error-prone: the sensory signal is often noisy; the relevant objects, whose appearance may not be entirely known in advance, are often embedded in irrelevant clutter, whose appearance and complex-

---

Disclaimer: *Most of the materials below are adapted from the following publication: B. Chen and P. Perona, Speed versus accuracy in visual search: Optimal performance and neural architecture. Journal of Vision 15(16), 9–9.*

B. Chen · P. Perona (✉)
Department of Electrical Engineering and of Computation and Neural Systems,
California Institute of Technology, Pasadena, CA, USA
e-mail: perona@caltech.Edu

Q. Zhao (ed.), *Computational and Cognitive Neuroscience of Vision*,
Cognitive Science and Technology, DOI 10.1007/978-981-10-0213-7_6

ity may also be unknown. Thus to reduce detection errors the visual system must account for the noise structure of the sensors and the uncertainty of the environment. Second, **time** is of the essence: the ability to detect quickly objects of interest is an evolutionary advantage. Speed comes at the cost of making more errors. Thus, it is critical that each piece of sensory information is used efficiently to produce a decision in the shortest amount of time while maintaining the probability of errors within an acceptable limit.

A note for readers who are engineering-minded: understanding how humans tradeoff speed versus accuracy in visual search could help us unveil the fundamental computational principles the brain uses in situations where information is scarce and when time is expensive. These principles could guide us in designing better machines for similar problems. For example, autonomous driving vehicles rely on object detectors to avoid obstacles and pedestrians. The quality of the detector's input images, and in turn its accuracy, is affected by the length of the exposure time, especially at night when the ambient light level is low. At the same time, the detector must maintain a low delay to be useful. The detector thus faces the same problem of speed (exposure time) versus accuracy tradeoff, and could benefit from the computational principles discussed in this chapter.

1 The Phenomenology of Visual Search

There are two crucial quantities in visual search: the **response time** (RT, how long after an observer is exposed to a scene before it generates a response) and the **error rates** (ER). The error rates include the **false positive rate** (FPR), which is the fraction of times when the observer claims to have found a target even though the scene does not contain any, and the **false negative rate** (FNR), which is the fraction of times when the observer claims no target when there is one. We are interested in how these quantities are affected by the structure of the search task.

The general **set-up** of a visual search task is as shown in Fig. 1a. An observer sits down in front of a computer monitor. The monitor display a series of images that consist of distractors and sometimes targets. The goal of the observer is to decide whether a target object is present in a cluttered image as quickly and accurately as possible while maintaining fixation at the center of the image. The decision is **binary**, and the two categories of stimuli are: target-present ($C = 1$) and target-absent ($C = 0$), as shown in Fig. 1b. When the target is present, its location is not known in advance; it may be one of L locations in the image. The observer only reports whether the target appears, but not where. For now, we limit the number of targets to be at most **one**.

In our experiments the target and distractor objects appear at M locations ($M \leq L$) in each image where M reflects the complexity of the image and is known as the **"set-size"**. The objects are simplified to be oriented bars, and the only feature in which the target and distractor differ is orientation. Target distinctiveness is controlled by the difference in orientation between target and distractors, the

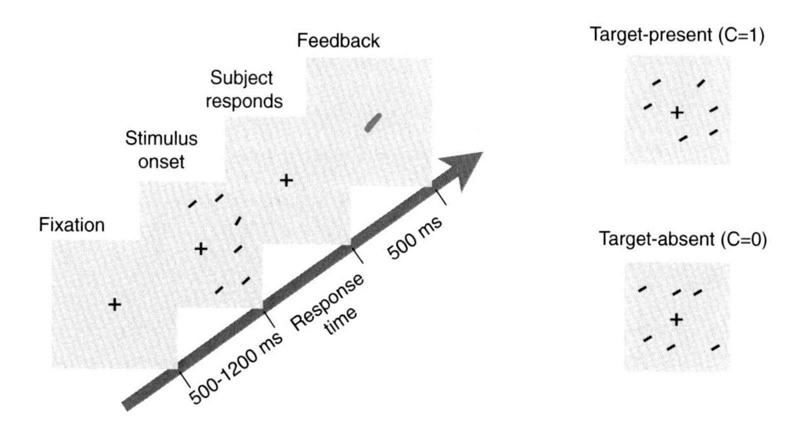

Fig. 1 Visual search setup **a** Each trial starts with a fixation screen. Next, the "stimulus" is displayed. The stimulus is an image containing M oriented bars that are positioned in M out of L possible display locations ($M = 6, L = 12$ in this example). One of the bars may be the target. The stimulus disappears as soon as the subject responds by pressing one of two keys, to indicate whether a target was detected or not. Feedback on whether the response was correct is the presented on the screen, which concludes the trial. The subject were instructed to maintain center-fixation at all times and respond as quickly and as accurately as possible. **b** Examples of a target-present trial stimulus and a target-absent trial stimulus

"orientation-difference" $\Delta\theta$. Prior to image presentation, the set of possible orientations for the target and the distractor is known, whereas the set-size and orientation-difference may be unknown, and may change from one image to the next.

In this design we strive to have the simplest experiment that captures all the relevant variables, namely the dependent variables RT and ERs, as well as the independent variables the set-size M and the orientation-difference $\Delta\theta$. To do so we first simplify the appearance of the stimuli so that we can focus on modeling search strategies instead of building classifiers. Second, we eliminate eye-movements by forcing fixation at the center of the image at all times because saccade planning is a rich phenomenon of its own that many are struggling to explain. Third, we have randomized the placement of the targets and the distractors (details in Sect. 4), duration between trials, and stimulus orientation etc. to eliminate potential biases.

The visual search literature records a rich set of phenomena that regarding the RT and ERs of human observers. We list three in Fig. 2. An intuitive phenomenon is the "set-size effect". As the amount of clutter increases in the display, the subject tends to take longer to respond. The slope of RT with respect to the set-size M depends on the distinctiveness between the target and the distractor $\Delta\theta$. The smaller is $\Delta\theta$, the more difficult is the task and the larger the slope. A less intuitive phenomenon is the "search asymmetry effect" that the slope for target-absent is roughly twice the slope for target-present (Many other dependent variables display the set-size effect and search asymmetry, interested reader is referred to Palmer (1994)). Lastly, the RT distributions is heavy-tailed: the log RTs roughly follow a Gaussian distribution. The list of phenomena goes on.

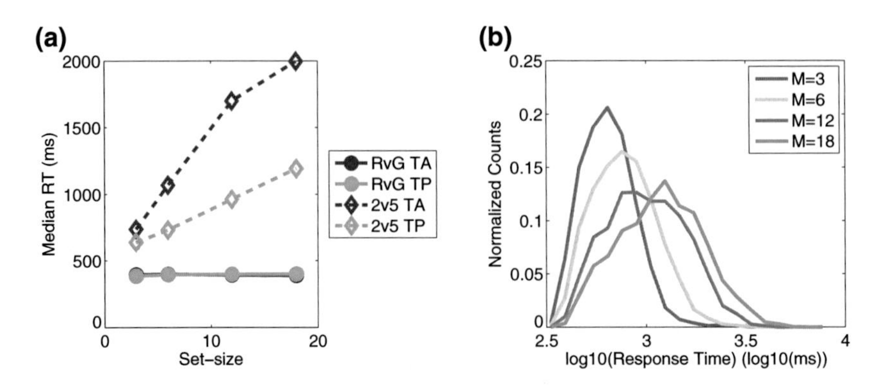

Fig. 2 Selected list of visual search phenomena **a** The "set-size" effect. Median RT increases linearly with set-size. The slope depends on the trial type (target-absent trials have roughly twice the slope) and task difficulty. The two tasks are searching for a red bar among green bars (easy) and searching for a "2" among "5"s (hard). **b** RT histograms for different set-sizes ($\{3, 6, 12, 18\}$), plotted in log domain based 10

Psychophysicists have put forward many models that can describe a subset of the phenomena fairly well, but most models fall short in accounting for all the phenomena. Describing all phenomena in one model is a challenging task. The model needs to be flexible enough to accommodate changes of the environment, e.g. different set-sizes, or different probability distributions on the set-sizes, etc. In addition, the model needs to be efficient enough so that it can be easily transferred from one environment to the next. Furthermore, there are countless unintended events, such as the subject blinking, getting fatigued or being distracted, that could pollute the behavioral data.

Example 1.1 One example of existing theories that can only partially describe visual search behaviors is the Feature-integration-theory (Treisman and Gelade 1980). It says that there are two stages of search: a pre-attentive stage where objects are analyzed for details in parallel, and a focused attention stage where each object is inspected one after another. With some revision this theory can explain both the set-size effect and search asymmetry. When the target is distinctive, search happens in the pre-attentive stage and thus RT is independent of set-size. When the target is barely separable from the distractor, each object requires focused attention and search enters the second stage, where RT is linear in set-size. When target is absent, an observer needs to scan all objects before declaring target-absent. When target is present, the observer only needs to scan to the target, and on average, half of the objects. Therefore, the RT slope for target-absent is twice that of target-present. However, this reasoning also implies that when target is absent, all the trials will take roughly the same amount of time, namely the time it takes to scan all objects, while RTs in the target-present trials will have more variation due to the uncertainty in the target location. However, as Wolfe et al. (2010) pointed out, the target-absent RTs follow a log-normal-like distribution and are as variable as the target-present trials. Therefore, the intuitive theory of two-stage search fails to explain the RT histograms.

Therefore, instead of describing human behaviors in a variety of visual search problems, we seek to study the **optimal** behavior on a per-situation basis. The optimal behavior can be used as a gold standard to measure human performance. The question we ask is, given input observations and prior knowledge about the task, what is the best ER versus RT tradeoff (optimality will be defined later) that an observer could achieve? The process of solving for the optimal behavior is called the "ideal observer analysis" (Sect. 2), which we see below.

2 Ideal Observers

2.1 *Sensory Input*

The first piece of the ideal observer analysis is to identify the input to the problem. We consider sensory input from the early stages of the visual system (retina, LGN and primary visual cortex), where raw images are processed and converted into a stochastic stream of observations (details below). The anatomy, as well as the physiology, of these stages are well characterized by Hubel and Wiesel (1962). These mechanisms compute local properties of the image, such as color contrast, orientation, spatial frequency, stereoscopic disparity and motion flow (Felleman and Essen 1991), and communicate these properties to downstream neurons for further processing. The communication takes the form of sequences of action potentials/spikes from orientation-selective neurons in V1 (Hubel and Wiesel 1962).

The firing patterns of the neurons are modeled with an homogeneous Poisson process (Sanger 1996). This means that each neuron fires at a fixed rate of λ spikes /second given the input image, and the timings of the spikes are independent of each other. More specifically, the number n of events (i.e. action potentials) that will be observed during one second is distributed as

$$P(n|\lambda) = \lambda^n e^{-\lambda}/n!$$

The firing patterns are produced over the time interval $[0, t]$ by a population of N neurons, also known as a **"hypercolumn"**, from each of the L display locations. Each neuron has a localized spatial receptive field and is tuned to local image properties (Hubel and Wiesel 1962), which in our case is the local stimulus orientation; the preferred orientations of neurons within a hypercolumn are distributed uniformly in $[0°, 180°)$. λ_θ^i, the expected firing rate of the i-th neuron is a function of the neuron's preferred orientation θ_i and the stimulus orientation $\theta \in [0°, 180°)$:

$$\lambda_\theta^i = (\lambda_{max} - \lambda_{min}) \exp\left(-\frac{1}{2\psi^2}\left(\min_{k=-1,0,1}(|\theta - \theta_i + k180°|)\right)^2\right) + \lambda_{min} \quad (1)$$

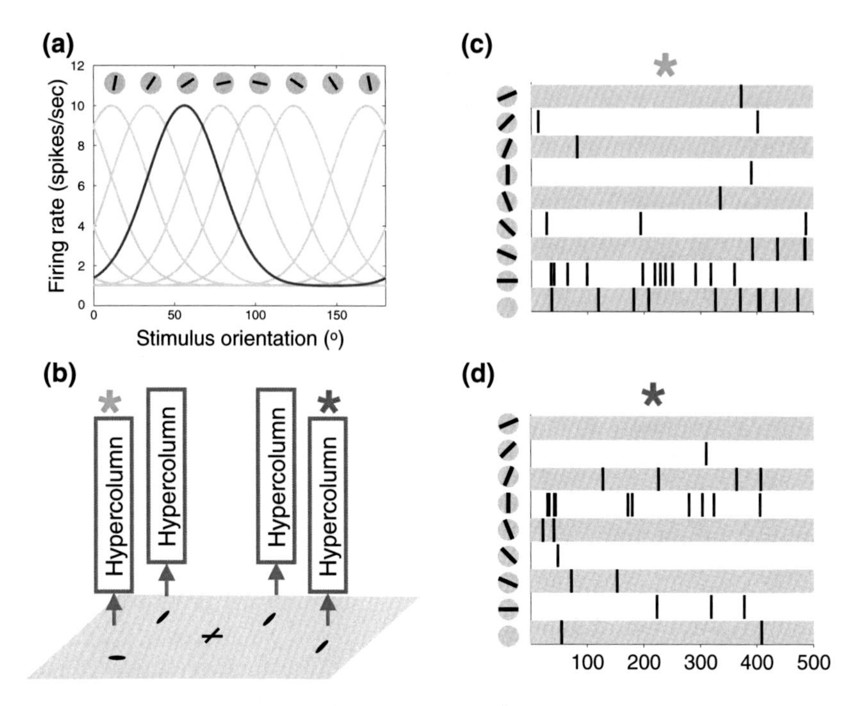

Fig. 3 V1 Hypercolumns **a** Orientation tuning curves λ_θ^i (Eq. 1) of a hypercolumn consisting of $N = 8$ neurons with half tuning width $\psi = 22°$, minimum firing rate $\lambda_{min} = 1$ Hz and maximum firing rate $\lambda_{max} = 10$ Hz. **b** V1 hypercolumns tessellate the input space, one for each visual location where an object (oriented bar) may appear. **c, d** Spike trains X_t^l at the target location (marked with *green star* in (**b**)) and a distractor location (*red star*)

(in spikes per second, or *Hz*) where λ_{min} and λ_{max} are a neuron's minimum and maximum firing rates, and $\psi \in (0°, 180°)$ is the half tuning width. Figure 3a shows the tuning functions of a hypercolumn of 8 neurons, Fig. 3b shows the spatial organization of the hypercolumns, and Fig. 3c, d shows the sample spike trains from two locations with different local stimulus orientations.

Why do we select the response of V1 hypercolumn neurons to be our input? Indeed there are multiple alternatives: the raw image, the response of the retina or LGN, and high-level signals that directly encode information regarding target presence. Our choice is based on flexibility and efficiency. Since the search problems considered here all involve just oriented bars placed distant apart, it would be redundant to model the neuronal hardware that gives rise to orientation-selectivity at this stage. Therefore, our level of abstraction should start at least from V1. On the other hand, although most visual search models assume high-level input signals, as in Palmer et al. (2000), Verghese (2001), Wolfe (2007), Purcell et al. (2012), they are not concerned with behaviors across multiple visual search tasks. As we see later, we will interpret the input from V1 neurons depending on the probabilistic structure of the task, which is a key to the generalizability of the ideal observer.

Why do we use a Poisson process to model the V1 spike trains? While Gaussian firing rate models (Verghese 2001) have also been used in the past, the Poisson model represents more faithfully the spiking nature of neurons (Sanger 1996; Beck et al. 2008; Graf et al. 2011). Second, we do not using electrophysiological recordings from V1 neurons (Graf et al. 2011) because large scale recordings from the entire V1 is not currently possible.

2.2 *Optimality*

Now that we have specified the assumptions regarding sensory input, we are ready to define optimality. Given the firing pattern of V1 neurons that is caused by the image, an observer faces a double decision. First, at each time instant it has to decide whether the information in the input collected so far is sufficient to detect the target. Second, once information is deemed sufficient, it has to decide whether the target is present or not. Moreover, as we have discussed before, RT, the amount of time the observer spends in collect information, trades off with ER, the number of errors.

An observer that optimally trades off ER versus RT is called an "ideal observer". Optimality is measured in terms of the Bayes risk (Wald and Wolfowitz 1948; Busemeyer and Rapoport 1988):

$$\text{BayesRisk} = \mathbb{E}[\text{RT}] + C_p\mathbb{E}[\text{Declare } C = 1|C = 0] + C_n\mathbb{E}[\text{Declare } C = 0|C = 1] \tag{2}$$

where $\mathbb{E}[\text{RT}]$ is the expected RT, $\mathbb{E}[\text{Declare } C = 1|C = 0]$ is the probability of the observer making a target-present decision when the target is absent, or the false positive rate, and likewise $\mathbb{E}[\text{Declare } C = 0|C = 1]$ is the false negative rate. C_p and C_n are two free parameters: the cost (in seconds) of false positive errors and the cost of false negative errors.

For example, C_p might be quantified in terms of the time wasted exploring an unproductive location while foraging for food, and C_n may be the time it takes to move to the next promising location. The relative cost of errors and time is determined by the circumstances in which the animal operates. For example, an animal searching for scarce food while competing with conspecifics will face a high cost of time (any delay in pecking a seed will mean that the seed is lost) and low cost of error (pecking on a pebble rather than a seed just means that the pebble can be spat out). Conversely, an airport luggage inspector faces high false reject error costs and comparatively lower time costs. C_p and C_n determine how often the observer is willing to make one type of error vs. the other, and vs. waiting for more evidence. Thus, the Bayes risk measures the combined RT and ER costs of a given search mechanism. Given a set of inputs, the ideal observer is the mechanism minimizing such cost (Fig. 4a).

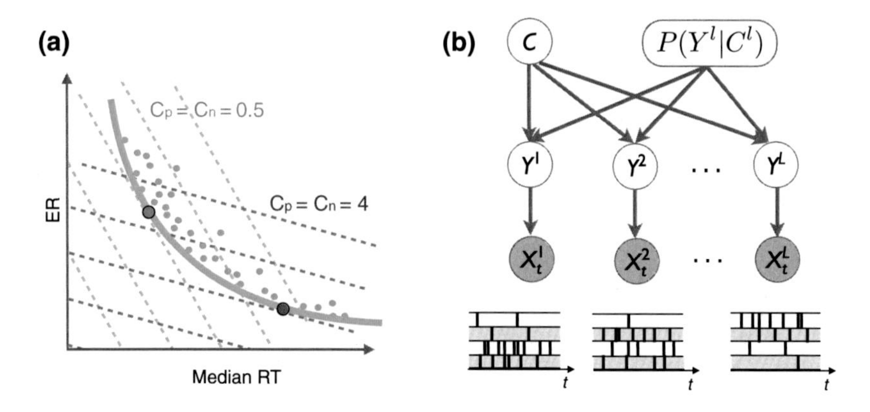

Fig. 4 Ideal Observer and graphical model for visual search **a** Generalization of ideal observer for studying the ER versus RT tradeoff. An ideal observer minimizes the Bayes risk (Eq. 2). An observer's performance is indicated by a *green dot* in the ER versus median RT plane. The *dashed lines* indicate equi-cost contours for a given ratio of the errors versus time cost. *Blue lines* correspond to relatively low error cost, while the *red lines* correspond to high error cost (for simplicity the two error costs are equal here). The *green curve* indicates the performance of the ideal observer (the lower envelope of all observers). The *blue* and *red dots* indicate the locus where the ideal observer curve is tangent to the *blue* and *red* equi-error lines. Such points correspond to the setting of the observer that minimizes the Bayes risk. **b** A generative model of the stimulus. The stimulus class C and a prior distribution on the stimulus orientation $P(Y^l|C^l)$ decide, for each display location l, the orientation Y^l (may be blank). The orientation Y^l determines in turn the observation X_t^l, which are firing patterns from a hypercolumn of V1 orientation-selective neurons at location l over the time window $[0, t]$ (The firing patterns of 4 neurons are shown at each location)

In visual psychophysics, "ideal observers" (Wilson 2011) are typically computed for tasks where the amount of signal is fixed (e.g. the stimulus is displayed for a fixed amount of time), and **only ER** must be minimized. By contrast, our ideal observers dynamically adjust the length of stimulus presentation based on the information collected so far to **jointly minimize RT and ER.**

We also see why the ideal observer definition is with respect to a specific set of assumptions regarding the input signal. First, the signal to noise ratio of the V1 front-end controls the rate at which we can receive information. Second, the correlation in the input affects how the ideal observer should decode information. Lastly, the input modality is restricted only to the orientation-tuned neurons in V1.

In general, the ideal observer of general visual search is often too expensive to compute exactly. Fortunately, in common settings **the performance of the ideal observer is indistinguishable from that of the Sequential Probability Ratio Test (SPRT)** (Wald 1945), which may be evaluated efficiently. Thus, while the SPRT is not strictly optimal for visual search (Chen and Perona 2014), we still refer to it as the ideal observer, bearing in mind that it is optimal for all practical purposes.

3 The Sequential Probability Ratio Test

In this section, we will use the term SPRT and the ideal observer interchangeably (see reasoning above). We will first present SPRT for a series of visual search task, and defer the optimality analysis to Sect. 5.

Call X_t the input observations, which is the collection of firing patterns of the V1 hypercolumn neurons from all display locations collected over the time window $[0, t]$, the SPRT takes the following form:

$$S(X_t) \triangleq \log \frac{P(C=1|X_t)}{P(C=0|X_t)} \begin{cases} \geq \tau_1 & \text{Declare target present} \\ \leq \tau_0 & \text{Declare target absent} \end{cases} \tag{3}$$

It considers $S(X_t)$, the log likelihood ratio of target-present ($C = 1$) vs. target-absent ($C = 0$) probability with respect to the observations X_t. A target-present decision is made as soon as $S(X_t)$ crosses an upper threshold τ_1, while a target-absent decision is made as soon as $S(X_t)$ crosses a lower threshold τ_0. Until either event takes place, the observer waits for further information. For convenience we use base 10 for all our logarithms and exponentials, i.e. $\log(x) \triangleq \log_{10}(x)$ and $\exp(x) \triangleq 10^x$.

Here we assume that target-present and target-absent share the same prior probability of 0.5, hence the log posterior ratio $\log P(C=1|X_t)/P(C=0|X_t)$ is identical to the log likelihood ratio $\log P(X_t|C=1)/P(X_t|C=0)$. If the prior probability is not uniform, one can obtain the log posterior ratio by adding the log prior ratio $\log P(C=1)/P(C=0)$, a simple application of Bayes rule. Thus for simplicity, it is sufficient to be concerned with computing the log likelihood ratio $S(X_t)$ only.

The thresholds $\tau_1 > 0$ and $\tau_0 < 0$ control the maximum tolerable error rates. For example, if $\tau_1 = 2$, i.e. a target-present decision is taken when the stimulus is $> 10^2$ times more likely to be a target than a distractor, then the maximum false positive rate is 1 %; Similarly If $\tau_0 = -3$ then target likelihood is $< 10^{-3}$ times the distractor's, and the false negative rate is at most 0.1 %. τ_1 and τ_0 are judiciously chosen by the observer to minimize the Bayes risk in Eq. 2, and hence are functions of the costs of errors. For example, if $C_p > C_n$, the observer should be less reluctant to make a false negative error, and thus should set $|\tau_0| < \tau_1$. In addition, if both C_p and C_n are large, the observer should increase $|\tau_0|$ and τ_1 so that fewer errors are made in general at the price of a longer RT. Given this relationship, we will parameterize the SPRT with the thresholds τ_0 and τ_1 instead of the costs of errors C_p and C_n.

Therefore, an ideal observer for trading off ER and RT computes decisions using the SPRT (Wald 1945), which compares the log likelihood ratio $S(X_t)$ between target-present and target-absent to a pair of thresholds τ_0 and τ_1. Next we explain how $S(X_t)$ is computed.

3.1 Notations

Let X_t^l denote the activity of the neurons at location l during the time interval $[0, t]$ in response to a stimulus presented at time 0. $X_t = \{X_t{}^l\}_{l=1}^L$ is the ensemble responses of all neurons from all locations. Let $\mathcal{L}_\theta(X_t{}^l) \triangleq \log P(X_t{}^l|Y^l = \theta)$ denote the log likelihood of the spike train data $X_t{}^l$ when the object orientation Y^l at location l is θ degrees. When there is only one location (as in visual discrimination as below), the location superscript is omitted. The target orientation and the distractor orientation are denoted respectively by θ_T and θ_D. In many cases, the target orientation is not unique, but sampled from a set $\Theta_T = \{\theta_1, \theta_2, \ldots\}$ of many possible values. Similarly Θ_D is the domain for the distractor orientation. $n_T = |\Theta_T|$ and $n_D = |\Theta_D|$ are the number of candidate target and distractor orientations, respectively.

3.2 $S(X_t)$ for Homogeneous Discrimination

Given a probabilistic structure of the task (in the form of a graphical model, as we will see later) $S(X_t)$ can be systematically constructed from the visual inputs. To see this, we start from the simplest example.

The problem of homogeneous discrimination is illustrated in Fig. 5a, and the corresponding graphical model in Fig. 5b. The target or the distractor can appear at only one display location ($L = M = 1$) known to the subject, and the target and distractor have distinct and unique orientations θ_T and θ_D, respectively. The visual system needs to determine whether the target or the distractor is present in the test image.

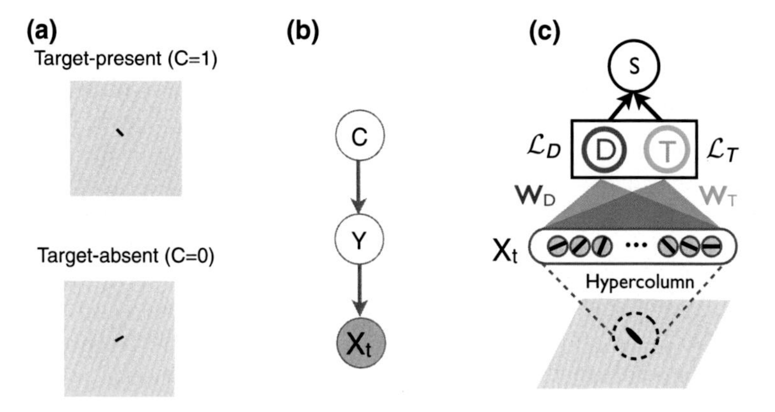

Fig. 5 Homogeneous visual discrimination **a** Example of a target-present and a target-absent stimulus for visual discrimination. **b** The corresponding graphical model. **c** The flow of computation of SPRT. The image generates spiking inputs in the hypercolumn according to Eq. 4. The spikes are combined to create the log likelihood $\mathcal{L}_D(X_t)$ for the distractor and $\mathcal{L}_T(X_t)$ for the target. The two log likelihoods are contrasted (Eq. 6) to compute the test statistics $S(X_t)$ for SPRT

The evidence X_t is a set of spike trains from N orientation-tuned neurons (which can be generalized to be sensitive to color, intensity, etc.) collected during the time interval $(0, t)$. Let $X_t^{(i)}$ be the set of spikes from neuron i in the time interval from 0 to t, K_t^i the number of spikes from neuron i in X_t^i, and K_t the total number of spikes, then the likelihood of $X_t^{(i)}$ when stimulus orientation is θ is given by a Poisson distribution:

$$P(X_t^{(i)}|Y=\theta) = \text{Poiss}(K_t^i|\lambda_\theta^i t) = (\lambda_\theta^i t)^{K_t^i}\frac{\exp(-\lambda_\theta^i t)}{K_t^i!} \tag{4}$$

where λ_θ^i is the firing rate of neuron i when the stimulus orientation is θ.

The observations from the hypercolumn neurons are independent from each other, thus the log likelihood of X_t is given by (Fig. 5c)

$$\begin{aligned}\mathcal{L}_\theta(X_t) &\triangleq \log P(X_t|Y=\theta) = \log\prod_{i=1}^{N} P(X_t^{(i)}|Y=\theta)\\ &= \sum_{i=1}^{N}\log\left((\lambda_\theta^i t)^{K_t^i}\frac{\exp(-\lambda_\theta^i t)}{K_t^i!}\right)\\ &= \sum_{s=1}^{K_t} W_\theta^{i(s)} - t\sum_{i=1}^{N}\lambda_\theta^i + \text{const}\end{aligned} \tag{5}$$

where $W_\theta^i = \log\lambda_\theta^i$ is the contribution of each action potential from neuron i to the log likelihood of orientation θ, and "const" is a term that does not depend on θ and is therefore irrelevant for the decision. The first term is the "diffusion" that introduces jumps in $\mathcal{L}_\theta(X_t)$ whenever a spike occurs. The second term is a "drift" term that moves $\mathcal{L}_\theta(X_t)$ gradually in time. When the tuning curves of the neurons tessellate regularly the circle of orientations, as is the case in our model (Fig. 3a), the average firing rate of the hypercolumn under different orientations is approximately the same, and the drift term may be safely omitted. A sample diffusion is shown in Fig. 6a.

Therefore, the log likelihood ratio can be computed by:

$$\begin{aligned}\text{Visual discrimination: } S(X_t) &\triangleq \log\frac{P(X_t|C=1)}{P(X_t|C=0)} = \log\frac{P(X_t|Y=\theta_T)}{P(X_t|Y=\theta_D)}\\ &= \mathcal{L}_{\theta_T}(X_t) - \mathcal{L}_{\theta_D}(X_t)\\ &= \sum_{s=1}^{K_t} W^{i(s)} + t(\sum_{i=1}^{N}\lambda_{\theta_D}^i - \lambda_{\theta_T}^i)\end{aligned} \tag{6}$$

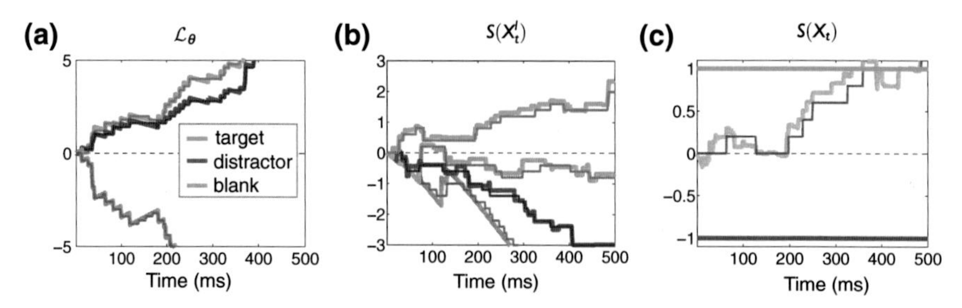

Fig. 6 Internal signals of SPRT. **a** Log likelihoods over time $\mathcal{L}_\theta(X_t)$ for the target, the distractor, and "blanks" (see Example 3.2), computed from the hypercolumn input in Fig. 3c according to Eq. 5. $\mathcal{L}_\theta(X_t)$ for blanks **b** Local log likelihoods $S^l(X_t{}^l)$ from four locations (color coded) computed using Eq. 8. **c** The log likelihood ratio $S(X_t)$ computed from the local log likelihood ratios in (**b**) according to Eq. 7. SPRT computes a decision by comparing $S(X_t)$ to a pair of *upper* (*green*) and *lower* (*red*) thresholds. This examples shows a target-present decision being reached at approximately 350 ms. Thick lines represent signals from the ideal observer, and thin lines from the spiking network (see Sect. 6)

Equation 6 is still a diffusion, with the jump weights $W^i = W^i_{\theta_T} - W^i_{\theta_D}$ and a drift rate of $\sum_{i=1}^{N}(\lambda^i_{\theta_D} - \lambda^i_{\theta_T})$. The constant terms from Eq. 5 cancel. Observing the fact that the total population response of the hypercolumn to θ_T and to θ_D should be identical, we see that the drift term can also be omitted.

Therefore, as first pointed out by Stone (1960), SPRT on homogeneous discrimination may be computed by a diffuse-to-bound mechanism (Ratcliff 1985). In addition, as shown by Wald (1945), SPRT is optimal in minimizing the Bayes risk in Eq. 2 and thus optimal. Later in Sect. 5 we provide empirical evidences of this optimality result.

3.3 $S(X_t)$ for Homogeneous Search

Now that we have analyzed the case of discrimination (one item visible at any time) we will explore the case of search (multiple items present simultaneously, one of which may be the target). Consider the case where all the L display locations are occupied by either a target or a distractor (i.e. $L = M > 1$) and the display either contains exactly one target or none. The target orientation θ_T and the distractor orientation θ_D are again unique and known (Fig. 1b, graphical model in Fig. 7a).

Call $l_T \in \{1, 2, \ldots, L\}$ the target location and assume uniform prior on l_T. The log likelihood ratio of target-present vs target-absent is given by Chen et al. (2011) (Fig. 7b):

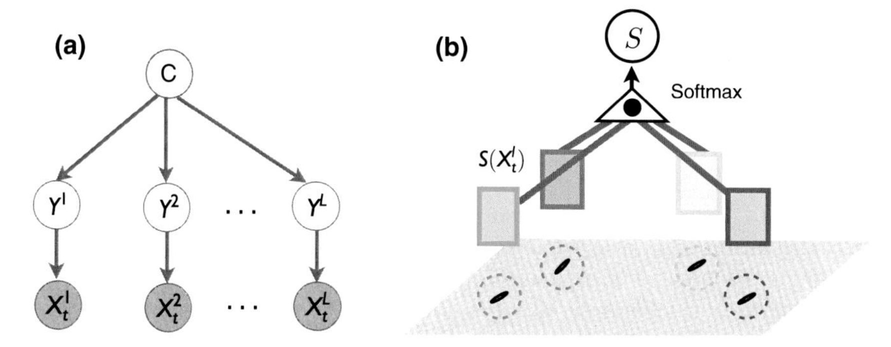

Fig. 7 Homogeneous visual search **a** The graphical model for homogeneous visual search. **c** The flow of computation of SPRT. Each rectangle contains a homogeneous discrimination network in Eq. 5c that computes the local log likelihood ratio $S^l(X_t)$. The $S^l(X_t)$s are combined using a softmax (Eq. 9) to compute the test statistics $S(X_t)$ for SPRT (Eq. 7)

$$\text{Homogeneous visual search: } S(X_t) = \log \frac{\sum_{l_T} P(X_t|l_T)P(l_T|C=1)}{P(X_t|C=0)} = \log \frac{1}{L} \sum_{l_T} \frac{P(X_t|l_T)}{P(X_t|C=0)}$$
$$= \log \frac{1}{L} \sum_{l_T} \frac{P(X_t^{l_T}|\theta_T) \prod_{l \neq l_T} P(X_t^l|\theta_D)}{\prod_l P(X_t^l|\theta_D)}$$
$$= \log \frac{1}{L} \sum_{l_T} \frac{P(X_t^{l_T}|\theta_T)}{P(X_t^{l_T}|\theta_D)} = \mathcal{S}\max_{l_T} \left(S^{l_T}(X_t^{l_T}) - \log(L) \right) \tag{7}$$

where

$$S^l(X_t^l) \triangleq \mathcal{L}_{\theta_T}(X_t^l) - \mathcal{L}_{\theta_D}(X_t^l) \tag{8}$$

is the log likelihood ratio for homogeneous discrimination at location l (see Fig. 6b for an example), and $\mathcal{S}\max(\cdot)$ is the "softmax" function. For any vector $\mathbf{v}$ and a set of indices $\mathcal{I}$:

$$\mathcal{S}\max_{i \in \mathcal{I}} (\mathbf{v}) \triangleq \log \sum_{i \in \mathcal{I}} \exp(v_i) \tag{9}$$

Softmax can be thought of as the marginalization operation in log probability space: it computes the log probability of a set of mutually-exclusive events from the log probabilities of the individual events. For example for two mutually-exclusive events A_1, A_2 we have $P(A_1 \bigcup A_2) = P(A_1) + P(A_2)$, then $\log P(A_1 \bigcup A_2) = \mathcal{S}\max_{i=1,2} \left(\log P(A_i) \right)$. Since the target may appear at one of L locations, and these events are mutually exclusive with respect to each other, the local log likelihoods should be combined using a softmax.

It is important to note that the log likelihood ratio for homogeneous search is **not a diffusion**. Rather, it combines diffusions in a non-linear fashion (via a softmax).

Diffusion is the dominant model for fitting behavioral data for decision making problems, so why bother with the ideal observer? At a conceptual level, the focus of an ideal observer is to study the optimal RT vs ER tradeoff, whereas the goal of most diffusion models is to fit the data. At a practical level, when the search environment (such as the set-size and orientation-difference) changes, a diffusion model may require additional parameters to specify how the diffusions should change accordingly (Palmer 2005; Drugowitsch et al. 2012). By contrast, the ideal observer is parameter-free as long as the search task is specified. This parsimonious formulation allows the ideal observer model to generalize to novel experimental settings (we will see this in Sect. 4.2 and Fig. 13c-f), which is non-trivial for diffusion models. Furthermore, our model connects the underlying cortical mechanisms and physiological parameters to the subjects' behavior. For example, in Sect. 4.1 we will see that our ideal observer analysis predicts that the visual system employs gain-control mechanisms for estimating the complexity of the scene. This prediction would not be possible if one just fit the data with diffusions. Therefore, while diffusion models remain great phenomenological models for decision-making mechanisms, we use the ideal observer model to study the optimal tradeoff of ER and RT across different experimental settings.

Unlike the case for homogeneous discrimination, SPRT is not the ideal observer for homogeneous search. In fact, as soon as either the target-present or target-absent hypotheses contains multiple sub-hypotheses (e.g. the target-present hypothesis here is comprised of L sub-hypotheses, one for each target location), SPRT is not optimal. In Sect. 5 we give details of the optimality analysis as well as empirical verification of the quasi-optimality of SPRT.

3.4 $S(X_t)$ for Heterogeneous Search

What happens when the set-size and/or orientation-difference are unknown to the subject in advance? Fortunately we have discussed the two basic components of SPRT, namely a diffusion process to compute the local log likelihood ratio, and a softmax operation to combine mutually exclusive hypotheses. By stacking these operations strategically, one can implement SPRT for a variety of visual search tasks.

As depicted in the graphical model in Fig. 4b, we introduce "conditional distractor distribution" (CDD) to reflect the uncertainty regarding the display property. CDD is the distribution of orientation Y^l at any non-target location l, i.e. $P(Y_l|C^l = 0)$. We denote CDD with ϕ where $\phi_\theta \triangleq P(Y_l = \theta|C^l = 0)$. Thus ϕ is a n_D-dimensional probability vector. i.e. each element of ϕ is non-negative, and all elements sum to one.

Example 3.1 **Unknown orientation-difference** Consider the case where orientation-difference is unknown and the set-size is known (Fig. 8b). Assume the

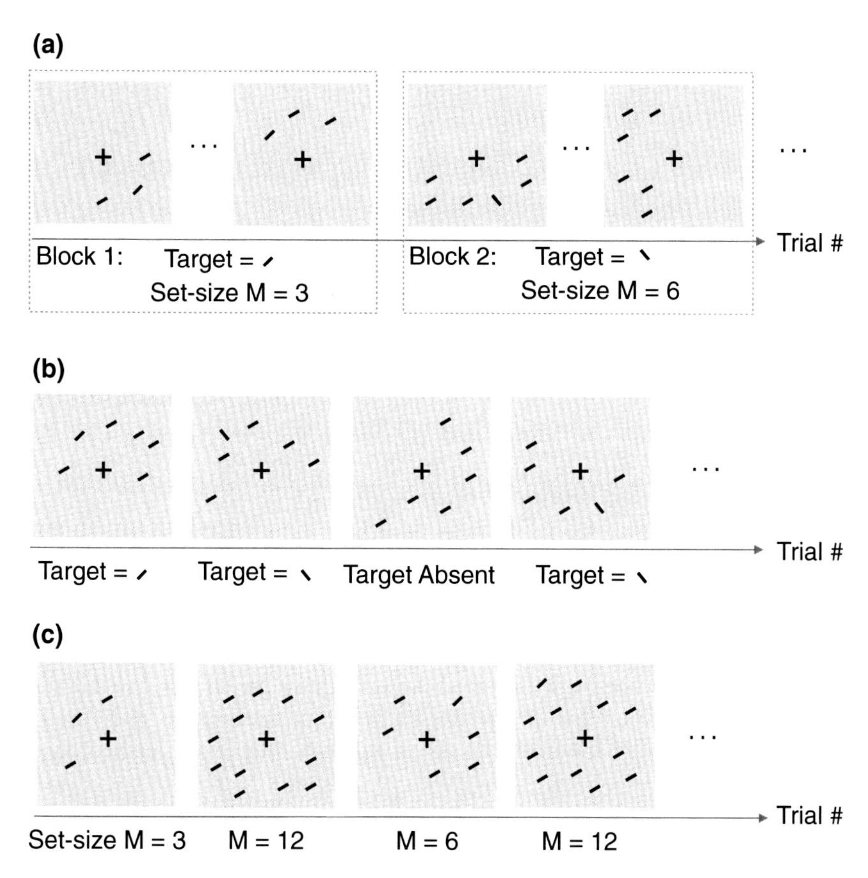

Fig. 8 Common visual search settings. **a** "Blocked": the orientation-difference and the set-size remain constant within a block of trials (outlined in dashed box) and vary only between blocks. **b** "Mix contrast": the target orientation varies independently from trial to trial while the distractor orientation and the set-size are held constant. **c** Mix set-size: the set-size is randomized between trials while the target and distractor orientations are fixed

distractor orientation is always $\theta_T = 0°$ but the distractor orientation is sampled uniformly at random from $\Theta_D = \{20°, 30°, 45°\}$, and identical across the visual field. If we can identify the distractor orientation θ_D, we can reduce this problem to a homogeneous search problem and use Eq. 7 to compute SPRT. To do so we will make use of three CDDs. Here a CDD is a three dimensional vector of

$$\phi = [P(Y^l = 20°|C^l = 0), P(Y^l = 30°|C^l = 0), P(Y^l = 45°|C^l = 0)]$$

The three CDDs are:

$$\phi^{(1)} = [1, 0, 0]; \phi^{(2)} = [0, 1, 0]; \phi^{(3)} = [0, 0, 1];$$

with equal prior probability $P(\phi^{(i)}) = 1/3, \forall i$.

Example 3.2 **Unknown set-size** Consider another example where the set-size M is sampled uniformly from $\{3, 6, 12\}$, and distractor orientation is fixed at 30° (Fig. 8c). The total number of display locations is $L = 12$.

In this case, denote $Y^l = \emptyset$ that a non-target location being blank. If there are M display items, then the probability of any non-target location being blank is $(L - M)/L$. A CDD is a two dimensional vector of

$$\phi = [P(Y^l = 20°|C^l = 0), P(Y^l = \emptyset|C^l = 0)]$$

and the three different set-sizes may be represented by three CDDs of equal probability

$$\phi^{(1)} = [3/12, 9/12], \phi^{(2)} = [6/12, 6/12], \phi^{(3)} = [1 - \epsilon, \epsilon] \quad (10)$$

where ϵ is a small number to prevent zero probability.

Note that the setup in Eq. 10 only approximates the probabilistic structure of **Exp. 3**. This is because the blank placements are not independent of one another. In other words, for a given set-size M, only M locations can contain a distractor. If we place a distractor at each location with probability M/L, we do not always observe M distractors. Instead, the actual set-size follows a Binomial distribution with mean M. However, this is a reasonable approximation because the human visual system can generalize to unseen set-sizes effortlessly. In addition, the values of M used in our experiments are often different enough $\{3, 6, 12\}$ that the i.i.d. model is equally effective in inferring M.

The log likelihood ratio may be computed as:

$$S(X_t) = \underset{l=1...L}{\mathcal{S}\max} \left(S(X_t^l) - \log(L) \right) \quad (11)$$

$$\text{where } S(X_t^l) = \underset{\theta \in \Theta_T}{\mathcal{S}\max} \left(\mathcal{L}_\theta(X_t^l) - \log(n_T) \right) + \underset{\phi \in \Phi}{\mathcal{S}\max} \left(-\underset{\theta \in \Theta_D}{\mathcal{S}\max} \left(\mathcal{L}_\theta(X_t^l) + \log \phi_\theta \right) + Q_\phi(X_t) \right) \quad (12)$$

where $Q_\phi(X_t) \triangleq \log P(\phi|X_t)$ is the log posterior of the CDDs given the observations X_t.

The equation set (Eq. 11–12) may be obtained by a direct application of Bayesian inference on the graphical model in Fig. 4b. While the equations appear complicated, there are clear structures. The log likelihood ratio is obtained by nesting appropriately the models of homogeneous search and heterogeneous discrimination. At the highest level is the softmax over locations as in Eq. 7. At each location l, $S^l(X_t^l)$ is obtained as the difference between the log likelihood of the target with that of the distractor (Eq. 12), which is reminiscent of Eq. 8. In Eq. 12, computing the target log likelihood requires marginalizing over the unknown target orientation with a soft-

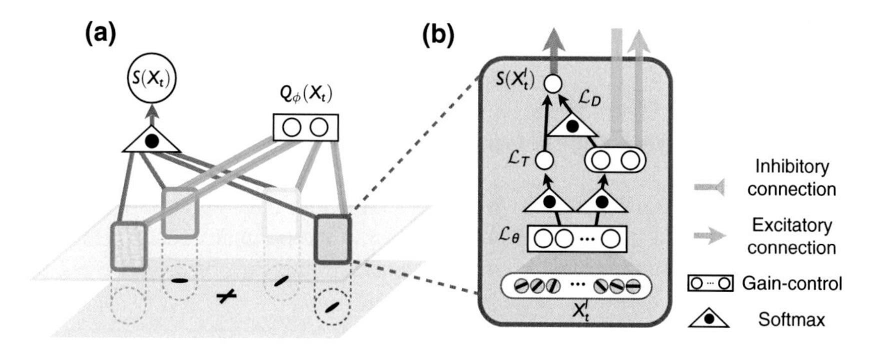

Fig. 9 Ideal observer for heterogeneous visual search and its spiking network implementation. **a** The ideal Bayesian observer for heterogeneous visual search is implemented by a five-layer network. It has two global circuits, one computes the global log likelihood ratio $S(X_t)$ (Eq. 11) from local circuits that compute log likelihood ratios $\{S^l(X_t^{\ l})\}_l$ (Eq. 12), and the other estimates scene complexity $Q_\phi(X_t)$ via gain-control. $Q_\phi(X_t)$ feeds back to the local circuit at each location. **b** The local circuit that computes the log likelihood ratio $S^l(X_t^{\ l})$. Spike trains X_t from V1/V2 orientation-selective neurons are converted to log likelihood for task-relevant orientations $\mathcal{L}_\theta$ (Eq. 5). The log likelihoods of the distractor $\mathcal{L}_D$ (second line of Eq. refeq:herospssearchspsdistractor) under every putative CDD are compiled together, sent (*blue* outgoing arrow) to the global circuit, and inhibited (*green* incoming arrow) by the CDD estimate Q_ϕ

max (again assuming uniform prior over possible target orientations in Θ_T). As for the distractor log likelihood, since both the CDD ϕ and the distractor orientation Y^l may be unknown and need to be marginalized, two softmaxes are necessary (the second line of Eq. 12). For simplicity, we do not describe the equations to compute $Q_\phi(X_t)$: it may be estimated simultaneously with the main computation by a scene complexity mechanism that is derived from first principles of Bayesian inference. This mechanism extends across the visual field and may be interpreted as wide-field gain-control (Fig. 9a). We will see the significance of this mechanism in Sect. 4.

A simpler alternative to inferring the CDD on a trial-by-trial basis is to ignore its variability completely by always using the same CDD obtained from the average complexity and target distinctiveness. More specifically, the approximated log likelihood ratio is:

$$\tilde{S}(X_t) \approx \underset{l=1,\ldots,L}{\mathcal{S}\max} \left(\underset{\theta\in\Theta_T}{\mathcal{S}\max} \left(\mathcal{L}_\theta(X_t^{\ l})\right) - \underset{\theta\in\Theta_D}{\mathcal{S}\max} \left(\mathcal{L}_\theta(X_t^{\ l}) + \log \bar{\phi}_\theta\right)\right) - \log(n_T L) \quad (13)$$

where $\bar{\phi}_\theta = \mathbb{E}(\phi_\theta)$ is the mean CDD for orientation θ with respect to the its prior distribution. This approach is suboptimal. Intuitively, if the visual scene switches randomly between being cluttered and sparse, then always treating the scene as if it had medium complexity would be either overly-optimistic or overly-pessimistic. Crucially, the predictions of this simple model are inconsistent with the behavior of human observers, as we shall see later in Fig. 11.

4 Model Prediction and Human Psychophysics

Now that we have seen how to implement SPRT given a visual search task, we show that it can predict existing phenomena in the literature and data collected by ourselves.

For the mathematically minded reader, a reasonable next step after presenting SPRT is to show that it is a good approximation to the optimal tradeoff. However, the procedure of computing the optimal tradeoff is rather involved, so we ask our readers to be patient and wait till the next section.

4.1 Qualitative Fits

A first test of our model is to explore its qualitative predictions of RT and ER in classical visual search experiments (Fig. 1a).

In a first simulation experiment (**Sim. 1**), we used a "blocked" design (Fig. 9a), where the orientation of targets and distractors as well as the number of items do not change from image to image within an experimental block. Thus, the observer knows the value of these parameters from experience. Accordingly, we held these parameters constant in the model. We assume that the costs of error are constant, hence we hold the decision thresholds constant as well. What changes from trial to trial is the presence and the location of the target, and the timing of individual action potentials in the simulated hypercolumns.

The model makes three qualitative predictions: (a) The RT distribution predicted by the model is heavy-tailed: it is approximately log-normal in time (Fig. 10b). (b) The median RT increases linearly, as a function of M, with a large slope for hard tasks (small orientation-difference between target and distractor), and almost flat for easy tasks (large orientation-difference) (Fig. 10a). The median RT is longer for target-absent than for target-present, with roughly twice the slope (Fig. 10a). The three predictions are in agreement with classical observations in human subjects (Fig. 2) (Treisman and Gelade 1980; Palmer et al. 2011).

In a second experiment (**Sim. 2**) we adopted a "mixed" design, where the distractors are known, but the orientation-difference is sampled from 10^0, 20^0 and 60^0, randomized from image to image (Fig. 9b). The subjects (and our model) do not know which orientation-difference is present before stimulus onset. The predictions of the model are shown in Fig. 10c. When the target is present both RT and ER are sensitive to the orientation-difference and will decrease as the orientation-difference increases, i.e. the model predicts that an observer will trade off errors in difficult trials (more errors) with errors in easy trials (fewer errors) to achieve an overall desired average performance, which is consistent with psychophysics data.

In **Sim. 3** we explored which one of two competing models best accounts for visual search when scene complexity is unknown in advance. Recall that in discussing heterogeneous search, we proposed two models, an ideal Bayesian observer

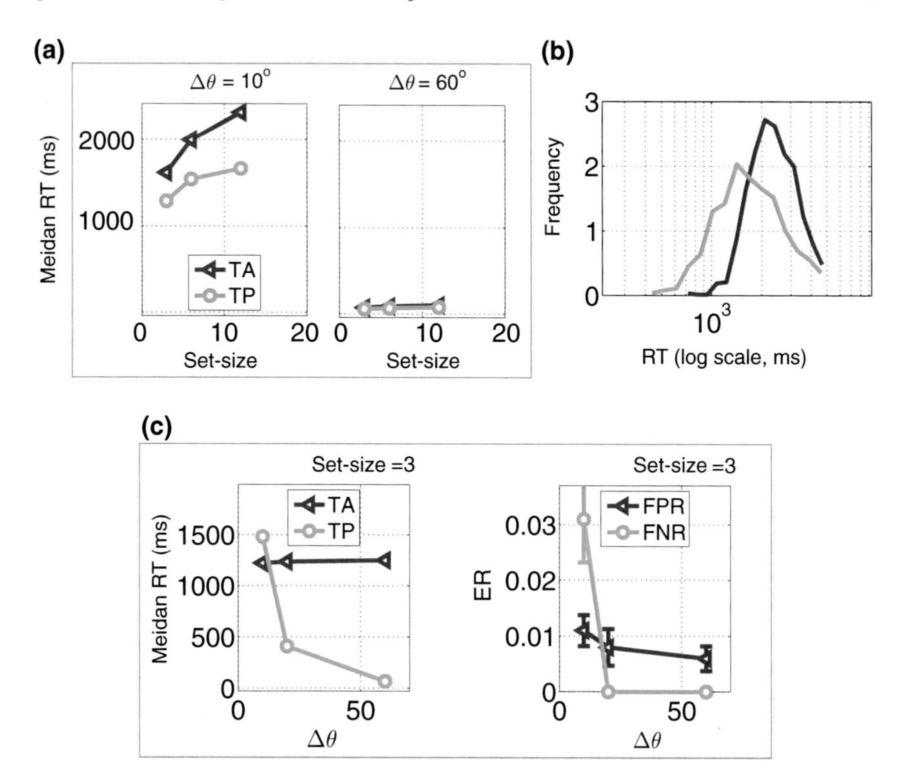

Fig. 10 Qualitative predictions of the ideal observer (**Sim. 1–3**). **a** Set-size effect on median RT under the blocked design (**Sim. 1**). The ideal observer predicts a linear RT increase with respect to set-size when the orientation-difference $\Delta\theta$ is low (10°, *left*) and a constant RT when the orientation-difference is low (60°, *right*). The target-absent (TA) RT slope is roughly twice that of target-present (TP). **b** RT histogram under the blocked design with a 10° orientation-difference and a set-size of 12 items. RT distributions are approximately log-normal. **c** Median RT (*upper*) and ER (*lower*) for visual search with heterogeneous target/distractor, mixed design (**Sim. 2**)

(Eq. 11) that is optimal and a simplified model (Eq. 13) that is sub-optimal. The ideal observer estimates the scene complexity parameter trial-by-trial (Eq. 11) and predicts that ERs are comparable for different set-sizes while RTs show strong dependency on set-size when the orientation-difference is small (Fig. 11a). The simplified model, where scene complexity is assumed constant (Eq. 13), predicts the opposite, i.e. that ER will depend strongly on set size, while RT will be almost constant when the target is present (Fig. 11b). Human psychophysics data (Wolfe et al. (2010), reproduced in Fig. 11c) show a positive correlation between RT and set-size and little dependency of ER on set-size, which favor the ideal model and suggest that the human visual system estimates scene complexity while it carries out visual search.

This example showcases one core reason why we study psychophysics. At a glance it seems like we are simply cooking up models to explain human behavior, and the explanation is the end-goal. In fact, model fitting serves a greater purpose of

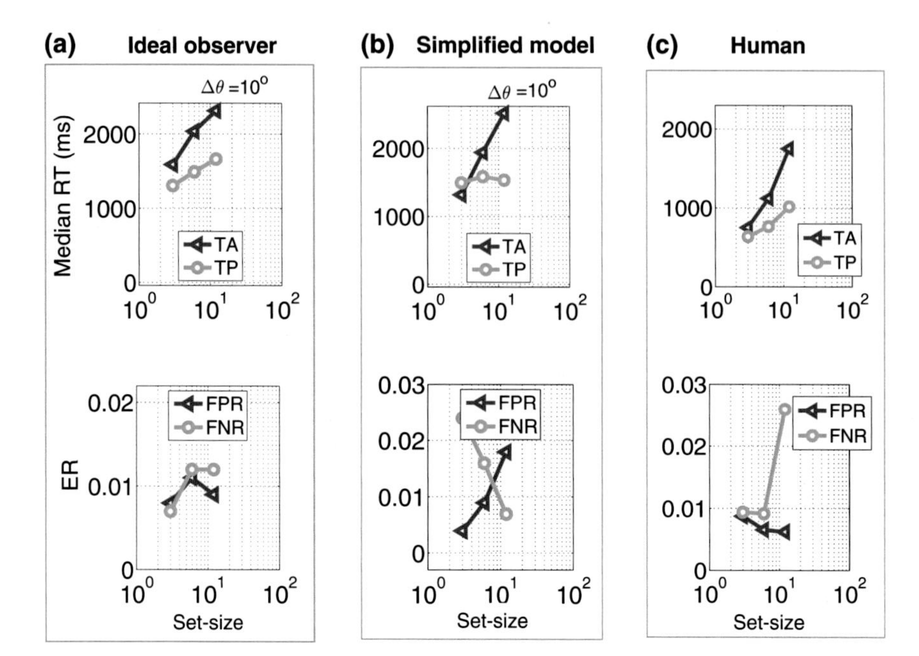

Fig. 11 Qualitative model predictions and psychophysics data on visual search with unknown set-size (**Sim. 3**). Median RT (*upper*) and ER (*lower*): false-positive-rate (FPR) and false-negative-rate (FNR), of visual search with homogenous target/distractor and unknown set-sizes (**Sim. 3**) under two models: the ideal observer (**a**) that estimates the scene complexity parameter ϕ (essentially the probability of a blank at any non-target location) on a trial-by-trial basis (Eq. 11) using a wide-field gain-control mechanism; a simplified observer (**b**) that uses average scene complexity $\bar{\phi}$ for all trials (Eq. 13). Psychophysical measurements on human observers (Wolfe et al. (2010), spatial configuration search in Figs. 2 and 3, reproduced here as (**c**)) are consistent with the optimal model (**a**). Simulation parameters are identical to those used in Fig. 10

studying the computational principles of the brain. The question we are interested in is whether the brain estimate scene complexity when conducting visual search. Since different answers generate different predictions, we can study this rather abstract question using quantifiable behavioral results.

4.2 *Quantitative Fits*

In order to assess our model quantitatively, we compared its predictions with data harvested from human observers who were engaged in visual search (Fig. 1a). Three experiments were conducted to test both the model and humans under different conditions. The conditions are parameterized by the orientation-difference chosen from $\{20°, 30°, 45°\}$ and the set-size from $\{3, 6, 12\}$. The blocked design were used in the first experiment (**Exp. 1**), where all $3 \times 3 = 9$ pairs of orientation-difference and set-size combinations were tested in blocks. The second experiment randomized orientation-difference from trial-to-trial while fixing the set-size at 12 (**Exp. 2**). The third randomized the set-size while holding the orientation-difference fixed at 30° (**Exp. 3**). The subjects were instructed to maintain eye-fixation at all times, and respond as quickly as possible and were rewarded based on accuracy.

We fit our model to explain the **full RT distributions and ERs** for each design separately. In order to minimize the number of free parameters, we held the number of hypercolumn neurons constant at $N = 16$, their minimum firing rate constant at $\lambda_{\min} = 1$ Hz, and the half-width of their orientation tuning curves at 22° (full width at half height: 52°) (Graf et al. 2011). Hence we were left with only three free parameters: the maximum firing rate of any orientation-selective neuron λ_{max} controls the signal-to-noise ratio of the hypercolumn; the upper and lower decision thresholds τ_0 and τ_1 control the frequency of false alarm and false reject errors. Once these parameters are given, all the other parameters of our model are analytically derived.

While our model takes care of the perceptual computational time, human response times also include a non-perceptual motor and neural conduction delay (Palmer et al. 2011). Therefore, we also use two additional free parameters per subject to account for the non-perceptual delay. We assume that the delay follows a log-normal distribution parameterized by its mean and variance.

In the blocked design experiment **Exp. 1**, the hypercolumn and the motor time parameters were fit jointly across all blocks (about 1620 trials); the decision thresholds were fit independently on each block (180 trials/block). In the mixed design experiments **Exp. 2-3**, all five parameters were fit jointly across all conditions for each subject because all conditions are mixed (440 trials/condition). See Fig. 12 for data and fits of a randomly selected individual, and Fig. 13a, b for all subjects in the blocked condition. In each experiment the model is able to fit the subjects' data well. The parameters that the model estimated (the maximum firing rate of the neurons λ_{max}, the decision thresholds τ_0, τ_1 are plausible (Vinje and Gallant 2000)).

It may be possible to model the inter-condition variability of the thresholds as the result of the subjects minimizing a global risk function (Drugowitsch et al. 2012). Therefore for each subject in the blocked design experiment **Exp. 1** we have tried fitting a common Bayes risk function (Eq. 2), parameterized by the two costs of errors C_p and C_n, across all blocks, and solving for the optimal thresholds for each block independently. This assumption reduces the number of free parameters for the blocked condition from 21 (2 thresholds × 9 conditions + 1 SNR + 2 motor parameters) to 5 (2 costs of errors + 1 SNR + 2 motor parameters), but at the cost of marked reduction in the quality of fits for some of the subjects. Therefore as far as our model is concerned, there was some block-to-block variability of the error costs.

Finally, we test the generalization ability of the ideal observer. We used the signal-to-noise ratio parameter (the maximum firing rate λ_{max}) and the two non-decision delay parameters estimated from the blocked experiment (**Exp. 1**) to predict the mixed experiments (**Exp. 2-3**). Thus for each mixed experiment, only two parameters, namely the decision thresholds τ_0 and τ_1, were fit. Despite the parsimony in parameterization, the model shows good cross-experiment fits (see Fig. 13c-f), suggesting that the parameters of the model refer to real characteristics of the subjects.

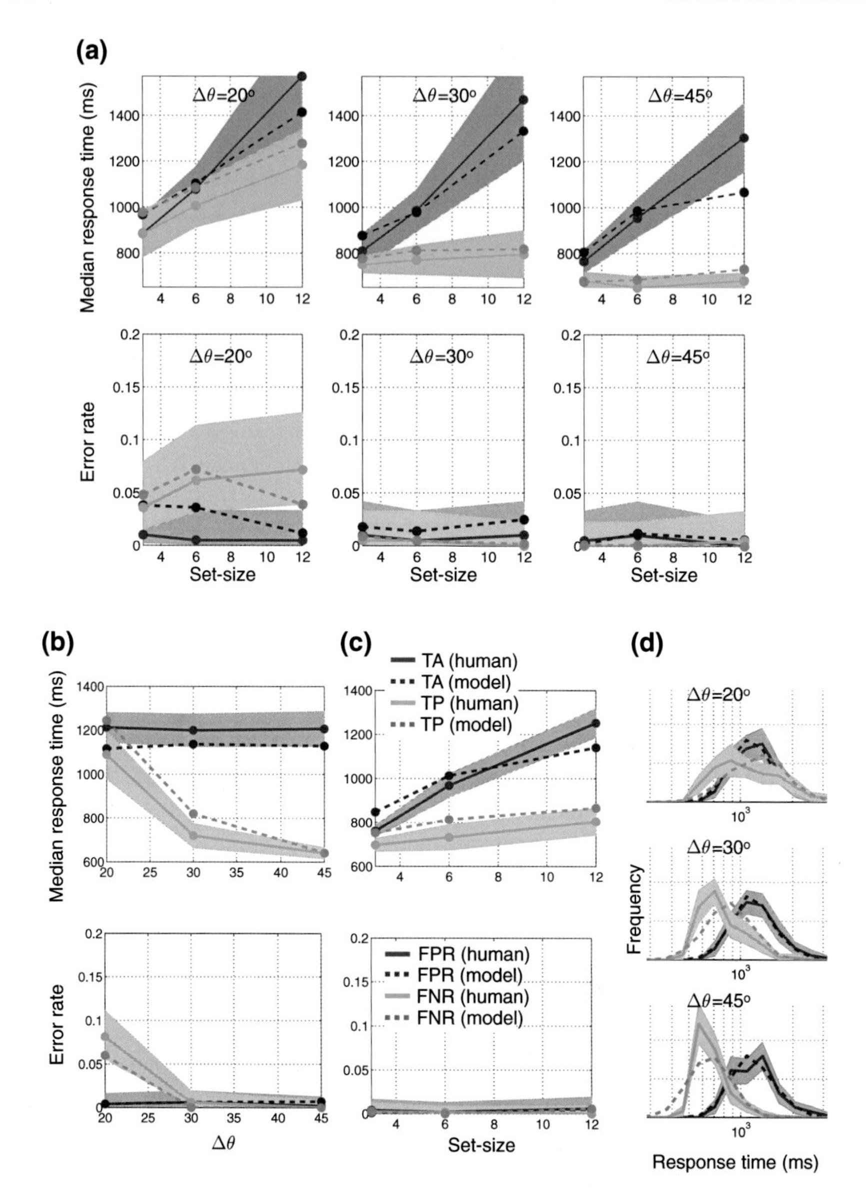

Fig. 12 Behavioral data of a randomly selected human subject and fits (ER, median RT and RT distributions) using the ideal observer (**a**) **Exp. 1**: "Blocked" design. All set-size M and orientation-difference $\Delta\theta$ combinations share the same hypercolumn and non-perceptual parameters; the decision thresholds are specific to each $\Delta\theta$-M pair. Fits are shown for RTs (*first row*) and ER (*second row*). **b-c** RT and ER for **Exp. 2**, the"mixed set size" (**b**) and **Exp. 3**, the "mixed contrast" design (**c**). **d** RT histogram for the "mixed contrast" design, grouped by orientation-difference

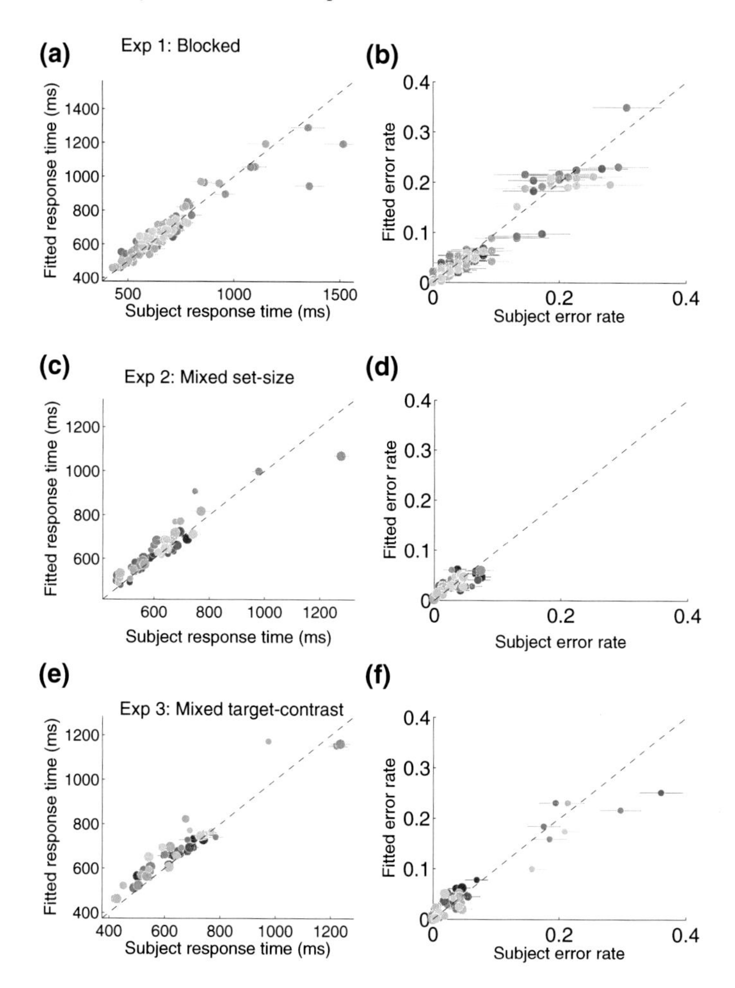

Fig. 13 Synopsis of fits to nine individual subjects. The rows correspond respectively to three designs: **Exp. 1** (blocked), **Exp. 2** (mixed contrast) and **Exp. 3** (mixed set size). The maximum firing rate of the hypercolumn λ_{max} and the two non-decision parameters for each subject are fitted using only the blocked design experiment, and used to predict median RT and ER for the two mixed design experiments. Colors are specific to subject. The small, medium and large dots correspond respectively to the orientation-differences 20°, 30°, and 45° in (**c-d**), and to the set-sizes 3, 6, and 12 in (**e-f**)

In conclusion, the ideal observer that prescribes the optimal behavior given task structure also predicts human visual search behavior. The ideal observer has a compact parameterization: on average, three parameters are needed to predict each experimental condition; many parameters (the signal-to-noise ratio of the hypercolumn and the motor time distribution) generalize across different experimental conditions.

The agreement between the optimal model predictions and the data collected from our subjects suggests that the human visual system may be optimal in visual search.

Our model uses $N = 16$ uncorrelated, orientation-tuning neurons per visual location, each with a half tuning width of 22° and a maximum firing rate (estimated from the subjects) of approximately 17 Hz. The tuning width agrees with V1 physiology in primates (Graf et al. 2011). While our model appears to have underestimated the maximum firing rate of cortical neurons, which ranges from 30 Hz to 70 Hz (Graf et al. 2011), and the population size N (which may be in the order of hundreds), actual V1 neurons are correlated, hence the equivalent number of independent neurons is smaller than the measured number. For example, take a population of $N = 16$ independent Poisson neurons, all with a maximum firing rate of 17 Hz, and combine every three of them into a new neuron. This will generate a population of 560 correlated neurons with a maximum firing rate of 51 Hz and a correlation coefficient of 0.19, which is close to the experimentally measured average of 0.17 (Graf et al. 2011) (see Vinje and Gallant (2000) for a detailed discussion on the effect of sparseness and correlation between neurons). Therefore, our estimates of the model parameters are consistent with primate cortical parameters. The parameters of different subjects are close but not identical, matching the known variability within the human population (Palmer et al. 2011; Essen et al. 1984). Finally, the fact that estimating model parameters from data collected in the blocked experiments allows the model to predict data collected in the mixed experiments does suggest that the model parameters mirror physiological parameters in our subjects.

5 Optimality Analysis

This is a self-contained technical section on the theoretical differences between SPRT and the ideal observer targeted towards the mathematically-minded readers. For practical visual search problems considered in this chapter, SPRT is indistinguishable from the optimal, and capable of explaining human behavior fairly well.

Now that we have seen how to use SPRT to generate ER and RT predictions for homogeneous and heterogeneous search tasks, the next question is, is SPRT optimal? As we have alluded to before, the SPRT is optimal only for homogeneous discrimination. Instead of showing the proof, we go for a more empirical solution.

Below, we provide an algorithm that solves for the ideal observer for a homogeneous visual search problem. This algorithm allows us to understand the optimal Bayes risk achievable by ANY observer, which serves as a gold standard to evaluate the performance of SPRT and of the humans. However, this algorithm is computationally tractable only for homogeneous discrimination and homogeneous search with 2 to 3 locations. We will see that problems of this scale is sufficient to reveal the difference between the ideal observer and SPRT.

5.1 Solving for the Ideal Observer

The ideal observer contains two components: a state space $\vec{Z}(t)$ over time and a decision strategy that associates each state and time with an action. One common constraint on the state space is that it must be Markov in time:

$$P(X_{t:t+\Delta t}|X_t, \vec{Z}(t)) = P(X_{t:t+\Delta t}|\vec{Z}(t))$$

in other words, $\vec{Z}(t)$ must be sufficient in summarizing past observations X_t so that given $\vec{Z}(t)$, future observations $X_{t:t+\Delta t}$ from time t to time $t + \Delta t$ become independent from the past. Once this constraint is satisfied, the problem may be formulated as a partially observable Markov decision process (POMDP) (Cassandra et al. 1994), and the optimal strategy may be solved exactly using dynamic programming.

What state formulation satisfies the Markov requirement? We choose $\vec{Z}(t)$ to be the local log likelihood ratios at all locations: $\vec{Z} : Z^l(t) = S^l(X_t{}^l) = \log \frac{P(X_t^l|C^l=1)}{P(X_t^l|C^l=0)}, l = 1 \ldots M$.

Example 5.1 To see why $\vec{Z}$ is a sufficient statistic to compute the likelihood of future observations, let $\Delta X = X_{t:t+\Delta t} \setminus X_t$ denotes new observations at all locations at time $t + \Delta t$, the likelihood of ΔX is obtained by marginalizing the target location l_T. Denote $l_T = 0$ the target-absent event:

$$P(l_T = 0|X_t) = P(C = 0|X) = \frac{1}{1 + \exp(S(t))} = \frac{1}{1 + \sum_l \exp(Z^l)/M} \tag{14}$$

For a target-present event where the target location is $l_T > 0$:

$$P(l_T|X_t) = \frac{\exp(Z_{l_T})/M}{1 + \sum_l \exp(Z^l)/M}$$

For notational convenience, define $Z^0 = \log(M)$, then the two equations above simplify to:

$$P(l_T|X_t) = \frac{\exp(Z_{l_T})}{\sum_{l=0}^{M} \exp(Z^l)}$$

The posterior on l_T is sufficient to compute the likelihood of ΔX:

$$\begin{aligned} P(\Delta X|X_t) &= \sum_{l_T=0}^{M} P(\Delta X, l_T|X_t) = \sum_{l_T=0}^{M} P(\Delta X|l_T)P(l_T|X_t) \\ &= \sum_{l_T=0}^{M} P(l_T|X_t) \prod_l P(\Delta X^l|C^l) \end{aligned}$$

where $C^l = 1$ iff $l = l_T$.

In addition, $\vec{Z}(t)$ is sufficient to compute the log likelihood ratio between target-present and target-absent as shown in Eq. 7:

$$S(X_t) = \log \frac{P(X_t|C=1)}{P(X_t|C=0)} = \underset{l=1\ldots M}{S\max}\left(Z^l(t)\right) - \log(M)$$

Recall from Eq. 6 that $Z^l(t) = S^l({X_t}^l)$ can be computed as a diffusion with no drift. For simplicity of analysis, we approximate the diffusions as gaussian random walks, as shown below.

At a target-present location, $Z^l(t)$ is computed by a summing a group of action potentials weighted by W^i according to the sending neuron i

$$\begin{aligned} Z^l(t) &= \sum_{s=1}^{K_t} W^{i(s)} \sim \sum_{i=1}^{N} Poisson(\lambda_{\theta_T} W^i) \\ &\approx \mathcal{N}\left(\mu_1 = \sum_{i=1}^{N} \lambda_{\theta_T} W^i, \sigma_1^2 = \sum_{i=1}^{N} \lambda_{\theta_T} (W^i)^2\right) \end{aligned} \tag{15}$$

Similarly at a distractor location:

$$Z^l(t) \approx \mathcal{N}\left(\mu_0 = \sum_{i=1}^{N} \lambda_{\theta_D} W^i, \sigma_0^2 = \sum_{i=1}^{N} \lambda_{\theta_D} (W^i)^2\right) \tag{16}$$

Therefore, we formulate $Z^l(t)$ as **a one-dimensional gaussian random walk** at each location (e.g. Thornton and Gilden 2007; Drugowitsch et al. 2012). $Z^l(t)$ is parameterize by the *drift-rate* μ_C which depends on the stimulus class C, and the variance σ^2 (Fig. 6b) (σ for the two classes are roughly identical due to symmetry). A larger drift-rate difference between the two classes $|\mu_1 - \mu_0|$ implies a higher signal-to-noise ratio, or equivalently, an easier discrimination problem. We only consider homogeneous search here so the drift rates per class are identical across locations.

The gaussian random walk approximation serves to simplify the presentation of the analysis. One can repeat all analyses in this chapter with Poisson diffusions Eq. 6 and obtain similar results regarding optimality.

5.2 Dynamic Programming

The optimal decision may be computed numerically using dynamic programming (Cassandra et al. 1994; Bellman 1956). Define $R(\vec{Z}(t), t)$ as the smallest Bayes risk an observer could incur starting from $\vec{Z}(t)$ at time t. For example, according to this definition the optimal Bayes risk is captured by $R(\vec{0}, 0)$, which is the total risk from time 0 onwards.

$R(\vec{Z}, t)$ is recursively given by:

$$R(\vec{Z}(t), t) = \min \begin{cases} C_n(1 - P_0(\vec{Z}(t))) & D = 0\text{: declare target absent} \\ C_p P_0(\vec{Z}(t)) & D = 1\text{: declare target present} \\ \Delta t + \mathbb{E}_{\vec{Z}(t+\Delta t)|\vec{Z}(t)} R(\vec{Z}(t + \Delta t), t + \Delta t) & D = \emptyset\text{: wait} \end{cases} \tag{17}$$

where $P_0(\vec{Z}(t))$ is the false rejection probability (explained later).

At any time t and any state $\vec{Z}(t)$, the ideal observer picks the action $D \in \{\emptyset, 0, 1\}$ that yields the lowest risk. If declaring target-absent, the observer makes a false rejection mistake and incurs a cost of C_n. The false reject probability, $P(C = 0|X_t)$ for a target-present stimulus X_t, can be computed from the state $\vec{Z}(t)$ and is denoted $P_0(\vec{Z}(t))$ (see Eq. 14). The option of declaring target-present is analyzed in the identical way. Last, if waiting for more evidence, the observer trades off the time cost Δt (time cost is 1) for a new observation, and access to the cumulative risk at $t + \Delta t$.

We use dynamic programming (Eq. 17) to solve for the optimal decision strategy for each $(\vec{Z}, t)$ pair. We set up the recurrence relationship in Eq. 17 by computing for each $\vec{Z}$:

$$\begin{aligned} P(\vec{Z}(t + \Delta t)|\vec{Z}(t)) &= \sum_{l_T=0}^{M} P(\vec{Z}(t + \Delta t)|l_T, \vec{Z}(t))P(l_T|\vec{Z}(t)) \\ &= \sum_{l_T=0}^{M} \mathcal{N}(\vec{Z}(t + \Delta t)|\vec{Z}(t) + \Delta\vec{Z}_{(l_T)}, V_{(l_T)}) \frac{\exp(Z^{l_T})}{\sum_{l=0}^{M} \exp(Z^l)} \end{aligned} \tag{18}$$

where $\Delta\vec{Z}_{(l_T)}$ and $V_{(l_T)}$ are the mean and variance of the change in $\vec{Z}$ when the target location is l_T. The notation $\Delta Z^l_{(l_T)}$ means the change of $\vec{Z}$ occurred within Δt at the l-th location where location l_T contains the target.

$$\begin{aligned} \Delta Z^l_{(l_T)} &= \mu_0 \Delta t & \forall l \neq l_T \\ \Delta Z^{l_T}_{(l_T)} &= \mu_1 \Delta t & \\ V^l_{(l_T)} &= \sigma^2 \Delta t & \forall l = 1, \ldots, M \end{aligned}$$

Having readily computed $P_1(\vec{Z})$ and $P(\vec{Z}(t + \Delta t)|\vec{Z}(t))$, we use backward induction to solve the dynamic programming equation (Eq. 17) by starting from an infinite horizon $t = \infty$, at which point the ideal observer is forced to declare either target-present or target-absent. We discretize the state space $\{\vec{Z}, t\}$ into a hypergrid. Since the state space grows exponentially with the length of $\vec{Z}$ (set-size M), dynamic programming can only be done for a small set-size (2 or 3).

The ideal observer is defined over a $M+1$ dimensional state space. The state space is separated by decision boundaries/surfaces into three different decision regions (Sobel et al. 1953). Furthermore, the recurrence Eq. 17 is time invariant. As a result, the optimal decision is constant in time (see Cassandra et al. 1994) and the decision surfaces have $M-1$ dimensions.

5.3 *Comparison with SPRT*

Is SPRT optimal in visual discrimination ($M = 1$)? Recall that the optimal decision boundaries are constant in time, and in one dimension, $\vec{Z}$ is the log likelihood ratio S and a scalar. As a result, the optimal decision strategy essentially applies a pair of constant thresholds over the log likelihood ratio. Figure 14a illustrates the decision boundaries. This is precisely SPRT. Therefore, for visual search, SPRT is optimal.

Is SPRT optimal in visual search ($M = L > 1$)? SPRT is a one dimensional test on the log likelihood ratio S whereas the optimal decision boundaries are in the two dimensions of $\vec{Z} = \{Z^1, Z^2\}$ (and constant in time). Since S can be computed from $\vec{Z}$ deterministically, we can project the decision boundaries of SPRT into two-dimensions (in terms of $\{Z^1, Z^2\}$) so that they can be compared against the optimal ones. Figure 14c, d shows that SPRT uses very different decision boundaries than the optimal ones.

Despite the disparity in decision boundaries, SPRT achieves almost the same Bayes risk as does the optimal. Figure 14b shows the Bayes risks for a variety of costs for errors.

For higher dimensional problems, the procedure for computing the optimal decision boundaries becomes exponentially more expensive, therefore it is impossible to conclude whether SPRT is optimal. However, one can restrict our search for the optimal decision strategy within a family of models that are compactly parameterized. Preliminary results suggest that SPRT is statistically indistinguishable from the optimal (within the restricted family) as long as the search is homogeneous. This is beyond the scope of this chapter. Interested readers are referred to Chen and Perona (2014).

In conclusion, using a computational procedure for computing the optimal decision strategy, we conclude that SPRT is near-optimal for visual search in low dimensions. Therefore, we use SPRT as a surrogate for the ideal observer.

6 Spiking Network Implementation

Finally, we explore the physical realization of the ideal observer and show that a simple network of spiking neurons may implement a close approximation to the decision strategy.

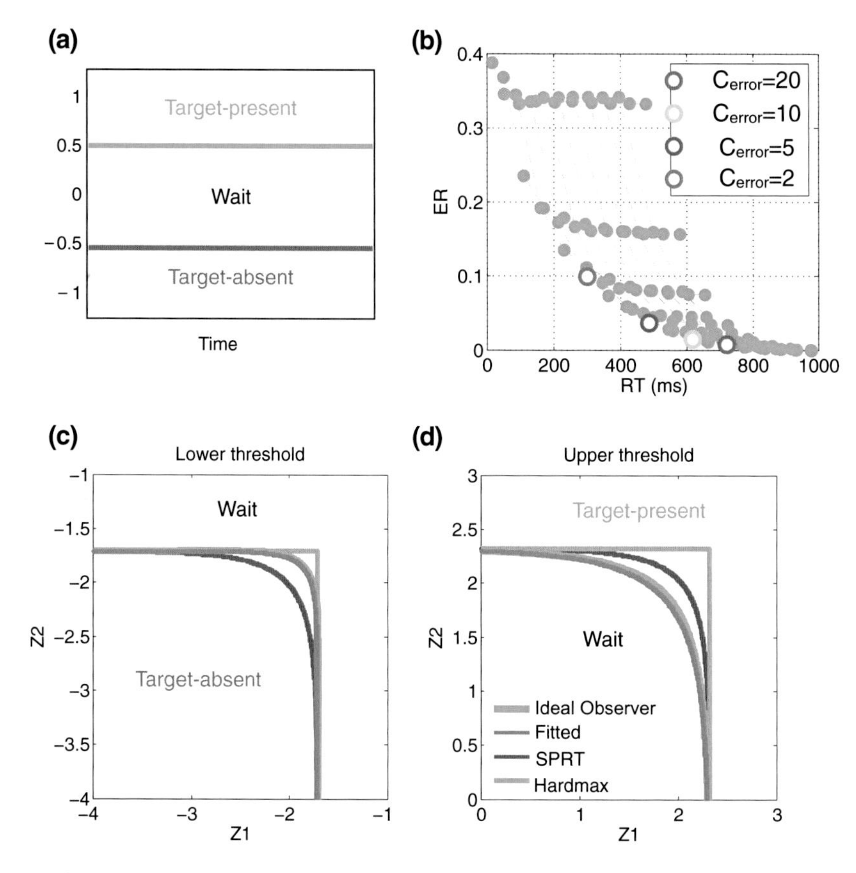

Fig. 14 Optimality analysis of SPRT **a** Optimal decision boundary obtained from the dynamic programming procedure Eq. 17 for visual discrimination ($M = 1$). The ideal observer compares $Z(t) = S(X_t)$ to a pair of thresholds, which is precisely SPRT. **b** Bayes risks (Eq. 2) of SPRT with different threshold pairs (τ_0, τ_1) (green dots) and the optimal decision strategy under a variety of costs of errors. The false positive and false negative errors have the same cost $C_p = C_n = C_{error}$ (in seconds per 100 % error). SPRT achieves the same ER vs RT tradeoff as the ideal observer in all cases provided that the SPRT thresholds are optimized. **c-d** The upper and lower thresholds of the ideal observer and the SPRT in a homogeneous visual search task with two locations ($M = 2$)

Local Log Likelihoods

We first explain how to compute $\mathcal{L}_\theta(X_t)$, the local log likelihood of the stimulus taking on orientation θ, from spiking inputs X_t from V1. $\mathcal{L}_\theta(X_t)$ is the building block of $S(X_t)$ (Eq. 6). Consider one spatial location, recall from Eq. 5 that the log likelihood is:

$$S_\theta(X_t) = \sum_{s=1}^{K_t} W_\theta^{i(s)} + \text{const}$$

The first term is a diffusion, where each spike causes a jump in $\mathcal{L}_\theta$. This term can be implemented by integrate-and-fire (Dayan and Abbott 2003) neurons, one for each relevant orientation $\theta \in \Theta_T \bigcup \Theta_D$, that receive afferent connections from all hypercolumn neurons with connection weights $w_\theta^i = \log \lambda_\theta^i$. The constant term is computationally irrelevant because it does not depend on the stimulus orientation θ; it may be removed by a gain-control mechanism to prevent the dynamic range of membrane potential from exceeding its physiological limits (Matteo et al. 1999). Specifically, one may subtract from each $\mathcal{L}_\theta$ a common quantity, e.g. the average value of the all the $\mathcal{L}_\theta$'s without changing $S(X_t^l)$ in Eq. 12.

Example 6.1 **Average gain-control** Average gain-control is the process of subtracting the mean from $\mathcal{L}_\theta$'s to remove unnecessary constants for decision and maintaining membrane potentials within physiological limits. Average gain-control may be conveniently done at the input using feedforward connections only. Specially, let $y_\theta(t)$ denote the mean-subtracted $\mathcal{L}_\theta$ signal, $w_\theta^i = \log \lambda_\theta^i$ denote the weights in Eq. 5, and $x_i(t) \in \{0, 1\}$ denote the instantaneous firing event during time t to $t + \Delta t$ from neuron i. The desired gain-controlled signal $y_\theta(t)$ may be computed by linear integration, as shown in Fig. 15a:

$$\dot{y}_\theta(t) = \sum_i \left(w_\theta^i - \frac{\sum_{\theta'} w_{\theta'}^i}{N} \right) x_i \tag{19}$$

Signal Transduction

The log likelihood $\mathcal{L}_\theta$ must be transmitted downstream for further processing. However, $\mathcal{L}_\theta$ is a continuous quantity whereas the majority of neurons in the central nervous system are believed to communicate via action potentials. We explored whether this communication may be implemented using action potentials (Gray and McCormick 1996) emitted from an integrate-and-fire neuron. Consider a sender neuron communicating its membrane potential to a receiver neuron. The sender may emit an action potential whenever its membrane potential surpasses a threshold τ_s. After firing, the membrane potential drops to its resting value, and the sender enters a brief refractory period whose duration (about 1*ms*) is assumed to be negligible. If the synaptic strength between the two neurons is also τ_s, the receiver may decode the signal by simply integrating such weighted action potentials over time. This coding scheme loses some information due to discretization. Varying the discretization threshold τ_s trades off the quality of transmission with the number of action potentials: a lower threshold will limit the information loss at the cost of producing more action potentials. Surprisingly, we find that the performance of the spiking network is very close to that of the Bayesian observer, even when τ_s is set high, so that a small number of action potentials is produced (see Fig. 15d,f for the quality of approximation for a toy signal and Fig. 6a-c for the quality of approximation for actual signals in the ideal observer.). The network behavior is quite insensitive to τ_s, thus we do not consider τ_s as a free parameter, and set its value to $\tau_s = 0.5$ in our experiments.

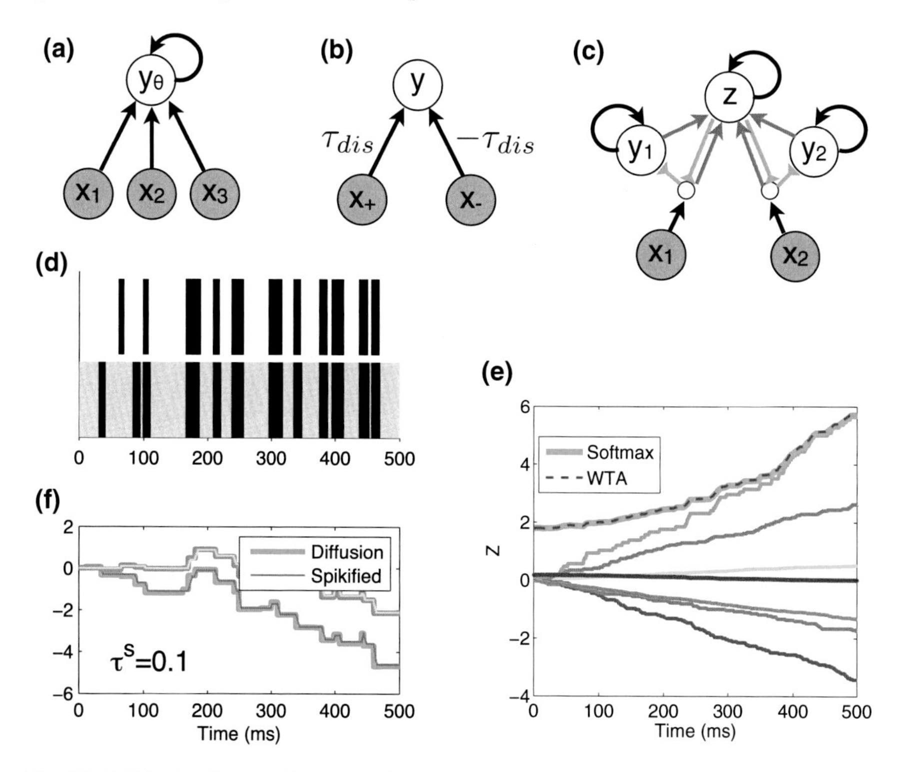

Fig. 15 Spiking implementation **a** A feedforward network implements the average gain-controlled network of Eq. 19. **b** Signal transduction. The positive and negative parts of the signal in x are encoded with integrate-and-fire neurons and transmitted to the receiver neuron y. **d** Two sender neurons communicate their membrane potentials using spike trains to a receiver neuron (only the negative neurons are shown). **f** The receiver reassemble the spike trains (*thick lines*) and reconstruct the senders' membrane potentials (*thin lines*). **c** Winner-take-all circuit for computing the softmax (Eqs. 20 and 21). **e** Comparison between the ground-truth and the WTA implementation of softmax of 7 neurons over time

Softmax

One of the fundamental computations in Eq. 11 is the softmax function (Eq. 9). It requires taking exponentials and logarithms, which have not yet been shown to be within a neuron's repertoire. Fortunately, it has been proposed that softmax may be approximated by a simple maximum (Chen et al. 2011; Ma et al. 2011), and implemented using a winner-take-all mechanism (Koch and Ullman 1987; Seung 2009) with spiking neurons (Oster et al. 2009). Through numerical experiments we find that this approximation results in almost no change to the network's behavior (see Fig. 15e). This suggests that an exact implementation of softmax is not critical, and other mechanisms that may be more neurally plausible have similar performances.

Example 6.2 One common implementation (Koch and Ullman 1987) of the softmax is described as follows. For a set of N spiking neurons, let $x_i(t) \in \{0, 1\}$ denote whether neuron i has spiked in time $t + \Delta t$. We introduce an additional N neurons $\{y_i\}_{i=1}^N$, where $y_i(t)$ denotes the membrane potential of the i-th additional neuron at time t. The desired quantity is the softmax over the cumulative signal in $x_i(t)$, denoted by $z(t)$. In other words:

$$z(t) \stackrel{\triangle}{=} \underset{i=1,\ldots,N}{\mathcal{S}\max}\left(w_i \sum_{\tau=0}^{t} x_i(\tau)\right)$$

$z(t)$ may be approximated by $\tilde{z}(t)$ using the following neuron equations:

$$\dot{\tilde{z}}(t) = \sum_i y_i(t) w_i x_i(t) \quad (20)$$

$$\dot{y_i}(t) = y_i(t)(w_i x_i(t) - \dot{\tilde{z}}(t)) \quad (21)$$

Figure 15e shows that $\tilde{z}(t)$ approximates $z(t)$ well in a simple setup of 10 neurons with a common and small incoming weights $w_i = 0.05$ across all neurons.

Equation set (20) and (21) appear complicated, but they are simplify derived using Taylor expansion of the softmax function around the origin.

The time it takes for the winner-take-all network to converge is typically small (on *ms* level for 10's of neurons, and scales logarithmically with the number of neurons (Koch and Ullman 1987) compared to the inter-spike-intervals of the input neurons (around 30*ms* per neuron, and 12 ms for a hypercolumn of $N = 16$ neurons per visual location Vinje and Gallant 2000).

Decision

Finally, the log likelihood ratio $S(X_t)$ is compared to a pair of thresholds to reach a decision (Eq. 3). The positive and negative parts of $S(X_t)$, $(S(X_t))^+$ and $(-S(X_t))^+$, may be represented separately by two mutually inhibiting neurons (Gabbiani and Koch 1996), where $(\cdot)^+$ denotes halfwave-rectification: $(x)^+ \stackrel{\triangle}{=} \max(0, x)$. We can implement Eq. 3 by simply setting the firing thresholds of these neurons to the decision threshold τ_1 and $-\tau_0$, respectively.

Alternatively, $S(X_t)$ may be computed by a mechanism akin to the ramping neural activity observed in decision-implicated areas such as the frontal eye field (Woodman et al. 2008; Heitz and Schall 2012; Purcell et al. 2012). $(S(X_t))^+$ and $(-S(X_t))^+$ could be converted to two trains of action potentials using the same encoding scheme described above in the Signal Transduction section. The resultant spike trains may be the input signal of an accumulator model (e.g. Bogacz et al. 2006). The model has been shown to be implementable as a biophysically realistic recurrent network (Wang 2002; Lo and Wang 2006; Wong et al. 2007) and capable of producing and thresholding ramping neural activity to trigger motor responses (Woodman et al. 2008; Heitz and Schall 2012; Purcell et al. 2012; Mazurek et al. 2003; Cassey et al. 2014).

While both neural implementations of $S(X_t)$ are viable options, in the simulations used in this study we opted for the first.

Network structure

If we combine the mechanisms discussed above, i.e. local gain-control, an approximation of softmax, a spike-based coding of analog log likelihood values as well as the decision mechanism, we see that the mathematical computations required by the ideal observer can be implemented by a deep recurrent network of spiking neurons (Fig. 9a).

The overall network structure is identical to the diagram (Fig. 9b) explained in Sect. 3.4. It is composed of local "hypercolumn readout" networks (Fig. 9b), and a central circuit that aggregates information over the visual field. The local network computes the local log likelihood ratio $S^l(X_t^l)$ (Eq. 12) and simultaneously computes the local log likelihood for each CDD. The CDD log likelihoods are aggregated over all locations and sent to a gain-control unit to estimate the posterior of the CDD $Q_\phi = \log P(\phi|X_t)$, which capture the most likely set-size and orientation-difference. At each time instant this estimate is fed back to the local networks to compute $S(X_t^l)$ (Eq. 12).

It is important to note that both the **structure** and **the synaptic weights** of the visual search network described above were derived **analytically** from the hypercolumn parameters (the shape of the orientation-tuning curves), the decision thresholds, and the probabilistic description of the task. The network designed for heterogeneous visual search could dynamically switch to simpler tasks by adjusting its priors (e.g. $P(\phi)$). The network has only three degrees of freedom, rather than a large number of network parameters (Ma et al. 2011; Alex et al. 2012).

7 Chapter Summary

Searching for objects amongst clutter is one of the most valuable functions of our sensory systems. Best performance is achieved with fast response time (RT) and low error rates (ER); however, response time and error rates are competing requirements which have to be traded off against each other. The faster one wishes to respond, the more errors one makes due to the limited rate at which information flows through the senses. Conversely, if one wishes to reduce error rates, decision times become longer. In order to study the nature of this trade-off we derived an ideal observer for visual search; the input signal to the model is action potentials from orientation-selective hypercolumn neurons in primate striate cortex V1, the output of the model is a binary decision (target-present versus target-absent) and a decision time.

Five free parameters uniquely characterize the model: the maximum firing rate of the input neurons and the maximum tolerable false-alarm and false-reject error rates, as well as two parameters characterizing response delays that are unrelated to decision. Once these parameters are set, RT histograms and ER may be computed for any experimental condition. Our model may be implemented by a deep neural network composed of integrate-and-fire and winner-take-all mechanisms. The network structure is completely deterministic given the probabilistic structure of the search task. Signals propagate from layer to layer mostly in a feed-forward fashion; however, we find that two feedback mechanisms are necessary: (i) gain control (lateral inhibition) that is local to each hypercolumn and has the function of maintaining signals within a small dynamic range, and (ii) global inhibition that estimates the complexity of the scene. Qualitative comparison of model predictions with human behavior suggests that the visual system of human observers indeed does estimate scene complexity as it carries out visual search, and that this estimate is used to control the gain of decision mechanisms.

Despite the parsimony, our model is able to quantitatively predict human behavior in a variety of visual search conditions. Without physiological measurements of the hypercolumn parameters (number of neurons, maximum firing rate, etc.) directly from human subjects, one can not assess optimality. After all, we may be overestimating the signal-to-noise ratio in the front-end while humans are sub-optimal. Nonetheless, the estimated hypercolumn parameters are plausible, suggesting that humans may employ an optimal strategy for visual search.

References

Alex K, Ilya S, Geoff H (2012) Imagenet classification with deep convolutional neural networks. Adv Neural Inf Process Syst 25:1106–1114

Beck JM, Ma WJ, Kiani R, Hanks T, Churchland AK, Roitman J, Shadlen MN, Latham PE, Pouget A (2008) Probabilistic population codes for Bayesian decision making. Neuron 60(6):1142–1152

Bogacz R, Brown E, Moehlis J, Holmes P, Cohen JD (2006) The physics of optimal decision making: a formal analysis of models of performance in two-alternative forced-choice tasks. Psychol Rev 113(4):700–765

Busemeyer JR, Rapoport A (1988) Psychological models of deferred decision making. J Math Psychol 32(2):91–134

Carandini M, Heeger DJ, Movshon JA (1999) Linearity and gain control in v1 simple cells. In: Models of cortical circuits. Springer, pp. 401–443

Cassandra AR, Kaelbling LP, Littman ML (1994) Acting optimally in partially observable stochastic domains. In: AAAI, vol 94, pp 1023–1028

Chen B, Navalpakkam V, Perona P (2011) Predicting response time and error rates in visual search. In: Adv Neural Inf Process Syst 2699–2707

Chen B, Perona P (2014) Towards an optimal decision strategy of visual search. arXiv:1411.1190. Accessed 1 Nov 2014

Chung-Chuan L, Xiao-Jing W (2006) Cortico-basal ganglia circuit mechanism for a decision threshold in reaction time tasks. Nature Neurosci 9(7):956–963

Drugowitsch J, Moreno-Bote R, Churchland AK, Shadlen MN, Pouget A (2012) The cost of accumulating evidence in perceptual decision making. J Neurosci 32(11):3612–3628

Drugowitsch J, Moreno-Bote R, Churchland AK, Shadlen MN, Pouget A (2012) The cost of accumulating evidence in perceptual decision making. J Neurosci 32(11):3612–3628

Fabrizio G, Christof K (1996) Coding of time-varying signals in spike trains of integrate-and-fire neurons with random threshold. Neural Comput 8(1):44–66

Felleman DJ, Van Essen DC (1991) Distributed hierarchical processing in the primate cerebral cortex. Cereb Cortex 1(1):1–47

Geisler WS (2011) Contributions of ideal observer theory to vision research. Vis Res 51(7):771–781

Graf ABA, Kohn A, Jazayeri M, Movshon JA (2011) Decoding the activity of neuronal populations in macaque primary visual cortex. Nature Neurosci 14(2):239–245

Gray CM, McCormick DA (1996) Chattering cells: superficial pyramidal neurons contributing to the generation of synchronous oscillations in the visual cortex. Science 274(5284):109–113

Heitz RP, Schall D (2012) Neural mechanisms of speed-accuracy tradeoff. Neuron 76(3):616–628

Hubel DH, Wiesel TN (1962) Receptive fields, binocular interaction and functional architecture in the cat's visual cortex. J Physiol 160(1):106–154

Koch C, Ullman S (1987) Shifts in selective visual attention: towards the underlying neural circuitry. In: Matters of intelligence. Springer, pp 115–141

Matthias O, Rodney D, Shih-Chii L (2009) Computation with spikes in a winner-take-all network. Neural Comput 21(9):2437–2465

Mazurek ME, Roitman JD, Ditterich J, Shadlen MN (2003) A role for neural integrators in perceptual decision making. Cereb. Cortex 13(11):1257–1269

Mervyn S (1960) Models for choice-reaction time. Psychometrika 25(3):251–260

Ma WJ, Navalpakkam V, Beck JM, van den Berg R, Pouget A (2011) Behavior and neural basis of near-optimal visual search. Nat Neurosci 14(6):783–790

Palmer J (1994) Set-size effects in visual search: the effect of attention is independent of the stimulus for simple tasks. Vis Res 34(13):1703–1721

Palmer J, Verghese P, Pavel M (2000) The psychophysics of visual search. Vis Res 40(10):1227–1268

Palmer EM, Horowitz TS, Torralba A, Wolfe JM (2011) What are the shapes of response time distributions in visual search? J Exp Psychol: Hum Percept Perform 37(1):58–71

Palmer J, Huk AC, Shadlen MN (2005) The effect of stimulus strength on the speed and accuracy of a perceptual decision. J vis 5(5):376–404

Peter C, Andrew H, Brown Scott D (2014) Brain and behavior in decision-making. PLoS Comput Biol 10(7):e1003700

Peter D, Abbott LF (2003) Theoretical neuroscience: computational and mathematical modeling of neural systems. J Cogn Neurosci 15(1):154–155

Purcell BA, Schall JD, Logan GD, Palmeri TJ (2012) From salience to saccades: multiple-alternative gated stochastic accumulator model of visual search. J Neurosci 32(10):3433–3446

Richard B (1956) Dynamic programming and lagrange multipliers. Proc Nat Acad Sci USA 42(10):767

Roger R (1985) Theoretical interpretations of the speed and accuracy of positive and negative responses. Psychol Rev 92(2):212–225

Sanger TD (1996) Probability density estimation for the interpretation of neural population codes. J Neurophysiol 76(4):2790–2793

Seung HS (2009) Reading the book of memory: sparse sampling versus dense mapping of connectomes. Neuron 62(1):17–29

Sobel M et al (1953) An essentially complete class of decision functions for certain standard sequential problems. Ann Math Stat 24

Thornton TL, Gilden DL (2007) Parallel and serial processes in visual search. Psychol Rev 114(1):71

Treisman AM, Gelade G (1980) A feature-integration theory of attention. Cogn Psychol 12(1):97–136

Treisman AM, Gelade G (1980) A feature-integration theory of attention. Cogn Psychol 12(1):97–136

Van Essen DC, Newsome WT, Maunsell JHR (1984) The visual field representation in striate cortex of the macaque monkey: asymmetries, anisotropies, and individual variability. Vis Res 24(5):429–448

Verghese P (2001) Visual search and attention: a signal detection theory approach. Neuron 31(4):523–535
Vinje WE, Gallant JL (2000) Sparse coding and decorrelation in primary visual cortex during natural vision. Science 287(5456):1273–1276
Wald A (1945) Sequential tests of statistical hypotheses. Ann Math Stat 16(2):117–186
Wald A, Wolfowitz J (1948) Optimum character of the sequential probability ratio test. Ann Math Stat 19(3):326–339
Wolfe JM (2007) Guided search 4.0. Integr Models Cogn Syst 99–119
Wolfe JM, Palmer EM, Horowitz TS (2010) Reaction time distributions constrain models of visual search. Vis Res 50(14):1304–1311
Wolfe JM, Palmer EM, Horowitz TS (2010) Reaction time distributions constrain models of visual search. Vis Res 50(14):1304–1311
Wong K-F, Huk AC, Shadlen MN, Wang X-J (2007) Neural circuit dynamics underlying accumulation of time-varying evidence during perceptual decision making. Front Comput Neurosci 1
Woodman GF, Kang M-S, Thompson K, Schall JD (2008) The effect of visual search efficiency on response preparation neurophysiological evidence for discrete flow. Psychol Sci 19(2):128–136
Xiao-Jing W (2002) Probabilistic decision making by slow reverberation in cortical circuits. Neuron 36(5):955–968

The Pupil as Marker of Cognitive Processes

Wolfgang Einhäuser

1 The Pupil Is a Readily Accessible Marker of Neural Processes

Of all peripheral physiological measures of neural activity, pupil size is arguably the easiest to access. In fact, it is one of the few markers of another's internal state that is available to the observer's naked eye. This property renders pupil size a cue that can be exploited in social interactions. This is no news to those renaissance women, who extracted atropine from the deadly nightshade (*atropa belladonna*) to dilate their pupils in order to increase their perceived attractiveness, eventually leading to the plant's commonly used name of "belladonna". Not surprisingly, pupillometry has been used rather widely in the early days of cognitive psychology and flourished in the 1960s. With the advent of other techniques, like EEG and later PET and fMRI, peripheral physiology, and thus pupillometry, became somewhat unfashionable. However, since the turn of the millennium, when video-based oculography started to become widespread, pupil size could be measured as a "by-product" of eye-tracking data. This led pupillometry to have its own renaissance with the number of pupillometry publications increasing each year. The present chapter will deal with those recent developments with a strong focus on the cognitive aspects controlling pupil size.

W. Einhäuser (✉)
Institut für Physik, Technische Universität Chemnitz, Chemnitz, Germany
e-mail: wolfgang.einhaeuser-treyer@physik.tu-chemnitz.de

Q. Zhao (ed.), *Computational and Cognitive Neuroscience of Vision*,
Cognitive Science and Technology, DOI 10.1007/978-981-10-0213-7_7

2 Modulation of the Pupil's Response to Light by Cognitive Factors

2.1 Awareness and Imaginary Light Sources Modulate the Pupil Light Reflex

Thinking of changes in pupil size, the pupil light reflex (PLR) readily comes to everyone's mind: when light levels incident on the eye increase, the pupil constricts. Even this seemingly simple and reflexive behavior has, however, cognitive components attached. Rather than responding to light physically entering the eye, the pupil at least to some extent reacts to the subjective perception of lightness. This somewhat counterintuitive phenomenon becomes particularly evident when two stimuli of different luminance are presented to the two eyes. At each instant, both pupils reflect the luminance of the stimulus that reaches awareness: when the brighter stimulus is consciously perceived, the pupil constricts more than during times when the darker stimulus is perceived (Naber et al. 2011; Fahle et al. 2011). Similarly, when one eye is perceptually suppressed from awareness, flashes presented to this eye elicit less constriction than during periods when the respective eye is dominant (Bárány and Halldén 1948; Lorber et al. 1965).

Such cognitive command over the pupil light reflex is not restricted to stimuli that are actually physically present. A substantial body of recent research indicates that the mere indication of the presence of a light source can elicit a pupillary constriction. For example, the pupil constricts when exposed to a picture depicting the sun, even if images that are identical in luminance and similar with respect to low-level properties yield no constriction or even a dilation (Binda et al. 2013). These responses to a depicted sun are sensitive to image rotation, but persist in cartoon pictures (Naber and Nakayama 2013). Along similar lines, brightness illusions trigger pupil constrictions (Laeng and Endestad 2012) and even merely imagining a situation that is intrinsically dark ("dark room") or bright ("sunny sky") yields differential pupil responses (Laeng and Sulutvedt 2014). In sum, even the seemingly simple pupil light reflex is under substantial cognitive control.

2.2 The Pupil Light Response Can Mark the Focus of Attention

There are several interactions between pupil size and selective attention, and some of them may even point to a common neural circuitry (see Sect. 5.4). Similar to imagining a light source, attention can modulate the pupil light reflex. If a display contains a dark and a bright region, the pupil becomes smaller when the bright region is attended than when the darker surface is attended (Binda et al. 2013; Binda and Murray 2015; Mathôt et al. 2013). Importantly, this differential effect on pupil size holds for covert attentional shifts; that is, shifts of attention without shifts of gaze. In this case again the visual input to the retina remains identical, only the cognitive factor, the locus of attention, distinguishes the two situations. This effect can be exploited to track attention by a procedure reminiscent of measuring attention to flickering stimuli by steady-state visually evoked potentials (SSVEPs) in EEG (Morgan et al. 1996), albeit at substantially lower frequencies: when different regions of the visual field alternate at different temporal frequencies between bright and dark, the attended region can be inferred from the dominant frequency in the pupil response (Naber et al. 2013b). The attentional modulation of the pupil response extends to feature-based attention: if the dark and the bright surface overlap, still attention to the bright surface yields a stronger constriction than attention to the dark one (Binda et al. 2014). Importantly, the attentional modulation of the pupil light reflex provides sufficient temporal resolution to track covert attentional shifts that accompany the preparation of an overt shift (i.e., of an eye movement, cf. Deubel and Schneider 1996): preparation of an eye movement to a dark surface is accompanied by a relative pupil dilation compared to the preparation of an eye movement to a bright surface (Mathôt et al. 2015a). Despite the substantial latency of the pupil response (about 500 ms) as compared to the coupling of overt and covert attention (on the order of 200 ms), Mathôt et al. (2015a) could elegantly isolate this preparatory response by occasionally swapping bright and dark surface around the initiation of the eye movement. These studies offer the possibility that the pupil light reflex could be used to achieve a temporally fine-grained tracking of attentional deployment in space.

3 Cognitive Factors Controlling Pupil Dilation

3.1 Pupil Dilation Signals Cognitive Load

Most of the cognitive effects on the pupil mentioned in Sect. 2 seem related to the perception of lightness and therefore—at least conceptually—tied to the pupil light reflex. Independent of any relation to lightness, however, the pupil has been implicated to signal a vast variety of cognitive and emotional processes and states. Already in the early 1960s it was demonstrated that pupil size relates to cognitive load. For example, when solving arithmetic problems, pupil size increases and the amount of increase relates to the difficulty of the problem (Hess and Polt 1964; Boersma et al. 1970). Similarly, pupil size scales with the number of items stored in memory (Kahneman and Beatty 1966; see also Sect. 3.8). The increase in pupil diameter with task difficulty extends to hierarchical processing. When comparing same-different judgements on physical stimulus properties (same letter in same case), on identity (same letter in a different case) or on category (vowel versus consonant), these judgements yield increasing pupil dilation (Beatty and Wagoner 1978). Similar results hold for picture-word matching, where dilations are larger for superordinate than for subordinate and basic categories, corresponding to the difficulty assessed by behavioral measures (van der Meer et al. 2003). The pupil as marker for cognitive load is not restricted to stimuli that are visual in nature: pupil dilation can be employed in monitoring task difficulty during listening. For example, pupil dilation increases with decreasing speech intelligibility, even though this effect plateaus once intelligibility becomes low and may even slightly reverse once conditions become too difficult (Zekveld and Kramer 2014). This (slight) reversal is also known for memory load: once memory capacity is exceeded the pupil constricts relative to the dilation at maximum capacity (Poock 1973). On an individual level, the amount of task-related pupil dilation differs between participants of high-fluid intelligence and normal individuals, but only for highly demanding tasks (van der Meer et al. 2010). This suggests that pupil dilation may indeed be indicative of available cognitive resources. In sum, pupil dilation is likely a signal of processing resources and their use, dilation continues to increase while the available processing capacity is filled up, eventually saturates and in some cases shows a decrease if the capacity is "overloaded".

3.2 Pupil Dilation and Arousal

Studying the dependence of pupil dilation on cognitive load, Bradshaw (1967) confirmed the dependence for solving arithmetic problems, but did not find a similar dependence for multi-solution "word-games". In addition, there was a rapid dip observed once the solution to a problem was found and reported. This dependence on the report for pupil size to shrink may signal the "end of a job" (Bradshaw), a notion also alluded to in Hess (1965). Together with a general decline of pupil size over the course of an experiment, Bradshaw interpreted these findings as evidence that pupil dilation signals arousal. "Arousal" in this line of research mostly refers to a general state of increased vigilance or task-awareness, and indeed the use of pupillometry for continuous alertness monitoring has been suggested (Kristjansson et al. 2009). Much more commonly, however, pupil dilation is associated with sexual or emotional arousal. Large pupils as signal of a (sexually) aroused state may provide the underpinning to the *belladonna* effect (see also Sect. 4.1) and it is tempting to speculate that this signal also contributes to the low-light levels of candle-light dinners being perceived as "romantic". In the scientific literature, the interest in such arousal dates back to early work by Hess and Polt (1960), who used potentially sexually arousing stimuli to modulate "interest value" (see Aboyoun and Dabbs 1998 for a more detailed account). Although many paradigms—sometimes implicitly—assume that emotional arousal is involved in mediating the pupil response, explicit manipulations of arousal in pupillometric studies are surprisingly rare. Interestingly, the pupil response seems to depend little on the valence of the stimulus (positive versus negative), but rather scales with the strength of the stimulus in either direction (Janisse 1974). This general finding was confirmed using affective pictures from the IAPS database: Again, the pupil response to affective pictures (pleasant or unpleasant) is larger as compared to neutral stimuli with little difference between pleasant and unpleasant (Bradley et al. 2008). These authors contrast this to other physiological measures, such as heart rate and electrodermal activity, which on the same stimuli more clearly distinguish between positive and negative valence. Even when visual stimuli are presented only extremely briefly, the subsequent pupil response is modulated by the affect participants associate with the depicted information (Naber et al. 2012). Like for cognitive load, the effect of arousal is not limited to visual stimuli. Instead, for emotional sounds the same pattern is observed: positive and negative sounds induce a larger pupil with comparably little difference between the two (Partala and Surakka 2003). The pupil response to visually or acoustically arousing stimuli is corroborated by the result that actual physical pain, for example an electric shock applied to the fingertips, results in pupil dilation that scales with stimulus intensity (Chapman et al. 1999; Geuter et al. 2014). In healthy observers, this pain response extends to pain empathy; that is, the pupil dilates when pain of other individuals is depicted (Paulus et al. 2015). In sum, the pupil seems to dilate to arousing stimuli. Since arousal is a comparably broad and unspecific signal, the challenge remains to identify the extent to which arousal per se dilates the pupil and to which extent the observed dilations result from other factors that yield the pupil to dilate and in addition are associated with increased levels of arousal.

3.3 *Pupil Dilation and Anxiety*

In addition to mere arousal, pupil dilation has been associated with a variety of emotions. Frequently, however, these emotions are supposed to increase arousal and thus indirectly yield pupil dilation. Anxiety, either induced by the expectation of pain or by showing threatening stimuli is one of the best studied emotions in the context of pupillometry. An early study showed that in a memory task, pupil dilation is larger for participants with a high audience anxiety as compared to a low anxiety control group (Simpson and Molloy 1971). In line with this result, the pupil light reflex towards a light flash is smaller (i.e., there is less constriction, the pupil remains larger) and the overall pupil size is larger for students scoring higher on a trait anxiety scale (Nagai et al. 2002). In this paradigm trait anxiety, however, was strongly correlated with state anxiety, and factoring out state-anxiety effects from trait-anxiety measures shows that the state-anxiety effect on the pupil is substantially larger than the trait-anxiety effect (Bitsios et al. 2002). The specific effects of anxiety in such paradigms can be dissociated from a general increase in alertness or vigilance: when observers anticipate an electric shock or having to detect a faint auditory signal, pupil size at trial onset increases as an effect of alertness. The pupil constriction to a visual stimulus that serves as indicator of the upcoming sound or electric shock (i.e., as unconditioned stimulus, see also Sect. 3.7) is only diminished for the anticipation of the shock, but not for the anticipation of the sound (Bitsios et al. 2004). This supports a specific effect of anxiety on pupil size that rides on top of a more general arousal or alertness signal.

3.4 Direct Effects of Attention and Salience on Pupil Dilation

Although attention and arousal are frequently related and sometimes mixed-up terminologically, they are distinct concepts. Unlike arousal, attention is selective to a particular spatial region, time point or feature-subset of the stimulus, and increasing attention to some aspects of the stimulus implies reduction elsewhere (James 1890). In contrast, arousal up- or down-regulates processing throughout the sensory space in an unspecific manner. In Sect. 2.2 effects of attention towards bright or dark surfaces were discussed, where pupil size marks attention through a modulation of the pupil light response. Complementary to this stimulus-related effect, attentional orienting in itself may yield dilations of the pupil.

If pupil dilation is indeed linked to selective attention, stimuli that are likely to attract attention ("salient" stimuli) should yield a larger pupil response. First evidence in this direction has already been provided by an early paper on the pupil response: when stimuli of higher "interest value" are presented to an observer the pupil dilates more (Hess and Polt 1960). However, this early study based "interest value" mainly on the subjective expectation which stimulus should be of interest to male and female observers, respectively. It therefore may simply reflect sex-specific arousal or preference patterns (Rieger and Savin-Williams 2012; but see: Aboyoun and Dabbs 1998); Libby et al. (1973) demonstrated that "attentionally interesting" pictures in general yield the pupil to dilate. More importantly, these authors showed that another common measure of arousal, heart rate, is not correlated to the pupil dilation on a trial-by-trial basis. This presented one of the earliest arguments that other factors than arousal may have a direct influence on pupil size. In more recent approaches, bottom-up salience measures (e.g., Itti et al. 1998) have been related to the pupil response. If observers are asked to perform a saccade to a peripheral target, the pupil response increases with contrast. Importantly, this adds to the effect of attending a bright or dark surface as described above, and may thus be interpreted as response to stimulus salience (Wang et al. 2014). This interpretation is supported by similar results in macaque monkeys, where the increase in visual salience has the same effect on the pupil as microstimulation of the superior colliculus (Wang et al. 2012, 2014). Contrary to an interpretation of pupil dilation signaling bottom-up salience, however, the pupil dilates *more* when areas of a natural scene are about to be attended that are *low* in bottom-up salience (Mathôt et al. 2015b). Importantly, this particular study controls for unspecific factors, such as task-related effort or arousal by showing that the effect reduces in a dual-task setting as compared to fixation guidance. One interpretation of this result is that less salient stimuli require more cognitive resources (more "top-down" signal) to attract attention. In this view, pupil size in these experiments again reflects cognitive load rather than attention per se. While these results are not conflicting on the level of the data, the extent to which the pupil signals (i) bottom-up low-level salience, (ii) higher-level, but still stimulus-driven interestingness, or (iii) task-modulated priority, which combines bottom-up and top-down cues, remains an exciting issue open for further experimental investigation. In any case, it will remain important to carefully control for effects of arousal or cognitive load that will often be associated with orienting attention.

3.5 The Pupil Signals Decisions

If participants have to signal a decision after a delay, their pupil during this period is larger than if no decision is required (Simpson and Hale 1969). While several other factors could be involved in this difference (e.g., an arousal difference between the task conditions), this result was one of the first indications that pupil dilation accompanies the process of decision making as such. Indeed, the timing of the pupil response is closely linked to the time the decision is made: if observers are asked to make a free decision to press a button once within a fixed interval—similar to the classical Libet experiment (Libet et al. 1983)—the pupil dilates with a fixed lag to the free choice, and the pupil response to the decision persists even if the motor act to signal the decision is delayed (Einhäuser et al. 2010). Such results, however, raise an interesting question: which are the decision variables to which the pupil responds? Is it still the mere arousal associated with the decision or does the pupil signal specific decision variables, such as reward or risk?

3.6 Pupil Dilation as Marker of Violated Expectations and Surprise

Pupil dilation has frequently been associated with the perception of rare events and the violation of expectation. In Friedmann et al. (1973) it has been argued that if the occurrence of a specific stimulus (a double click among single clicks) is unknown at trial onset ("is uncertain") the pupil is larger the smaller the probability of occurrence is. At least at low probability, this inverse relation of pupil size to stimulus probability is independent of whether the stimulus is a target or a distractor (Qivuan et al. 1985). This relation of pupil size to probability is also evident in "oddball" experiments: similar to the P3 ERP component, pupil dilation is associated with detection of the oddball (Gilzenrat et al. 2010), though P3 and pupil dilation may be related to different aspects of processing (Kamp and Donchin 2015). A similar pupil dilation is also found for targets during rapid serial visual presentation that deviate from non-targets only in their categorical properties, such as the semantic category (Privitera et al. 2010). Importantly, this pupil response is still observed—albeit smaller in magnitude—when the target is missed or no immediate response is required. Whether these specific findings relate to the deviation of targets from the distractors as such or to a more general response to oddity is therefore unclear. Larger pupil dilation for the unexpected extends to complex cognitive judgments, such as the temporal order of events inferred from language: a sequentially presented pair of words that is consistent with the temporal order yields a weaker pupil dilation than similar word pairs that imply an incorrect temporal order (Nuthmann and van der Meer 2005). This result also illustrates a possible link between pupil responses to unexpected events to those to increased cognitive load: the unexpected temporal order—as any unpredictable event—may require more processing resources, thereby impose a higher cognitive load and thus dilate the pupil more. Similar results for a pupil response to violation of expectation in language are also found for series of temporal order judgements (Raisig et al. 2010) as well as for the non-occurrence of an expected rhyme (Scheepers et al. 2013). In sum, there is converging visual, auditory, and linguistic evidence that pupil dilation accompanies events that are improbable or violate expectations.

To get more quantitative predictions regarding the violation of expectations across these domains, a solid framework from (neuro-)economic decision making theories is required and indeed has recently entered the pupillometry field. A challenge for such quantification of distinct decision variables to the pupil response rests on the observation that many of the relevant candidates are confounded in everyday decision making and also in typical paradigms. For example, obtaining reward is frequently linked to arousal, vigilance, effort or the risk associated with the decision. The notion that the pupil specifically signals surprise has found support by an experiment involving a simple card game that dissociates changes in decision variables from the decision process as such. It reveals that pupil dilation is more tightly linked to risk prediction error than to reward, reward prediction error or risk per se (Preuschoff et al. 2011) and the effect of risk prediction error can add to unspecific effects like arousal or task demands. Since risk prediction error is one possible quantification of surprise, these results support the notion of the pupil as indicator of surprise.

3.7 Pupil Dilation and Learning

In the aforementioned card game, outcome is independent across trials and actually independent of the observer's choices. In many realistic situations, however, outcome is determined by choices and by the history of previous trials. To maximize reward in such situations, observers should learn from their experience to optimize their choices. If the outcome related to a particular choice option is uncertain, frequently an exploitation-exploration trade-off is faced. Shall one stay with an option for which outcome is less uncertain or should one explore alternative options with more uncertain outcome? Using a four-armed bandit game, Jepma and Nieuwenhuis (2011) introduced this issue in the pupillometry field. Their observers were faced with four options in each trial and the reward for each option varied smoothly across trials. Using a reinforcement-learning model (Daw et al. 2006), the authors could classify trials along the exploration-exploitation axes. Larger baseline pupil size (i.e., pupil size prior to making the choice) was associated with more explorative behavior both across trials within an observer as well as between individuals.

The notions of uncertainty and the exploitation-exploration trade-off are combined in a number prediction task (Nassar et al. 2012). Observers had to predict numbers drawn from a Gaussian distribution, whose mean occasionally changed at time points unbeknownst to the observer ("change points"). The authors distinguish the average pupil diameter in a period 2 s after the outcome (the true number) is revealed from the change in pupil diameter (2nd minus 1st second) in this interval. The probability of a change point correlated positively with pupil diameter change, in line with the surprise interpretation given above. In contrast, the uncertainty about the underlying probability distribution was coupled to the average pupil dilation. Since exploration behavior is linked to increased uncertainty, this result is in line with the aforementioned exploitation/exploration tradeoff being reflected in average pupil size. The same study also makes a step towards causality. An unpredictable (surprising) but task-irrelevant auditory cue increased pupils in both analyzed phases and at the same time learning rate (as measure of the exploitation-exploration tradeoff) was affected. Obviously, it remains open whether both the pupil dilation and the change in behavior result from a common upstream source, or whether the circuits involved in pupil control (Sect. 5) indeed are identical to those controlling the exploration-exploitation trade-off and the resulting adaptation of learning rate. In any case, the pupil reflects measures of uncertainty and as such is related to a relevant modulator of learning behavior.

The relation of pupil dilation and learning also pertains to the arguably simplest form of learning—classical (i.e., Pavlovian) conditioning (Pavlov 1927). In classical conditioning a conditioned stimulus (CS+) is consistently followed by an unconditioned stimulus (US), while another stimulus (CS−) is not. When the US is an electric shock occurring 8 s after CS onset, the pupil response differentiates between CS + and CS− about half a second after CS onset (Reinhard et al. 2006). Hence, pupil dilation can be used as marker of successful conditioning with high temporal resolution. Moreover, pupil dilation as marker of reward-based strategies are even sufficiently robust to monitor participants' use of reward information, if they are unaware of a subliminally presented reward (Bijleveld et al. 2009).

In sum, the pupil provides a comparably robust marker of variables and states during learning. Besides the implications for the neural circuitry underlying both the control of learning and pupil dilation (Sect. 5), this also opens opportunities to continuously monitor internal variables that otherwise can only be probed at a few discrete time points if interference with the learning process as such shall be minimized.

3.8 Pupil and Memory

The notion of learning is tightly interwoven with concept of memory encoding and retrieval. Already the early studies on cognitive load and pupil size included working-memory load as one dominant example: pupil size scales with the number of items encoded into working memory (Beatty and Kahneman 1966; Beatty 1982 for a review). If memory capacity is exceeded, however, the pupil might again become smaller (Poock 1973).

When either related or identical words are repeated during an encoding phase for a subsequent memory test, pupil size depends on the spacing between the repetitions (Magliero 1983): smaller pupils are found for small spacing (0 or 1 item interleaved), whereas at larger spacing (4 or 8 items interleaved) pupil size is similar to the first presentation of the item. Words that have been learnt in a previous encoding phase ("old" words) elicit larger pupils during retrieval than words that have not been learnt ("new" words) and this effect is strongest for neutral words as compared to words of positive or negative valence ("pupil old/new effect", Võ et al. 2008). Pupil size scales with the "depth" by which an item is encoded: an item that is correctly remembered elicits a larger response than a merely "known" (i.e., familiar) item, and in turn these familiar items elicit larger responses than novel items; similarly, the response during retrieval depends on the attention given during encoding (Otero et al. 2011). Moreover, the pupil old/new effect is also observed when observers are instructed to conceal their memorization (Heaver and Hutton 2011). Together, this argues against a merely "effort" or "arousal" based explanation of the pupil old/new effect, but suggests that the pupil indeed genuinely signals a difference between familiarity and novelty. The old/new effect extends to complex natural scenes: during retrieval scenes that have been seen before elicit a larger pupil than novel scenes (Naber et al. 2013c). Similar to the findings with words, this natural scene old/new effect persists in the absence of a qualified response and scales with the subjective confidence as measure of the memory depth.

Whether pupil dilation at the time of *encoding* is predictive of later retrieval success is an issue of debate. In Magliero (1983) the effect of spacing between repetitions on pupil was virtually identical when the same or a similar word was repeated, whereas later retrieval performance showed opposing spacing dependence between these experiments. This provided some evidence that pupil dilation—at least across different conditions—might not be predictive of retrieval success. Similarly, for the "old/new" effect with words, no relation between pupil dilation during encoding and later retrieval was found (Võ et al. 2008). Using auditory stimuli, Papesh et al. (2012) found that *increased* pupil dilation during encoding relates to an item being remembered more likely, whereas Naber et al. (2013c) as well as Kafkas and Montaldi (2011) reported stronger *constriction* (or smaller dilation) to be predictive of better memory encoding for visual stimuli. In addition to obvious differences in stimulus material, modality and task that distinguish the four studies, it is possible that both effects are real: a larger pupil baseline during encoding may reflect a higher level of alertness and vigilance that broadly enhances cognitive processes including memory encoding. The stronger constriction, in turn, can be related to a novelty signal that fosters an item to be remembered (cf. Knight 1996), an interpretation in line with the pupillometric data at retrieval. With an appropriate paradigm and possibly additional measures (cf. Sect. 7) there may therefore be a straightforward way to reconcile the apparently conflicting findings.

Such reconciliation could have practical consequences: knowing the probability that an item will be remembered at the time of encoding could be used to repeat or prolong the presentation of items for which the pupil predicts bad memory. In this way, a feedback of pupillometric data onto item presentation may allow more efficient encoding and thus better overall learning success.

3.9 Pupil Dilation Accompanies Changes in Perception

The relation of pupil to novelty on the one hand and consolidation of a decision on the other hand, also manifests itself in a perceptual effect. When an ambiguous stimulus is presented or two distinct stimuli are presented to the eyes, perception alternates between perceptual alternatives over time, a phenomenon referred to as rivalry. The change of perception is accompanied by a dilation of the pupil that peaks about 1 s after the perceptual change and the stability of the percept is to some extent predicted by the pupil dilation at its onset (Einhäuser et al. 2008). This finding extends to many forms of rivalry (Hupe et al. 2009), to the first experience of a specific perceptual interpretation (Kietzmann et al. 2011), and to binocular rivalry (Naber et al. 2011). One interpretation of this finding states that the mechanisms that consolidate a decision in the presence of equal outcome estimates for an alternative also yield perception to consolidate when two or more stimulus interpretations are equally likely. Alternatively, the dilation after the change of the percept could be related to a notion of surprise in the event a stable percept changes into something else (Kloosterman et al. 2015). Reconciling this views, surprise might actually be a trigger signal that allows consolidation. In addition, there frequently is a small dip observed at the onset of the new perceptual state and it would be interesting to see whether this is related to a brief instability reflected by less dilation that enables the perceptual switch or whether the dip relates to a novelty signal related to increased pupillary constriction. In any case, the relation between changes in pupil size and in perceptual interpretation seems a robust phenomenon with plenty, as of now largely unexplored, possible applications.

3.10 The Challenge to Dissociate Distinct Cognitive Factors that Dilate the Pupil

Since the 1960s, many cognitive phenomena and processes have been related to pupil size. The present section should have illustrated that a dissociation between these factors is not always straightforward. In many paradigms the variable of interest, such as for example novelty or violation of expectancy, may be associated with increased task difficulty and thus with an increased cognitive load. Task difficulty and cognitive load, in turn, may be associated with increased arousal.

Nonetheless, it seems likely that many of the discussed factors have a unique contribution to pupil size, at least at some point in the complex time course of the pupil response. Many of the aforementioned recent studies have begun to dissociate two or more of the factors. To follow this lead and develop paradigms that carefully dissociate the many potential contributors to pupil dilation and constriction will remain one of the key challenges for pupillometric studies in the near future to eventually decide which cognitive processes indeed drive pupil dilation directly and which do so by modulating comparably global variables, such as vigilance and arousal. Only with such an understanding through appropriate experimental design, questions on the neural substrate and the ecological relevance of the pupil response —beyond a reflex to increased illuminance on the eye—can be fully appreciated.

4 The Pupil in Social Interactions

The function of the pupil light reflex seems obvious, to regulate the amount of light incident on the retina. Indeed, pupil size seems to optimize visual acuity for a given luminance level (Campbell and Gregory 1960). In contrast, whether there is a functional role of pupil-size changes due to cognitive factors or whether it is just an epiphenomenon of neurotransmitter release is open. One hypothesis for a functional role of pupil dilation under constant illumination is a social function. There are two components to investigating social effects of the pupil. First, it needs to be demonstrated that the pupil reacts to the social cue of interest or reveals information about the beholder's emotional state (Sects. 3.2, 3.3); second, it needs to be tested whether the information conveyed by a person's pupil indeed influences judgements by others about the internal state or intentions of the pupil's beholder.

4.1 Belladonna—Observed Pupil Size and Attractiveness

Owing to the "belladonna" effect described at the beginning of this chapter, the relation of pupil size to perceived attractiveness has drawn arguably most attention among all social effects of pupil size. In typical paradigms, photographs of males and/or females are manipulated to increase or decrease the depicted pupil size while keeping other facial features constant. Then observers are asked to judge attractiveness of the respective pictures and frequently the observers' own pupil dilation is measured to assess to what extent one's own pupil mimics the observed pupil size.

In an early scientific inquiry into this question, Simms (1967) demonstrated that for pictures depicting a person of the opposite sex, larger pupils in the picture evoke a larger pupil response, while the reverse is true for pictures depicting a person of the same sex. Hess and Petrovich (1987) report a pilot experiment, which confirms the finding for males watching a picture of a female and that the subjects reported "greater positive feelings" (p. 329) for the photograph with the larger pupil, even though the majority of subjects was apparently unaware of the pupil size manipulation in the picture. Bull and Shead (1979) find an asymmetry between men and women: in their study all pupil effects on attractiveness were limited to pictures of females. In line with Simms (1967), they found reversed patterns for female observers, who rated images with larger pupils as less good looking, as compared to adult males, who attributed higher attractiveness ratings to pictures with larger pupils. Interestingly, male children (age 10) match the pattern of women observers. In a large sample of female participants, attractiveness ratings for pictures depicting females decrease with increasing pupil size, while the ratings show a quadratic dependence for pictures depicting males, with largest ratings at medium sized pupils (Tomlinson et al. 1978). Female preference for medium as compared to large pupil sizes in male pictures may in addition be modulated by selection preferences (Tombs and Silverman 2004). Consequently, the results on females rating pictures of males are somewhat mixed. However, the results on pictures of females seem more or less unanimous: adult male observers perceive females with larger pupils as more attractive, while the reverse is true for female observers. This lends at least some scientific support for belladonna's intended effect.

4.2 Observed Pupil Size Modulates the Perception of Emotions

Beyond attractiveness, perceived pupil size also seems to enhance the emotion perceived in another person's face. Pupil size in a picture affects observer's sadness judgement, in that faces with smaller pupils are rated more negative (i.e., as more sad, Harrison et al. 2006). The same study finds a correlation between the observer's own pupil size and the perceived pupil size. Interestingly, the effect in this study with a small sample seems restricted to sadness stimuli, and does not extend to neutral, happy or angry faces. Pupil size during sadness processing is also reflected in a variety of brain structures associated with emotion processing, including the amygdala (Harrison et al. 2006). The exact role of the amygdala in processing perceived pupil size has, however, remained somewhat controversial. Similar to the early studies by Simms (1967), Demos et al. (2008) presented pictures of females with manipulated pupil size to a group of male observers and found stronger amygdala activity when images with larger pupils were presented, again with observers being unaware of the pupil size manipulation. Unlike the results in (Harrison et al. 2006), this implies that amygdala responds stronger to more dilated pupils, consistent with the notion that amygdala stimulation would lead to dilated pupils, at least in animals (Ursin and Kaada 1960). Amemiya and Ohtomo (2012) confirm that amygdala mirrors the observed pupil size, but find their effect to be independent of gender and attractiveness ratings, though limited to human as compared to cat faces. In any case, the monitoring of another's pupil size appears to be a continuous process: there is differential brain activation if the observed pupil size is correlated with one's own pupil size as compared to a situation in which the observed size is negatively related to one's own pupil size (Harrison et al. 2009). In this view, the conflict between observed and expressed pupil size prompts social attention.

4.3 The Pupil in Competitive Scenarios

Since the pupil size signals decisions (Sect. 3.5) and is related to the level of deception employed (Wang et al. 2006; see Sect. 8.1), it seems plausible to assume that observing a competitor's pupil could be used to one's own advantage. While it is frequently purported that this might be one reason for poker players to wear sunglasses and it seems that the information can in principle be read off another one's pupil at average interaction distance, explicit knowledge is needed for the exploitation of such information (Naber et al. 2013a). Taken together with the emotional and attractiveness signals, of which the observers in the aforementioned studies are typically also unaware, this highlights that observers typically draw inferences about another person's state based on their pupil size without the observers' explicit awareness.

5 The Physiological Basis of Pupil Control

5.1 Interplay of Pathways Involving Sympathetic and Parasympathetic System

Since the focus of the present chapter is on cognitive aspects, the basic physiology and anatomy are only covered briefly, and the reader is referred to review articles and books on this topic (e.g., Loewenfeld 1993; Samuels and Szabadi 2008a, b) for more detail. In brief, pupil size is controlled by two muscles, the iris sphincter muscle (yielding constriction) and the iris dilator muscle (yielding dilation). The sphincter is innervated by the parasympathetic system, and receives input from the Edinger-Westphal (EW) nucleus via a disynaptic cholinergic projection. The EW nucleus receives direct input from the pretectal nucleus, which receives input directly from the retina, thus controlling the pupil light response. The iris dilator is innervated by the sympathetic system via an all excitatory projection from the Locus Coerulus (LC) involving a noradrenergic (NA) projection from LC to pre-ganglionic sympathetic neurons that disynatptically project to the muscle via cholinergic and another α1-NA synapse. Importantly, LC also projects via an inhibitory α2-NA synapse to the EW, thereby inhibiting parasympathetic con-striction. Consequently, LC plays an important role in the control of the pupil. NA-release both dilates the pupil directly via the sympathetic pathway and indirectly through inhibition of the parasympathetic pathway (Samuels and Szabadi 2008a, b). Consequently, both autonomic systems, sympathetic and parasympathetic, are involved in mediating cognitive effects on the pupil.

5.2 *Dissociating Less Constriction from More Dilation*

Since inhibition of the parasympathetic pathway has the same net dilatory effect as excitation of the sympathetic pathway, the two effects are hard to dissociate by pupillometry alone. Even though it seems conceivable to image the differential muscle activity of sphincter and dilator non-invasively, such attempts to the best of the author's knowledge have not yet succeeded. Consequently, separation of the two pathways seems only possible by pharmacological intervention. Steinhauer et al. (2004) performed a set of two experiments to address this issue first by indirect and then by direct means. They had participants perform continuous arithmetic tasks of different difficulty (subtraction of 7, addition of 1) under light or dark conditions. In line with the studies on cognitive load (Sect. 3.1), before verbalization was started (i.e., during task preparation) pupils dilated more for the difficult task. On top of this baseline, differential task effects were only observed under comparably higher illumination, suggesting that the baseline effect was mainly sympathetic and the add-on mostly parasympathetic. Selective pharmacological blockade of either muscle then confirmed the comparably mild effect of the sympathetic pathway, as compared to the stronger modulation—both for the baseline and the add-on dilation—through inhibition of the parasympathetic pathway. In combination these data suggest that conducting experiments under different illumination levels allows some dissociation between parasympathetic and sympathetic pathway, even without pharmacological intervention.

5.3 *The Role of the Locus Coerulus (LC)*

Besides anatomical considerations, the role of the LC in pupil control has two further lines of support. First, models of LC function are remarkably well matched by pupillometric findings; second, there is evidence from physiology and imaging that converges to a tight coupling between pupil dilation and LC activity. Rather independent from pupillometric studies, it has been suggested that LC relates to behaviour in a similar way that is observed for pupil dynamics. A link between LC activity and arousal has been suggested at about the same time as the earliest pupillometry studies (see Berridge 2008 for review). However, similar to the pupillometric references to arousal (Sect. 3.2), arousal is a very broad term. More specifically, Aston-Jones and Cohen (2005) as well as Bouret and Sara (2005) identified two modes of LC activity: a phasic mode that corresponds to the consolidation of behaviour and a tonic mode that rather reflects task disengagement.

Exactly this exploitation/exploration tradeoff and its link to an adaptive gain theory of LC function motivated Jepma and Nieuwenhuis' (2011) and Gilzenrat et al.'s (2010) pupillometric studies, which found this tradeoff reflected in pupil dynamics (see Sects. 3.6, 3.7). Moreover, it has been suggested that the LC-NA system signals surprising events, in particular unexpected uncertainty (Yu and Dayan 2005; Dayan and Yu 2006). Again, this result parallels and precedes the observed effects of surprise and unexpected uncertainty on pupil size (Sect. 3.6). Consequently, several functions that have been suggested to be controlled by LC based on physiological evidence in non-human animals, have also been linked to pupil size in human observers. Despite such wealth of converging indirect evidence for the involvement of the LC in the control of cognitive effects on pupil dilation, direct electrophysiological evidence for the link between LC activity and pupil size is surprisingly scarce. To date, there is only a single set of published electrophysiological results in monkey (Rajkowski et al. 1993). Localization of LC in standard functional imaging is difficult (Astafiev et al. 2010). Hence, only recently the combination of simultaneous functional MRI and pupillometry during an oddball task with neuromelanin-sensitive structural imaging to localize LC could provide the first direct evidence for the co-variation of LC activity and pupil size in humans (Murphy et al. 2014). Together with the electrophysiological data in monkey, the pharmacological and anatomical data, this provides converging evidence for a strong connection between LC-NA activity and pupil size.

5.4 The Superior Colliculus (SC) Hypothesis

Recently, an alternative hypothesis to LC as predominant driver of cognitive effects on the pupil has emerged. Noting the tight link between attentional orienting and pupil size (Sects. 2.2 and 3.4), the superior colliculus (SC), a structure that integrates information from many cortical and subcortical sources and controls attentional deployment in space as well as gaze shifts, has been implicated to play an important role in the control of pupil size (see Wang and Munoz 2015 for detailed review). In monkey, microstimulation of the LC triggers pupil dilation even if no overt shift of attention (i.e., no gaze shift) is evoked (Wang et al. 2012). Similar pupil dilation is seen in humans when directing attention to a salient target (Wang et al. 2014). Moreover, pupils are more dilated for faster and for more difficult (anti- versus prosaccade) saccadic eye movements (Wang et al. 2015) in humans, which is in line with SC activity patterns in monkeys. The SC and the LC hypotheses are, however, not mutually exclusive. Anatomically, SC could yield increases in pupil size in a similar way as LC: inhibition of parasympathetic constriction or activation of sympathetic dilation. It is possible that LC and SC subserve complementary function: for example, LC activity is apparently involved in consolidating a decision to saccade by signaling the final commitment to saccade execution (Kalwani and Gold 2008). In this view, SC may guide the orienting process and LC then eventually "protect" the chosen action against competing signals. As such, SC would be more involved in judging the sensory aspect of the decision, and LC would track and consolidate the behavioral response (see also Clayton et al. 2004). In any case, dissociating the roles and determining the interplay between LC and SC in controlling the pupil will be a most exciting topic for combined neurophysiological and pupillometric research in the near future.

6 Technical Issues

6.1 *Spectral Sensitivity of the Pupil Light Response and "Isoluminance"*

When using visual stimuli to study cognitive influences on pupil size, experimental design should minimize the effects of the stimuli as such. Ideally, one would like to operate in a regime where the pupil is neither too large nor too small to avoid ceiling and floor effects, respectively. An obvious approach seems to render stimuli across conditions isoluminant. Stimulus design is, however, complicated by the fact that the relation between light incident on the eye and the resulting size of the pupil is not straightforward. Besides the integrated luminous flux incident on the eye, i.e., the illuminance, factors like stimulus size and the age of the observer modulate pupil size (Watson and Yellott 2012). Functions that provide pupil size as function of adapting light level in addition extrapolate into the mesopic and scotopic range, even if their luminance definition assumes photopic light levels and thus a cone-dominated spectral sensitivity. To complicate the issue of the spectral sensitivity of the pupil size, the pupil's adaptation to light levels is not only controlled by the so-called image-forming pathway of vision (i.e., by rods and cones), but also by the "non-image-forming" pathway, based on intrinsically photo-sensitive retinal ganglion cells (ipRGC) with melanopsin as photopigment. This is evidenced by Melanopsin knock-out rodents showing a diminished pupil light reflex (Lucas et al. 2003), together with data showing that ipRGCs seem to play a similar functional role in primates, including humans (Dacey et al. 2005). When psychophysically isolating pupil responses to these three classes of photosensitive cells (rods, cones, ipRGCs), the pupil light reflex to comparably brief flashes turns out to be different for the three regimes: rod and ipRGC mediated responses are mostly linear in the flux density of light incident on the cornea, whereas cone-mediated responses are comparably insensitive to stimulus size and mostly depend on stimulus luminance (Park and McAnay 2015). While the exact dependence of pupil responses to light is an active field of research, the dependence on 3 different classes of photosensitive cells indicates a complication for stimulus design when studying cognitive effects on pupil size. Luminance by definition weights the irradiant power with the absorption sprectra of the relevant photoreceptor; that is, for the photopic range, with the $V(\lambda)$ curve. With the spectral sensitivity of melanopsin being different from the $V(\lambda)$ curve, design of stimuli that are "isoluminant" from the perspective of the pupil (in the sense that they trigger an identical response of the relevant photoreceptor class) is a non-trivial task. Even under clearly photopic conditions, truly isoluminant will not necessarily imply an unchanged signal for the pupil control circuits. Consequently, even if stimuli are isoluminant between different conditions (say differently colored cues for different attention tasks), a careful balancing of stimuli across conditions is certainly advisable.

6.2 Foreshortening

Many experimental paradigms ask for observers being free to shift their gaze. Besides the many *true* effects on pupil size that are associated with attentional and gaze shifts outlined above, shifts in gaze may also yield an *apparent* effect on the pupil size measurement, especially when a camera or a video-based eye-tracker are used to measure pupil size. It is important to realize that any camera measures the projection of an area on its 2D receptor surface (i.e., the film or in most cases its CCD chip). Consequently, any rotation of the eye relative to the camera's line of sight will change the size of its projection, with the full pupil area only measured if the area is perpendicular to the camera's line of sight (i.e., if the observer looks straight ahead into the camera). If this so-called "foreshortening" of area is not considered by the eye-tracker's software and not corrected for either, eye position may confound pupil measurements. Ideally, the foreshortening error should be corrected for, and several recent papers lay out procedures to do so in detail for commonly used eye-tracking systems (e.g., Gagl et al. 2011; Brisson et al. 2013; Hayes and Petrov 2015). If the eye-tracker or camera geometry is not sufficiently well known or if the estimation of the gaze direction contains large errors (e.g., because only the pupil was tracked), for small deviations from the straight ahead and large pupillometric effects, the foreshortening effect might be negligible in some cases. Nonetheless, even in these cases it is highly advisable to verify that any apparent pupillometric difference between any two conditions is not in fact a difference in gaze patterns.

7 Combination of Pupillometry with Other Techniques

7.1 Combination with EEG

The combination of pupillometry and EEG dates back into the early days of pupillometric research. A prototypical example in this respect is the relation between pupil dilation and the P3 component of the ERP. In their paper reporting the pupil probability effect, Friedmann et al. (1973) simultaneously record EEG and pupil response and use the similarity in the response of the pupil and the P3 component to support the notion of the P3 as oddball detector related to attentional orienting. Only recently, it has been shown that the amplitudes of the pupil response and the P3 are not linearly correlated, but instead large P3 amplitudes correlate with medium pre-trial pupil sizes (Murphy et al. 2011). This suggests that P3 and pupil size may have complementary relations to LC-NA activity and thus together may index phasic and tonic LC activation during continuous task performance. Even when dissociating different subcomponents in the P3, such as responses to deviance (rare events on a categorical level) and novelty (of a particular item), the pupil response does not correlate with either on a trial-by-trial basis and thus seems to signal complementary information (Kamp and Donchin 2015). Interestingly, the relation between performance and pupil size, especially prior to trial onset, seems to depend on the difficulty of the specific task at hand. Consequently, the precise relation between ERP components, such as the P3, and pupil responses for a broader variety of tasks seems an interesting issue for further investigation. Not only can remote video-based pupillometry and EEG be combined rather simply from a methodological and technological point of view, but also do paradigms that are particularly well suited for EEG (e.g., requiring no eye movements, motor responses outside the time window of interest, etc.) also bear advantages for pupillometry. The results that ERP and pupil response are not redundant but rather complementary make such a combination of the two techniques therefore seem most promising for all areas in which either of them has been widely used.

7.2 Combination with FMRI

Since most eye-tracking manufacturers started to offer high-resolution devices that can be used inside an MRI device, the simultaneous combination of functional imaging and pupillometry has become feasible. The potential usefulness of this combination is three-fold. First, functional imaging may help unveiling the neural circuitry underlying pupil control (see Sect. 5); second, provided the difficulty to localize pupil control structures like the LC in standard fMRI techniques (Astafiev et al. 2010), the relation of BOLD activity to pupil dilation can been used to verify and improve the localization of these structures (Sterpenich et al. 2006); third, for paradigms for which a relation between a cognitive factor and pupil size are well established, pupil size can serve as continuous regressor in lieu of an overt discrete response (e.g., a button press). Such an approach has first been suggested, examined and validated in the context of a working memory task (Siegle et al. 2003), highlighting that pupil size can capture inter- and intra-subject variability to improve fMRI sensitivity. More recent examples include pupil size as index of listening load (Zekveld et al. 2014) or as marker of pain empathy in patients and healthy controls (Paulus et al. 2015) during simultaneous fMRI. Especially in the context of resting state fMRI, where no task is being performed and thus any overt report on one's own state would interfere with the resting-state logic, the correlation between pupil size and BOLD signal may provide an additional tool (Yellin et al. 2015). In sum, using pupil size as continuous marker of internal state and as regressor to brain-activity data not only potentially increases power, but also may decrease confounds related to an overt response during imaging.

7.3 Combination with Other Peripheral Measures

Throughout the history of pupillometry it has frequently been combined with other peripheral physiological measures, such as heart rate or electrodermal activity. However, even if these measures were obtained simultaneously, they frequently have been first analyzed in isolation and then summary measures, such as peaks, latencies or average time courses have been compared. Early studies, even if they reported differences—for example in latency—frequently focused on the similarity of pupil size to the other peripheral measures (e.g., Tursky et al. 1969). Only recently, the question how the information contained in the raw time series could be combined has come more into focus. For example, in the context of pain, the appropriate combination of electrodermal and pupillometric measurements over time provided better predictions of individual pain ratings than indices derived from either measure alone (Geuter et al. 2014). Given the ease of access to all of these measures, and provided the analytic tools to analyze complex time series available to date, it will be exciting to see to which other fields this complementarity extends to.

8 Applications

8.1 Lie Detection

The tight link between arousal and pupil size as well as the link to decision making have spurred the hope to use pupil size as marker for lie detection, similar to the electrodermal activity used by means of a "polygraph" in some jurisdictions. In this logic, lying and deception require more mental resources or yield more arousal and consequently a larger pupil. Indeed, early attempts (Berrien and Huntington 1943) showed more variation (dilation followed by constriction) in pupil size during deceptive than during non-deceptive responses, though the information about the deception in the pupil did not exceed information contained in heart rate. More recently, it was confirmed that sending deceptive messages yields pupil dilation, and found that on average pupil size scales with the degree of deception (Wang et al. 2006). While this may be beneficial in situations where the average outcome shall be optimized over a series of many individual games, the sensitivity and specificity is unlikely to allow identification of an individual lie or deceptive measure. In addition, since increasing mental load or arousal allows some indirect volitional control over the pupil, performing some math or thinking of a dark arousing situation during control (baseline) trials, may present an effective countermeasure against such efforts. Consequently, it is exceedingly unlikely that pupillometry can provide a "lie-detection" tool that is sufficiently robust to reach the standards required for court and should not be relied upon in any interrogation.

8.2 *Diagnostics and Treatment Monitoring*

In first-responder practice and intensive care, the pupil light reflex is a widely used tool to assess normal brain function. In the intact brain, shining a bright light into one eye will yield an immediate constriction of both pupils. Provided the ease of access to the pupil, this raises the question as to whether more sophisticated pupil responses can similarly be applied to more complex diagnostic tasks. Since monitoring the pupil can readily be achieved in clinical settings, the potential use for clinical pupillometric applications is large. It becomes especially interesting in patient groups who are incapable of providing reliable information about their internal cognitive state, such as young children or psychiatric patients.

Pupil as marker of schizophrenia dates back at least to a study in the mid-1940s (May 1948). Interestingly, this study already reports both an abnormal pupil light reflex as well as a reduced pupil dilation to applying a painful stimulus (a twitch to the skin at the neck) in schizophrenics as compared to healthy controls. With the technological advance in pupillometric measurements, it became feasible to use pupillometry as a tool of treatment monitoring and disease progression in schizophrenia (see Steinhauer and Hakerem 1992 and Steinhauer 2002 for detailed accounts). Besides generally reduced pupil responsiveness and longer latencies, abnormalities of the pupil response in schizophrenia include a smaller dilation to rare events (Steinhauer and Zubin 1982), and have been related to working memory deficits (Granholm et al. 1997), inappropriate timing of attentional allocation (Granholm and Verney 2004), and overall limitation of processing resources (Fish and Granholm 2008). Provided that eye-movement abnormalities also present an important marker for schizophrenia (e.g., Kojima et al. 2001) and the relation between orienting circuitries and the pupil response (Sects. 3.4, 5.4), concurrent assessments of pupil and eye-movement abnormalities seem an interesting issue for future research in schizophrenia.

The relation of pupil dilation to anxiety that is observed in subclinical populations (Sect. 3.3) extends to anxiety disorders and phobias. For example, the pupil reaction to pictures of slithering snakes compared to control stimuli is increased only for snake phobics but not for controls, while pictures of attacking snakes affect the pupils of both populations alike (Schaefer et al. 2014). Patients suffering from posttraumatic stress disorder show increased pupil dilation as compared to controls during task performance, which is not specific to traumatic stimuli (Felmingham et al. 2011). Pupil as measure of anxiety may also be used to assess vulnerability in children: children of anxious mothers show increased pupil responses to angry faces (Burkhouse et al. 2014). Similarly, children of depressed mothers show increased pupil responses to sad faces. In turn, pupil dilation also shows abnormalities in depressed adults, such as decreased dilation for rare events as compared to controls, though to a lesser extent than in schizophrenics (Steinhauer and Zubin 1982). Such examples illustrate the potential usefulness of pupillometry as objective tool to diagnose and monitor psychiatric diseases or vulnerability to such.

Of the many neurological diseases that may affect pupil size more or less directly, Parkinson's disease (PD) has recently received particular interest. Patients suffering from PD show a slowing of the pupil light reflex that is reflected in an increased latency, a decreased amplitude as well as decreased velocities and accelerations of the PLR (Micieli et al. 1991; Giza et al. 2011). Since PD patients show specific impairments for tasks involving anti-saccades (Briand et al. 1999), it will also be interesting to test the results relating to the SC hypothesis of pupil control (Sect. 5.4) in PD patients. Moreover, it will be an interesting issue to test, whether pupilloemetric abnormalities extend to other diseases that have been suggested as prodromal forms for PD, such as restless leg syndrome or REM sleep behavioral disorder. The latter could be of particular interest, provided that an impairment of the LC-NA system is a likely cause of this disorder, especially if it co-occurs with PD (Garcia-Lorenzo et al. 2013).

Finally, provided the relation of pupil size to memory and cognitive load, a relation to dementia and memory impairment is apparent. Indeed patients suffering from Alzheimer's disease (AD) show less pupillary constriction as compared to normals (Fountoulakis et al. 2004; Prettyman et al. 1997), which likely results from reduced levels of ACh in AD (Francis et al. 1999). In sum, the aforementioned examples illustrate how a large variety of psychiatric and neurological disorders show abnormalities in pupil responses. Given that pupillometry can be performed with comparably inexpensive and easy to maintain equipment, the vast potential for pupillometry as tool for early diagnosis as well as the monitoring of treatment and disease progression has only been started to be explored.

8.3 Communication

Increasing and decreasing mental load, for example by conducting mental arithmetic, gives some volitional control over the pupil, which can be used to conduct binary tasks (such as replying to yes/no questions) with potential application to communication with the otherwise immobilized and possibly as diagnostic aid for patients whose state of consciousness is in question (Stoll et al. 2013a). Importantly, this procedure does not require any patient-specific training and is thus a potential means of *newly establishing* communication. Even though for this group of patients communication at essentially any speed is a considerable improvement over no communication at all, the current throughput of about 1 yes/no question per 30 s (i.e., 2 bits per minute) is clearly suboptimal. It will be interesting to see whether effects of covertly shifting attention to black or white surfaces or to surfaces that flicker at a fixed frequency (Sect. 2.2) and/or combination with other brain-computer-interfacing techniques (such as EEG-based methods) will allow a substantial improvement and make pupillometric systems not only feasible in day-to-day clinical and rehabilitation practice, but also of interest to less restricted patients and possibly even healthy users.

9 Summary

The examples laid out in the present chapter demonstrate that changes in pupil size extend far beyond a purely reflexive behavior. High-level cognitive demands and processes both modulate the response of the pupil to incident light and by themselves affect pupil dilation directly. Consequently, the dynamics of the pupil presents a time-varying signal that seems almost as rich as event-related potentials in EEG. Unlike electrical activity, however, the pupil is a robust signal that can be read out with comparably inexpensive techniques and is available even to an observer's naked eye. Despite the substantial work in the 1960s, the modern pupillometric research is still a young field in a rapidly growing phase. With gaze-tracking technology becoming increasingly wide-spread, with high-speed cameras affordable, and pupil-tracking algorithms overcoming the technological hurdles associated with gaze-shifts, it seems well conceivable that pupillometry will increase its role towards becoming a standard tool to assess human cognitive and (neuro-)physiological function, and may serve as basis to a manifold of clinical and technical applications in the near future.

References

Aboyoun DC, Dabbs JN (1998) The hess pupil dilation findings: sex or novelty? Soc Behav Pers 26(4):415–419

Amemiya S, Ohtomo K (2012) Effect of the observed pupil size on the amygdala of the beholders. Soc Cogn Affect Neurosci 7:332–341. doi:10.1093/scan/nsr013

Astafiev SV, Snyder AZ, Shulman GL, Corbetta M (2010) Comment on "Modafinil shifts human locus coeruleus to low-tonic, high-phasic activity during functional MRI" and "Homeostatic sleep pressure and responses to sustained attention in the suprachiasmatic area". Science 328 (5976):309. doi:10.1126/science.1177200

Bárány H, Halldén U (1948) Phasic inhibition of the light reflex of the pupil during retinal rivlary. J Neurophysiol 11(1):25–30

Berrien FK, Huntington GH (1943) An exploratory study of pupillary responses during deception. J Exp Psychol 32(5):443–449

Beatty J, Wagoner BL (1978) Pupillometric signs of brain activation vary with level of cognitive processing. Science 199(4334):1216–1218

Beatty J (1982) Task-evoked pupillary responses, processing load, and the structure of processing resources. Psychol Bull 91(2):276–292

Berridge CW (2008) Noradrenergic modulation of arousal. Brain Res Rev 58(1):1–17. doi:10.1016/j.brainresrev.2007.10.013

Binda P, Pereverzeva M, Murray SO (2013) Attention to bright surfaces enhances the pupillary light reflex. J Neurosci 33:2199–2204
Binda P, Murray SO (2015) Spatial attention increases the pupillary response to light changes. J Vis 15(2):1. doi:10.1167/15.2.1
Binda P, Pereverzeva M, Murray SO (2014) Pupil size reflects the focus of feature-based attention. J Neurophysiol 112(12):3046–3052. http://dx.doi.org/10.1152/jn.00502.2014
Bijleveld E, Custers R, Aarts H (2009) The unconscious eye opener: pupil dilation reveals strategic recruitment of resources upon presentation of subliminal reward. cues. Psychol Sci 20 (11):1313–1315. doi:10.1111/j.1467-9280.2009.02443.x
Bitsios P, Szabadi E, Bradshaw CM (2002) Relationship of the 'fear-inhibited light reflex' to the level of state/trait anxiety in healthy subjects. Int J Psychophysiol 43(2):177–184
Bitsios P, Szabadi E, Bradshaw CM (2004) The fear-inhibited light reflex: importance of the anticipation of an aversive event. Int J Psychophysiol 52(1):87–95
Boersma F, Wilton K, Barham R, Muir W (1970) Effects of arithmetic problem difficulty on pupillary dilation in normals and educable retardates. J Exp Child Psychol 9(2):142–155
Bradley MM, Miccoli L, Escrig MA, Lang PJ (2008) The pupil as a measure of emotional arousal and autonomic activation. Psychophysiology 45:602–607
Bradshaw J (1967) Pupil size as a measure of arousal during information processing. Nature 216 (5114):515–516
Briand KA, Strallow D, Hening W, Poizner H, Sereno AB (1999) Control of voluntary and reflexive saccades in Parkinson's disease. Exp Brain Res 129:38–48
Brisson J, Mainville M, Mailloux D, Beaulieu C, Serres J, Sirois S (2013) Pupil diameter measurement errors as a function of gaze direction in corneal reflection eyetrackers. Behav Res Meth 45(4):1322–31. doi:10.3758/s13428-013-0327-0
Bull R, Shead G (1979) Pupil dilation, sex of stimulus, and age and sex of observer. Percept Mot Skills 49(1):27–30
Burkhouse KL, Siegle GJ, Gibb BE (2014) Pupillary reactivity to emotional stimuli in children of depressed and anxious mothers. J Child Psychol Psychiatry 55(9):1009–1016. doi:10.1111/jcpp.12225
Campbell FW, Gregory AH (1960) Effect of size of pupil on visual acuity. Nature 187:1121–1123
Chapman CR, Oka S, Bradshaw DH, Jacobson RC, Donaldson GW (1999) Phasic pupil dilation response to noxious stimulation in normal volunteers: relationship to brain evoked potentials and pain report. Psychophysiology 36(1):44–52
Clayton EC, Rajkowski J, Cohen JD, Aston-Jones G (2004) Phasic activation of monkey locus ceruleus neurons by simple decisions in a forced-choice task. J Neurosci 24(44):9914–9920
Dacey DM, Liao HW, Peterson BB, Robinson FR, Smith VC, Pokorny J, Yau KW, Gamlin PD (2005) Melanopsin-expressing ganglion cells in primate retina signal colour and irradiance and project to the LGN. Nature 433(7027):749–754
Daw ND, O'Doherty JP, Dayan P, Seymour B, Dolan RJ (2006) Cortical substrates for exploratory decisions in humans. Nature 441(7095):876–879
Dayan P, Yu AJ (2006) Phasic norepinephrine: a neural interrupt signal for unexpected events. Network 17(4):335–350
Demos KE, Kelley WM, Ryan SL, Davis FC, Whalen PJ (2008) Human amygdala sensitivity to the pupil size of others. Cereb Cortex 18:2729–2734
Deubel H, Schneider WX (1996) Saccade target selection and object recognition: evidence for a common attentional mechanism. Vis Res 36(12):1827–1837
Einhäuser W, Stout J, Koch C, Carter O (2008) Pupil dilation reflects perceptual selection and predicts subsequent stability in perceptual rivalry. Proc Natl Acad Sci USA 105(5):1704–1709. doi:10.1073/pnas.0707727105
Einhäuser W, Koch C, Carter OL (2010) Pupil dilation betrays the timing of decisions. Front Hum Neurosci 4:18. doi:10.3389/fnhum.2010.00018
Fahle MW, Stemmler T, Spang KM (2011) How much of the "unconscious" is just pre -threshold? Front Hum Neurosci 5:120. doi:10.3389/fnhum.2011.00120

Felmingham KL, Rennie C, Manor B, Bryant RA (2011) Eye tracking and physiological reactivity to threatening stimuli in posttraumatic stress disorder. J Anxiety Disord 25(5):668–673. doi:10.1016/j.janxdis.2011.02.010
Fish SC, Granholm E (2008) Easier tasks can have higher processing loads: task difficulty and cognitive resource limitations in schizophrenia. J Abnorm Psychol 117(2):355–363
Fountoulakis KN, St Kaprinis G, Fotiou F (2004) Is there a role for pupillometry in the diagnostic approach of Alzheimer's disease? A review of the data. J Am Geriatr Soc 52(1):166–168
Francis PT, Palmer AM, Snape M, Wilcock GK (1999) The cholinergic hypothesis of Alzheimer's disease: a review of progress. J Neurol Neurosurg Psychiatry 66(2):137–147
Friedman D, Hakerem G, Sutton S, Fleiss JL (1973) Effect of stimulus uncertainty on the pupillary dilation response and the vertex evoked potential. Electroencephalogr Clin Neurophysiol 34 (5):475–484
Gagl B, Hawelka S, Hutzler F (2011) Systematic influence of gaze position on pupil size measurement: analysis and correction. Behav Res Meth 43(4):1171–1181. doi:10.3758/s13428-011-0109-5
Garcia-Lorenzo D, Longo-Dos Santos C, Ewenczyk C, Leu-Semenescu S, Gallea C, Quattrocchi G, Pita Lobo P, Poupon C, Benali H, Arnulf I, Vidailhet M, Lehericy S (2013) The coeruleus/subcoeruleus complex in rapid eye movement sleep behaviour disorders in Parkinson's disease. Brain 136:2120–2129
Geuter S, Gamer M, Onat S, Büchel C (2014) Parametric trial-by-trial prediction of pain by easily available physiological measures. Pain 155(5):994–1001
Gilzenrat MS, Nieuwenhuis S, Jepma M, Cohen JD (2010) Pupil diameter tracks changes in control state predicted by the adaptive gain theory of locus coeruleus function. Cogn Affect Behav Neurosci 10(2):252–269. doi:10.3758/CABN.10.2.252
Giza E, Fotiou D, Bostantjopoulou S, Katsarou Z, Karlovasitou A (2011) Pupil light reflex in Parkinson's disease: evaluation with pupillometry. Int J Neurosci 121(1):37–43. doi:10.3109/00207454.2010.526730
Granholm E, Morris SK, Sarkin AJ, Asarnow RF, Jeste DV (1997) Pupillary responses index overload of working memory resources in schizophrenia. J Abnorm Psychol 106(3):458–467
Granholm E, Verney SP (2004) Pupillary responses and attentional allocation problems on the backward masking task in schizophrenia. Int J Psychophysiol 52(1):37–51
Harrison NA, Singer T, Rotshtein P, Dolan RJ, Critchley HD (2006) Pupillary contagion: central mechanisms engaged in sadness processing. Soc Cogn Affect Neurosci 1:5–17
Harrison NA, Gray MA, Critchley HD (2009) Dynamic pupillary exchange engages brain regions encoding social salience. Soc Neurosci 4:233–243. doi:10.1080/17470910802553508
Hayes TR, Petrov AA (2015) Mapping and correcting the influence of gaze position on pupil size measurements. Behav Res Meth
Heaver B, Hutton SB (2011) Keeping an eye on the truth? Pupil size changes associated with recognition memory. Memory 19(4):398–405. doi:10.1080/09658211.2011.575788
Hess EH, Polt JM (1960) Pupil size as related to interest value of visual stimuli. Science 132 (3423):349–350
Hess EH, Polt JM (1964) Pupil size in relation to mental activity during simple problem-solving. Science 143:1190–1192. doi:10.1126/science.143.3611.1190
Hess EH (1965) Attitude and pupil size. Sci Am 212:46–54
Hess EH, Petrovich SB (1987) Pupillary behavior in communication. In: Siegman AW, Feldstein S (eds) Nonverbal behavior and communication. Hillsdale, NJ, Lawrence Erlbaum
Hupé JM, Lamirel C, Lorenceau J (2009) Pupil dynamics during bistable motion perception. J Vis 9(7):10. doi:10.1167/9.7.10
Itti L, Koch C, Niebur E (1998) A model of saliency-based visual-attention for rapid scene analysis. IEEE Trans Pattern Anal Mach Intell 20:1254–1259
Janisse MP (1974) Pupil size, affect and exposure frequency. Soc Behav Pers 2:125–146
James W (1890) The principles of psychology. Holt, NewYork

Jepma M, Nieuwenhuis S (2011) Pupil diameter predicts changes in the exploration-exploitation trade-off: evidence for the adaptive gain theory. J Cogn Neurosci 23(7):1587–1596. doi:10.1162/jocn.2010.21548

Kahneman D, Beatty J (1966) Pupil diameter and load on memory. Science 154(3756):1583–1585

Kalwani RM, Gold JI (2008) The role of the locus coeruleus in motor commitment using the countermanding task. Soc Neurosci Abstr 165–169

Kamp SM, Donchin E (2015) ERP and pupil responses to deviance in an oddball paradigm. Psychophysiology 52(4):460–471. doi:10.1111/psyp.12378

Kafkas A, Montaldi D (2011) Recognition memory strength is predicted by pupillary responses at encoding while fixation patterns distinguish recollection from familiarity. Q J Exp Psychol (Hove) 64(10):1971–1989

Kietzmann TC, Geuter S, König P (2011) Overt visual attention as a causal factor of perceptual awareness. PLoS ONE 6(7):e22614. doi:10.1371/journal.pone.0022614

Knight R (1996) Contribution of human hippocampal region to novelty detection. Nature 383 (6597):256–259

Kojima T, Matsushima E, Ohta K, Toru M, Han YH, Shen YC, Moussaoui D, David I, Sato K, Yamashita I, Kathmann N, Hippius H, Thavundayil JX, Lal S, Vasavan Nair NP, Potkin SG, Prilipko L (2001) Stability of exploratory eye movements as a marker of schizophrenia—a WHO multi-center study. Schizophr Res 52:203–213. doi:10.1016/S0920-9964(00)00181-X

Kloosterman NA, Meindertsma T, van Loon AM, Lamme VA, Bonneh YS, Donner TH (2015) Pupil size tracks perceptual content and surprise. Eur J Neurosci 41(8):1068–1078. doi:10.1111/ejn.12859

Kristjansson SD, Stern JA, Brown TB, Rohrbaugh JW (2009) Detecting phasic lapses in alertness using pupillometric measures. Appl Ergon 40(6):978–986. doi:10.1016/j.apergo.2009.04.007

Laeng B, Endestad T (2012) Bright illusions reduce the eye's pupil. Proc Natl Acad Sci USA 109 (6):2162–2167. doi:10.1073/pnas.1118298109

Laeng B, Sulutvedt U (2014) The eye pupil adjusts to imaginary light. Psychol Sci 25(1):188–197. doi:10.1177/0956797613503556

Libet B, Gleason CA, Wright EW, Pearl DK (1983) Time of conscious intention to act in relation to onset of cerebral activity (readiness-potential). The unconscious initiation of a freely voluntary act. Brain 106 (Pt 3):623–642

Libby WL Jr, Lacey BC, Lacey JI (1973) Pupillary and cardiac activity during visual attention. Psychophysiology 10(3):270–294

Loewenfeld I (1993) The pupil: Anatomy, physiology, and clinical applications. Wayne State University Press, Detroit, MI

Lorber M, Zuber BL, Stark L (1965) Suppression of pupillary light reflex in binocular rivalry and saccadic suppression. Nature 208:558

Lucas RJ, Hattar S, Takao M, Berson DM, Foster RG, Yau KW (2003) Diminished pupillary light reflex at high irradiances in melanopsin-knockout mice. Science 299(5604):245–247

Magliero A (1983) Pupil dilations following pairs of identical and related to-be-remembered words. Mem Cognit 11(6):609–615

Mathôt S, van der Linden L, Grainger J, Vitu F (2015a) The pupillary light response reflects eye-movement preparation. J Exp Psychol Hum Percept Perform 41(1):28–35. doi:10.1037/a0038653

Mathôt S, van der Linden L, Grainger J, Vitu F (2013) The pupillary light response reveals the focus of covert visual attention. PLoS One 8(10):e78168. doi:10.1371/journal.pone.0078168. eCollection 2013

Mathôt S, Siebold A, Donk M, Vitu F (2015b) Large pupils predict goal-driven eye movements. J Exp Psychol Gen 144(3):513–521

May PR (1948) Pupillary abnormalities in schizophrenia and during muscular effort. J Ment Sci 94:89–98

van der Meer E, Friedrich M, Nuthmann A, Stelzel C, Kuchinke L (2003) Picture-word matching: flexibility in conceptual memory and pupillary responses. Psychophysiology 40(6):904–913 van der Meer E, Beyer R, Horn J, Foth M, Bornemann B, Ries J, Kramer J, Warmuth E, Heekeren HR, Wartenburger I (2010) Resource allocation and fluidintelligence: insights from pupillometry. Psychophysiology 47(1):158–169. doi:10.1111/j.1469-8986.2009.00884.x

Micieli G, Tassorelli C, Martignoni E, Pacchetti C, Bruggi P, Magri M, Nappi G (1991) Disordered pupil reactivity in Parkinson's disease. Clin Auton Res 1(1):55–58

Morgan ST, Hansen JC, Hillyard SA (1996) Selective attention to stimulus location modulates the steady-state visual evoked potential. Proc Natl Acad Sci USA May 14; 93(10):4770–4774

Murphy PR, O'Connell RG, O'Sullivan M, Robertson IH, Balsters JH (2014) Pupil diameter covaries with BOLD activity in human locus coeruleus. Hum Brain Mapp 35(8):4140–4154. doi:10.1002/hbm.22466

Murphy PR, Robertson IH, Balsters JH, O'Connell RG (2011) Pupillometry and P3 index the locus coeruleus-noradrenergic arousal function in humans. Psychophysiology 48 (11):1532–1543. doi:10.1111/j.1469-8986.2011.01226.x

Naber M, Frässle S, Einhäuser W (2011) Perceptual rivalry: reflexes reveal the gradual nature of visual awareness. PLoS ONE 6(6):e20910

Naber M, Hilger M, Einhäuser W (2012) Animal detection and identification in natural scenes: image statistics and emotional valence. J Vis 12(1):25. doi:10.1167/12.1.25

Naber M, Stoll J, Einhäuser W, Carter O (2013a) How to become a mentalist: reading decisions from a competitor's pupil can be achieved without training but requires instruction. PLoS ONE 8(8):e73302

Naber M, Nakayama K (2013) Pupil responses to high-level image content. J Vis 13(6):7. doi:10.1167/13.6.7

Naber M, Alvarez GA, Nakayama K (2013b) Tracking the allocation of attention using human pupillary oscillations. Front Psychol 4:919. doi:10.3389/fpsyg.2013.00919

Naber M, Frässle S, Rutishauser U, Einhäuser W (2013c) Pupil size signals novelty and predicts later retrieval success for declarative memories of natural scenes. J Vis 13(2):11. doi:10.1167/13.2.11

Nagai M, Wada M, Sunaga N (2002) Trait anxiety affects the pupillary light reflex in college students. Neurosci Lett 328(1):68–70

Nassar MR, Rumsey KM, Wilson RC, Parikh K, Heasly B, Gold JI (2012) Rational regulation of learning dynamics by pupil-linked arousal systems. Nat Neurosci 15(7):1040–1046. doi:10.1038/nn.3130

Nuthmann A, van der Meer E (2005) Time's arrow and pupillary response. Psychophysiology 42 (3):306–317

Otero SC, Weekes BS, Hutton SB (2011) Pupil size changes during recognition memory. Psychophysiology 4:1346–1353

Papesh MH, Goldinger SD, Hout MC (2012) Memory strength and specificity revealed by pupillometry. Int J Psychophysiol 83(1):56–64. doi:10.1016/j.ijpsycho.2011.10.002

Park JC, McAnany JJ (2015) Effect of stimulus size and luminance on the rod-, cone-, and melanopsin-mediated pupillary light reflex. J Vis 15(3):13. doi:10.1167/15.3.13

Partala T, Surakka V (2003) Pupil size variation as an indication of affective processing. Int J Human-Comput Stud 59:185–198

Paulus FM, Krach S, Blanke M, Roth C, Belke M, Sommer J, Müller-Pinzler L, Menzler K, Jansen A, Rosenow F, Bremmer F, Einhäuser W, Knake S (2015) Fronto-insula network activity explains emotional dysfunctions in juvenile myoclonic epilepsy: combined evidence from pupillometry and fMRI. Cortex 65C:219–231

Pavlov IP (1927) Conditioned reflexes: an investigation of the physiological activity of the cerebral cortex. Oxford University Press, London

Poock GK (1973) Information processing versus pupil diameter. Percept Mot Skills 37 (3):1000–1002

Prettyman R, Bitsios P, Szabadi E (1997) Altered pupillary size and darkness and light reflexes in Alzheimer's disease. J Neurol Neurosurg Psychiatry 62(6):665–668

Preuschoff K, 't Hart BM, Einhäuser W (2011) Pupil dilation signals surprise: evidence for noradrenaline's role in decision making. Front Neurosci 5:115. doi:10.3389/fnins.2011.00115

Privitera CM, Renninger LW, Carney T, Klein S, Aguilar M (2010) Pupil dilation during visual target detection. J Vis 10(10):3. doi:10.1167/10.10.3
Qiyuan J, Richer F, Wagoner BL, Beatty J (1985) The pupil and stimulus probability. Psychophysiology 22(5):530–534
Raisig S, Welke T, Hagendorf H, van der Meer E (2010) I spy with my little eye: detection of temporal violations in event sequences and the pupillary response. Int J Psychophysiol 76 (1):1–8. doi:10.1016/j.ijpsycho.2010.01.006
Rajkowski J, Kubiak P, Aston-Jones G (1993) Correlations between locus coeruleus (LC) neural activity, pupil diameter and behavior in monkey support a role of LC in attention. Soc Neurosci Abstr 19:974
Reinhard G, Lachnit H, König S (2006) Tracking stimulus processing in Pavlovian pupillary conditioning. Psychophysiology 43(1):73–83
Rieger G, Savin-Williams RC (2012) The eyes have it: sex and sexual orientation differences in pupil dilation patterns. PLoS ONE 7(8):e40256. doi:10.1371/journal.pone.0040256
Scheepers C, Mohr S, Fischer MH, Roberts AM (2013) Listening to limericks: a pupillometry investigation of perceivers' expectancy. PLoS ONE 8(9):e74986. doi:10.1371/journal.pone.0074986
Samuels ER, Szabadi E (2008a) Functional neuroanatomy of the noradrenergic locus coeruleus: its roles in the regulation of arousal and autonomic function part I: principles of functional organisation. Curr Neuropharmacol 6(3):235–253. doi:10.2174/157015908785777229
Samuels ER, Szabadi E (2008b) Functional neuroanatomy of the noradrenergic locus coeruleus: its roles in the regulation of arousal and autonomic function part II: physiological and pharmacological manipulations and pathological alterations of locus coeruleus activity in humans. Curr Neuropharmacol 6(3):254–285. doi:10.2174/157015908785777193
Schaefer HS, Larson CL, Davidson RJ, Coan JA (2014) Brain, body, and cognition: neural, physiological and self-report correlates of phobic and normative fear. Biol Psychol 98:59–69. doi:10.1016/j.biopsycho.2013.12.011
Siegle GJ, Steinhauer SR, Stenger VA, Konecky R, Carter CS (2003) Use of concurrent pupil dilation assessment to inform interpretation and analysis of fMRI data. Neuroimage 20 (1):114–124
Simms TM (1967) Pupillary response of male and female subjects to pupillary differences in male and female picture stimuli
Simpson HM, Hale SM (1969) Pupillary changes during a decision-making task. Percept Mot Skills 29(2):495–498
Simpson HM, Molloy FM (1971) Effects of audience anxiety on pupil size. Psychophysiology 8:491–496. doi:10.1111/j.1469-8986.1971.tb00481.x
Steinhauer SR, Zubin J (1982) Vulnerability to schizophrenia: information processing in the pupil and event-related potential. In: Usdin E, Hanin I (eds) Biological markers in psvchiatrv and neurology. Pergamon Press, Oxford, pp 371–385
Steinhauer SR, Hakerem G (1992) The pupillary response in cognitive psychophysiology and schizophrenia. Ann N Y Acad Sci 658:182–204
Steinhauer SR (2002) Cognition, psychopathology, and recent pupil studies. www.wpic.pitt.edu/research/biometrics/Publications/PupilWeb.htm. Retrieved 12 Aug 2016
Steinhauer SR, Siegle GJ, Condray J, Pless M (2004) Sympathetic and parasympathetic innervation of pupillary dilation during sustained processing. Int Psychophysiol 53:77–86
Sterpenich V, D'Argembeau A, Desseilles M, Balteau E, Albouy G, Vandewalle G, Degueldre C, Luxen A, Collette F, Maquet P (2006) The locus ceruleus is involved in the successful retrieval of emotional memories in humans. J Neurosci 26:7416–7423
Tombs S, Silverman I (2004) Pupillometry—a sexual selection approach. Evol Human Behav 25:221–228
Tomlinson N, Hicks RA, Pellegrini (1978) Attributions of female college students to variations in pupil size. Bull Psychon Soc 12(6):477–478
Tursky B, Shapiro D, Crider A, Kahneman D (1969) Pupillary, heart rate, and skin resistance changes during a mental task. J Exp Psychol 79(1):164–167

Ursin H, Kaada BR (1960) Functional localization within the amygdaloid complex in the cat. Electroencephalogr Clin Neurophysiol 12:1–20

Võ ML, Jacobs AM, Kuchinke L, Hofmann M, Conrad M, Schacht A, Hutzler F (2008) The coupling of emotion and cognition in the eye: introducing the pupil old/new effect. Psychophysiology 45(1):130–140

Wang JT, Spezio M, Camerer CF (2006) Pinocchio's pupil: using eyetracking and pupil dilation to understand truth-telling and deception in games. Am Econ Rev 100:984–1007

Wang CA, Boehnke SE, White BJ, Munoz DP (2012) Microstimulation of the monkey superior colliculus induces pupil dilation without evoking saccades. J Neurosci 14 32(11):3629–3636. doi:10.1523/JNEUROSCI.5512-11.2012

Wang CA, Boehnke SE, Itti L, Munoz DP (2014) Transient pupil response is modulated by contrast-based saliency. J Neurosci 34(2):408–417. doi:10.1523/JNEUROSCI.3550-13.2014

Wang CA, Brien DC, Munoz DP (2015) Pupil size reveals preparatory processes in the generation of pro-saccades and anti-saccades. Eur J Neurosci 41(8):1102–1110. doi:10.1111/ejn.12883

Wang CA, Munoz DP (2015) A circuit for pupil orienting responses: implications for cognitive modulation of pupil size. Curr Opin Neurobiol 33:134–140. doi:10.1016/j.conb.2015.03.018

Watson AB, Yellott JI (2012) A unified formula for light-adapted pupil size. J Vis 12(10):12. doi:10.1167/12.10.12

Yellin D, Berkovich-Ohana A, Malach R (2015) Coupling between pupil fluctuations and resting-state fMRI uncovers a slow build-up of antagonistic responses in the human cortex. Neuroimage 1(106):414–427. doi:10.1016/j.neuroimage.2014.11.034

Yu AJ, Dayan P (2005) Uncertainty, neuromodulation, and attention. Neuron 46(4):681–692

Zekveld AA, Kramer SE (2014) Cognitive processing load across a wide range of listening conditions: insights from pupillometry. Psychophysiology 51(3):277–284

Zekveld AA, Heslenfeld DJ, Johnsrude IS, Versfeld NJ, Kramer SE (2014) The eye as a window to the listening brain: neural correlates of pupil size as a measure of cognitive listening load. Neuroimage 101:76–86. doi:10.1016/j.neuroimage.2014.06.069

Social Saliency

Shuo Wang and Ralph Adolphs

Saliency historically refers to the bottom-up visual properties of an object that automatically drive attention. It is an ordinal property that depends on the relative saliency of one object with respect to others in the scene. Simple examples are a red spot on a green background, a horizontal bar among vertical bars, or a sudden onset of motion. Researchers have introduced the idea of a saliency map, an abstract and featureless map of the 'winners' of attention competition, to model the dynamics of visual attention. The standard saliency map involves channels like color, orientation, size, shape, movement or unique onset. But how do complex stimuli, especially stimuli with social meaning such as faces, pop out and attract attention? Suppose you are attending a big party: your attention might be captured by someone in a fancy dress, someone looking at you, someone who is attractive, familiar, or distinctive in some way. This happens essentially automatically, and encompasses a huge number of different stimuli that are all competing for your attention. What determines which is the most salient, and how can we best measure this?

Humans are social animals. We constantly interact with other people and the environment and we are unceasingly bombarded with various socially salient stimuli: faces, gestures, emoticons, and socially relevant pieces of text. These capture our attention, are encoded preferentially into memory, and influences our thoughts and actions. What is it about such stimuli that accomplishes these multiple effects? In particular, are there mechanisms analogous to those known to operate for low-level (non-social) saliency (such as visual motion and contrast)? Is there a finite vocabulary of social attention-grabbing cues, or a small set of dimensions that render social stimuli especially salient? Finally, how does the brain figure out what is important and what are the underlying neural mechanisms?

S. Wang (✉) · R. Adolphs
Computation and Neural Systems, California Institute of Technology, Pasadena, CA, USA
e-mail: wangshuo45@gmail.com

Q. Zhao (ed.), *Computational and Cognitive Neuroscience of Vision*,
Cognitive Science and Technology, DOI 10.1007/978-981-10-0213-7_8

Here, we discuss the emerging "social saliency"—social stimuli that attract attention, as well as the neural substrates of the attentional "social brain". This chapter is organized as follows. First, we begin by discussing the socially salient cues, including eye gazes, faces, head directions, finger gestures, postures, actions, biological motion, personal distance, social touch, and social rewards. We underscore the neural substrates underlying processing of such social stimuli. Second, we discuss the role of the amygdala in encoding saliency. Third, we discuss the deficits in processing socially salient stimuli in autism, and lastly, the possible involvement of the amygdala in these deficits.

1 Socially Salient Cues

In everyday life, we are constantly bombarded by social cues. Rich information can be derived through social cues. But how does the brain figure out the message conveyed by the social cues? How are social cues represented and integrated in the brain? In particular, is social interaction mediated through an mechanism that relies on saliency?

In this section, we discuss the socially salient cues, which include eye gazes, faces, head directions, finger gestures, postures, actions, biological motion, personal distance, social touch, and social rewards. We also discuss what is known about the functions and the neural underpinnings of each social cue, as well as the developmental ontogeny and comparative neurobiology of the social cues.

1.1 Eye Gazes

Eye gaze plays important roles in social communication. It functions for information seeking, signaling, controlling the synchronizing of speech, cueing for intimacy, avoiding undue intimacy, and avoiding excess input of information (Argyle et al. 1973). Especially, eye gaze directs attention and provides important sources of social information. Behavioral studies show that chimpanzees spontaneously follow human gaze direction and share joint visual attention (Povinelli and Eddy 1996). Rhesus monkeys can also follow gazes and use the attentional cues of other monkeys to orient their own attention to objects (Emery et al. 1997). Human studies show that newborns prefer faces with eyes open versus eyes closed (Batki et al. 2000), and that infants as early as 10 weeks of age follow the gaze of others (Hood et al. 1998), suggesting for the innateness of gaze cognition. Deficiencies in processing eye gaze can lead to complex disabilities such as autism (Pelphrey et al. 2002).

Friesen and Kingstone reported that humans attend reflexively to locations and objects that are being looked at by other people (Friesen and Kingstone 1998). Humans infer other people's movement trajectories from their gaze direction and use this information to guide their own visual scanning of the environment (Nummenmaa et al. 2009). However, the specificity of eye gazes in orienting attention has been questioned and it has been reported that other non-social stimuli, such as arrows, trigger reflexive shifts in attention in a manner behaviorally identical to those triggered by eyes (Ristic et al. 2002). Further investigation has differentiated these two types of attentional cues neuronally, showing that the neural systems subserving eye gazes and arrows are not equivalent—with the superior temporal sulcus (STS) being engaged exceptionally when the fixation stimulus is perceived as eyes (Kingstone et al. 2004). This is further supported by a study using a similar spatial cueing paradigm on a patient with circumscribed superior temporal gyrus damage who showed detection advantage only when cued by non-biological arrows but not gazes (Akiyama et al. 2006). The specificity of STS in cueing by gazes is also illustrated through patients with parietal damage—perceived gaze in faces can still trigger automatic shifts of attention in the contralesional direction, even though parietal damage causes spatial neglect and impairs the representation of location on the contralesional side (Vuilleumier 2002). This suggests a specific and anatomically distinct attentional mechanism through STS. Interestingly, attentional shift by gaze can be triggered without awareness (Sato et al. 2007).

STS activation has consistently been demonstrated in the normal brain when viewing eyes and this brain region been implicated in gaze processing. Recordings from single cells in awake, behaving monkeys have shown that this region of the temporal lobe is sensitive to face view and gaze direction (Perrett et al. 1985). Similarly, STS lesions in the rhesus monkey impair gaze direction discrimination (Campbell et al. 1990). Human neuroimaging studies have demonstrated that a superior temporal region centered in the STS is activated when a subject views a face in which the eyes shift their gaze (Puce et al. 1998). In a spatial cueing paradigm, the STS has been demonstrated to be sensitive to social context in which a gaze shift occurs (Pelphrey et al. 2003). Moreover, in a virtual reality environment, mutual gaze evokes greater activity in the STS than averted gaze, suggesting that the STS is involved in processing social information conveyed by gaze shifts within an overtly social context (Pelphrey et al. 2004). Multivariate pattern analysis of human functional imaging data has shown that anterior STS encodes the direction of another's attention regardless of how this information is conveyed (Carlin et al. 2011).

Perception of eye gaze also recruits the spatial cognition system in the intraparietal sulcus to encode and pay attention to the direction of another's gaze (Hoffman and Haxby 2000). Research with split-brain patients suggests that lateralized cortical connections between temporal lobe subsystems specialized for processing gazes and parietal lobe subsystems specialized for orienting spatial attention underlie the reflexive joint attention elicited by gazes (Kingstone et al. 2000). Besides the STS, contrasting between directed versus averted gaze produces a tight cluster of activation corresponding to and restricted to the central nucleus and the bed nucleus of the stria terminalis (termed the lateral extended amygdala) (Hoffman et al. 2007).

1.2 Faces

People often form judgments of others based on purely facial features. People are able to pick up subtle changes in facial structures from faces varying along one dimension to another (Oosterhof and Todorov 2008). Trait evaluations from faces can predict important social outcomes. Inferences of competence based solely on facial appearance predict the outcomes of elections (Todorov et al. 2005). Facial features can also influence sentencing decisions—inmates with more Afrocentric features received harsher sentences than those with less Afrocentric features (Blair et al. 2004). Remarkably, impressions and judgments of unfamiliar people can be formed by a very brief exposure to faces as short as 100 ms (Willis and Todorov 2006).

Primates have a dedicated visual system to process faces (Tsao et al. 2006). Electrophysiological recordings in monkeys (Rolls 1984; Leonard et al. 1985) and humans (Kreiman et al. 2000; Rutishauser et al. 2011) have found single neurons that respond not only to faces, but also to face identities, facial expressions and gaze directions (Gothard et al. 2007; Hoffman et al. 2007). Neuroimaging studies have revealed neural substrates for emotional attention, which might supplement but also compete with other sources of top-down control on perception (Vuilleumier 2005). In particular, the amygdala plays a crucial role in emotional attention and is required for accurate social judgments of other individuals on the basis of their facial appearance. The amygdala also shares parallel roles in humans and other animals in emotional influences on attention and social behavior (see Phelps and LeDoux 2005 for a review). Lesion studies showed that patients with complete bilateral amygdala damage judge unfamiliar individuals to be more approachable and more trustworthy than do control subjects (Adolphs et al. 1998). The impairment is most striking for faces to which normal subjects judge most unapproachable and untrustworthy. Besides, recent findings show that orbitofrontal cortex lesions result in abnormal social judgments to emotional faces (Willis et al. 2010). The relationship between the amygdala and saliency is reviewed in more detail in the next section.

Moreover, as reviewed in more detail below, people with autism are indicated to have altered saliency representations towards faces compared to non-face objects, as shown by reduced attention to faces compared to inanimate objects (Dawson et al. 2005; Sasson 2006), as well as circumscribed interests to a narrow range of inanimate subjects (e.g. gadgets, devices, electronics and Japanese animation, etc.) (Kanner 1943; Lewis and Bodfish 1998; South et al. 2005). It is even shown in children and adolescents (Sasson et al. 2008), as well as in 2–5 year-olds (Sasson et al. 2011), that people with autism fixated faces or people less than controls when freely viewing the arrays containing both faces and non-face objects. When looking within faces, the relative saliency of facial features is also altered in autism, as evidenced from both behavioral and neuronal findings: people with autism have an increased tendency to saccade away from (Spezio et al. 2007b) and actively avoid the eyes (Kliemann et al. 2010), but have an increased preference to fixate

(Neumann et al. 2006) and rely on information from the mouth (Spezio et al. 2007a). The behavioral abnormality is supported by neuronal evidence of abnormal processing of information from the eye region of faces in single cells recorded from the amygdala in neurosurgical patients with ASD (Rutishauser et al. 2013).

Interestingly, gaze cues interplay with facial emotion cues. It has been shown that direct gaze facilitates the processing of facially communicated approach-oriented emotions (e.g., anger and joy), whereas averted gaze would facilitate the processing of facially communicated avoidance-oriented emotions (e.g., fear and sadness), suggesting that gaze cues combine with facial emotion cues in the processing of emotionally relevant facial information (Adams and Kleck 2003).

1.3 Head Directions

Comparative research with non-human primates suggests that the orientation of the head might provide a stronger cue to another individual's attentional direction than eye-gaze alone (Langton et al. 2000). In humans, by manipulating the face directions of emotional expressions in the unilateral visual fields, it allows us to alter the emotional significance of the facial expression for the observer without affecting the physical features of the expression. It has been shown that the left amygdala increases activity for angry expressions looking toward the subjects than angry expressions looking away from them, suggesting that the amygdala is involved in emotional but not basic sensory processing for facial expressions (Sato et al. 2004). In infants, the emergence of the tendency to look where another person looks is a fundamental landmark in the development of referential communication. It has been found that normal infants 10–12 months old reliably look in the direction toward which adults turn their heads and eyes (Scaife and Bruner 1975).

There is often interplay between gaze direction and head direction. Averted gaze elicited fast orienting of visual attention occurred both when the head is seen from the front, and also when seen in profile even though the gaze is directed at the observer. Instead, a profile head with a compatible gaze direction does not elicit an attentional orienting, suggesting that the observer does not use information of the other person's gaze direction referenced to the observer, but referenced to the other person's head orientation (Hietanen 1999). However, human neuroimaging studies suggest that right anterior STS is invariant to head view and physical image features (Carlin et al. 2011). Furthermore, head direction cues also interact with body direction cues—only when the head is rotated in the cuing person's reference frame but not the observer's frame, the head direction cue can shift the observer's attention to the same direction (Hietanen 2002).

1.4 Finger Gestures

Finger pointing provides social information and captures attention. Many animals can use experimenter-given cues in the experimental setup and are sensitive to human gestural communication. In object-choice tasks, lowland gorillas complete the task better when the experimenter taps on or points at an object that contains a reward. Performance remains good when the experimenter gazes with eyes and head orients toward the correct object without manual gestures. In contrast, when only the experimenter's eye orientation serves as the cue, the gorillas do not appropriately complete the task (Peignot and Anderson 1999) (but also see Byrnit 2009). In a similar task, capuchin monkeys do not use the experimenter's gazing as a cue to find the correct baited object. In contrast, they do use gazing plus pointing, and it has been shown that pointing is necessary and sufficient under the conditions of the study (Anderson et al. 1995). Dogs are able to utilize pointing, bowing, nodding, head turning and glancing gestures of humans as cues for finding hidden food. Interestingly, this ability can be generalized from one person (owner) to another familiar person (experimenter) in using the same gestures as cues (Miklösi et al. 1998). Even fur seals were found to be able to follow human gestures—they are able to use cues involving a fully exposed arm or a head direction, but fail to use glance only, suggesting that a domestication process is not necessary to develop receptive skills to cues given by humans (Scheumann and Call 2004). In humans, finger pointing gestures could interfere with speech in a Stroop-type paradigm, suggesting that verbal and non-verbal dimensions are integrated prior to the response selection stage of processing (Langton et al. 1996).

1.5 Postures, Actions and Biological Motion

People not only pay attention to their only motion in the environment (Wang et al. 2012a), but also pay attention to and automatically infer other people's mental states such as intention from their motion and actions. Psychophysical and functional neuroimaging evidence shows that biological motion is processed as a special category, and the mechanism underlying the attribution of intentions to actions might rely on simulating the observed action and mapping it onto representations of our own intentions (Blakemore and Decety 2001). In a visual search paradigm with point-light animations, differentiating between actions requires attention in general and there are search asymmetries between actions (van Boxtel and Lu 2011). Particularly, animated threatening boxer targets pop out from emotionally-neutral walker distractors in a crowd, whereas walkers do not, showing that body cues signal important social information related to threat and survival (van Boxtel and Lu 2012). It has been suggested that body cues rather than facial expressions discriminate between intense positive and negative emotions (Aviezer et al. 2012).

Even laboratory rodents, the rat, the mouse, the guinea pig, and the golden hamster, use postures and acts to signal social information (Grant and Mackintosh 1963).

1.6 Personal Distance

When interpersonal space is invaded, people feel uncomfortable and aroused. People automatically and reliably regulate the distance between each other during social interaction (Hall 1966). The amygdala may be required to trigger strong emotional and arousal reactions when personal space is invaded, as evidenced by neuroimaging data showing amygdala activation in healthy individuals upon close personal proximity, and the lack of a personal space in an individual with complete amygdala lesions (Kennedy et al. 2009). This is consistent with monkey studies showing that amygdalectomized monkeys demonstrated increased social affiliation, decreased anxiety, and increased confidence compared with control monkeys (Emery et al. 2001). These effects might arise from a lack of saliency signal mediated by the amygdala to personal space violation. Furthermore, the amygdala mediates the approach and avoidance to ambiguous or threatening novel situations and people (Mason et al. 2006).

1.7 Social Touch

Social touch is a salient cue in our everyday social interactions since it plays a particularly important role in social bonding which, in turn, has a major impact on an individual's lifetime reproductive fitness (Dunbar 2010). Interpersonal touch provides an effective means of influencing people's various social behaviors (see Gallace and Spence 2010 for a review) and plays an important role in governing our emotional well-being (Field 2003). Human orbitofrontal cortex represents affectively positive and negative touch in different areas (Rolls et al. 2003) (also see Rolls 2010 for a review).

In addition to the fast-conducting myelinated afferent fibers responsible for tactile sensation, a system of slow-conducting unmyelinated tactile (CT) afferents is responsible for affective sensation, as supported by neuroimaging studies in a unique patient lacking large myelinated afferents showing that touch activates brain regions implicated in emotional and social processing such as insula, but not of somatosensory areas (Olausson et al. 2002). Further electrophysiological studies in healthy individuals have shown that soft brush stroking activates CT afferents but not myelinated afferents, suggesting that CT afferents constitute a privileged peripheral pathway for pleasant tactile stimulation that is likely to signal affiliative social body contact (Loken et al. 2009). However, recent neuroimaging studies in humans have shown that the response in primary somatosensory cortices to a sensual caress is modified by the perceived sex of the caresser, arguing for a more important role that somatosensory areas might play in affective processing than previously thought (Gazzola et al. 2012). Indeed, the different components of social touch, such as somatosensory experience, the proximity to the person, and an attribution of the somatosensory experience to the person, have been teased apart (Schirmer et al. 2011). It also illustrates that touch is a special and salient social cue that enhances visual attention and sensitizes ongoing cognitive and emotional processes.

1.8 Social Rewards

People not only pay attention to concrete, physical social cues delivered by direct body-body interaction, but also pay attention to more abstract social cues such as rewards. Attentional control is at the center of the function of dopamine in reinforcement learning and animal approach behavior (Ikemoto and Panksepp 1999; Wise 2004). Dopamine systems mediate the incentive saliency of rewards by specifically changing the perceptual representation of reward-conditioned stimuli such that they become salient and draw attention (Berridge and Robinson 1998). It has been shown that sensory and perceptual processing of reward-associated visual features is facilitated such that attention is deployed to objects characterized by these features, even when a strategic decision to attend to reward-associated features will be counterproductive and result in suboptimal performance (Hickey et al. 2010). Visual search for a salient target is slowed by the presence of an inconspicuous, task irrelevant distractor previously associated with monetary reward through learning, showing that valuable stimuli can modulate voluntary attention allocation (Anderson et al. 2011). Reward can even create oculomotor saliency and modulate the saccadic trajectories, suggesting low-level and non-strategic mechanisms that operate automatically (Hickey and van Zoest 2012).

Complex social behavior and decision making may share the same neural basis as the simple monetary evaluation and learning. The acquisition of one's good reputation robustly activated striatum as monetary rewards, suggesting a 'common neural currency' for rewards (Izuma et al. 2008). In healthy individuals, social and monetary reward learning share overlapping neural substrates (Lin et al. 2012b). In monkeys, neurons from orbitofrontal cortex signal both social values and juice rewards, and far more neurons signal social category than fluid value, despite the stronger impact of fluid reward on monkeys' choices (Watson and Platt 2012). Interestingly, people can make optimal reward choices without being fully aware of the basis of their decision (Wang et al. 2012b). Moreover, people with autism show various impairments in social decision making and reward learning (see below) (Izuma et al. 2011; Lin et al. 2012a, c).

2 Neural Representation of Saliency

A distributed network of visuomotor areas are proposed to encode a representation of saliency that combines bottom-up and top-down influences to identify locations for further processing. Neurons in the primate frontal eye field (FEF) exhibit the characteristics of a visual saliency map—they are not sensitive to specific features of visual stimuli, but their activity evolves over time to select the target of the search array (Thompson et al. 1996; Thompson and Bichot 2005). Visual activity in the FEF not only signals location of targets for orienting, but also signals movement-related saccade preparation (Murthy et al. 2009). However, in an adjacent area, supplementary eye field (SEF), only very few neurons selected the location of the search target (Purcell et al. 2012), showing a very limited role of SEF in encoding target saliency. Moreover, it has also been shown that neurons in lateral intraparietal (LIP) area reflect selection to salient stimuli defined by target when animals have to make a saccade toward the salient stimulus (Thomas and Paré 2007). Activity in the LIP correlates with monkey's planning of memory-guided saccades to goal-directed salient locations in visual search tasks (Ipata et al. 2006) and these neurons only respond to stimuli that are behaviorally significant (Gottlieb et al. 1998). Studies have even found pure bottom-up saliency signals in LIP in a passive fixation task without any top-down instructions (Arcizet et al. 2011). Furthermore, individual neurons in monkey area 7a of the posterior parietal cortex encode the location of the salient stimulus and can thus provide spatial information required for orienting to a salient spot in a complex scene (Constantinidis and Steinmetz 2001).

Besides cortical areas, subcortical superior colliculus (SC) encodes both stimulus identity and saccade goals during visual conjunction search (Shen and Paré 2007). Neuronal activity in the SC signals selection or increased saliency of subsequent saccade goals even before the initial saccade has ended (McPeek and Keller 2002), and a recent report has shown that the process can encompass at least two future saccade targets (Shen and Paré 2014), suggesting parallel processing of visual saliency in the SC. The causal functional role of the SC in target selection has been revealed by focal reversible inactivation in monkeys (McPeek and Keller 2004). Interestingly, even substantia nigra pars reticulata (SNr) has been shown to change activities with target selection and saccade initiation, which in turn may make substantial and direct contributions to the SC (Basso and Wurtz 2002).

On the other hand, neurons in the inferior temporal cortex have been suggested to play a critical role in representing and processing visual objects (Gross 1994; Logothetis and Sheinberg 1996; Tanaka 1997), which holds for both isolated objects and objects in complex natural scenes (Sheinberg and Logothetis 2001). Compared to visual areas in the temporal lobe, early visual areas such as V1 and V2 are generally accepted to represent low-level visual features, and V4 has been reported to show convergence of bottom-up and top-down processing streams that facilitate oculomotor planning for visual search (Mazer and Gallant 2003). V4 neurons not only enhance responses to a preferred stimulus in their receptive field matched a feature of the target, but also enhance responses to candidate targets selected for saccades (Bichot et al. 2005). Comparing pop-out and conjunctive stimuli, V4 neurons encode pop-out saliency in an top-down attention-dependent manner (Burrows and Moore 2009). In both single saccade tasks (Chelazzi et al. 1993; Tolias et al. 2001; Ogawa and Komatsu 2004) and tasks with naturalistic free-viewing (Sheinberg and Logothetis 2001; Mazer and Gallant 2003; Bichot et al. 2005), several studies have reported that temporal cortical neurons enhanced responses to visual stimuli when the stimulus in receptive field becomes the target, suggesting that task-relevant target saliency is encoded by these neurons.

3 Saliency and the Amygdala

The human amygdala is critical to process emotionally salient and socially relevant stimuli (Kling and Brothers 1992; Adolphs 2010). Earlier view of the amygdala in representing saliency lay in the fear-related domain and the amygdala has generally been conceptualized as a fear-processing module. This view was supported by animal models of fear conditioning (LeDoux 1993) and impairment of fear conditioning after amygdala damage in humans (Bechara et al. 1995; LaBar et al. 1995). Human studies demonstrated a selective impairment in recognizing fearful faces in subjects that lack a functional amygdala (Adolphs et al. 1994), mirrored by neuroimaging studies showing significant activation difference of the amygdala to fearful faces compared to happy faces (Morris et al. 1996). Interestingly, increased amygdala BOLD-fMRI to fearful stimuli was linked to serotonin transporter genes, which has been associated with anxiety-related behaviors (Hariri et al. 2002).

Recently, however, the amygdala has been proposed to respond to a broader spectrum of social attributes such as facial emotions in general and regulating a person's personal space (Kennedy et al. 2009), rather than being specific for fearful faces (Fitzgerald et al. 2006). Electrophysiological recordings in monkeys have found single neurons that respond not only to faces (Rolls 1984; Leonard et al. 1985), but also to face identities, facial expressions and gaze directions (Gothard et al. 2007; Hoffman et al. 2007). In humans, it has been reported that amygdala neurons are selective for a variety of visual stimuli (Kreiman et al. 2000). Single neurons in the human amygdala have been found to encode whole faces selectively (Rutishauser et al. 2011) and account for the abnormal face processing in autism (Rutishauser et al. 2013). A recent study has shown that neurons in the human amygdala encode subjective judgment of facial emotions, rather than simply their stimulus features (Wang et al. 2014a).

Values are salient social cues and the amygdala responds to values and rewards that are important to the organism (Baxter and Murray 2002). The primate amygdala represents the positive and negative value of visual stimuli during learning (Paton et al. 2006) and is sensitive to temporal reward structure (Bermudez et al. 2012). Monkeys with amygdala lesions showed impaired devaluation in selectively satiated food, indicating a failure to respond to the changing value of food rewards

(Malkova et al. 1997; Baxter et al. 2000). Furthermore, the amygdala neurons could predict the monkeys' save–spend choices while monkeys chose between saving liquid reward with interest and spending the accumulated reward (Grabenhorst et al. 2012). In rats, rapid strengthening of thalamo-amygdala synapses mediates cue–-reward learning (Tye et al. 2008).

Lastly, the amygdala processes more abstract attributes such as stimulus unpredictability (Herry et al. 2007). Amygdala lesions result in an absence or reduction of fixations on novel objects observed in monkeys (Bagshaw et al. 1972). It has also been shown that the amygdala mediates emotion-enhanced vividness (Todd et al. 2012), responds more to animate entities compared to inanimate ones (Yang et al. 2012), and is even selective to animals (Mormann et al. 2011). The amygdala has also been reported in modulating consolidation of object recognition memory, especially for highly emotionally arousing tasks (McGaugh 2000, 2004; Roozendaal et al. 2008).

However, there are also many examples showing no obvious corresponding behavioral impairment when the amygdala is lesioned. Recent findings have shown that preferential attention to animals and people is independent of the amygdala (Wang et al. 2015), and amygdala lesions do not lead to deficits in social attention as observed in people with autism (Wang et al. 2014b). These findings are consistent with preserved attentional capture by emotional stimuli and intact emotion-guided visual search in patients with acute amygdala lesions due to neurosurgical resection (piech et al. 2010, 2011). Besides compensatory circuits that might account for the intact social attention in amygdala lesion patients (Becker et al. 2012), these findings leave open the question of what are the essential structures mediating social saliency and to what extent the amygdala contributes to social saliency. These questions remain important topics for future studies.

Overall, the amygdala might act as a detector of perceptual saliency and biological relevance (Sander et al. 2005; Adolphs 2008). The functional role of the amygdala is supported by its connection with the visual cortices specialized for face processing (Vuilleumier et al. 2004; Moeller et al. 2008; Hadj-Bouziane et al. 2012) as well as reciprocal connections with multiple visually responsive areas in the temporal (Desimone and Gross 1979; Amaral et al. 2003; Freese and Amaral 2006) and frontal lobes (Ghashghaei and Barbas 2002).

4 Saliency and Autism

Autism is a disorder characterized by impairments in social and communicative behavior and a restricted range of interests and behaviors (DSM-5 2013). Individuals with autism show reduced attention to faces as well as to all other social stimuli such as the human voice and hand gestures, but pay more attention to inanimate objects (Dawson et al. 2005; Sasson 2006). Some characteristics, such as preference for inanimate objects and a lack of interest in social objects, are often evident very early in infancy (Kanner 1943; Osterling and Dawson 1994). Children with autism displayed significantly fewer social and joint attention behaviors, including pointing, showing objects, looking at others, and orienting to name (Osterling and Dawson 1994). People with autism also show circumscribed interests to a narrow range of inanimate subjects, a type of repetitive behavior occurring commonly in autism, and are fascinated with gadgets, devices, vehicles, electronics, Japanese animation and dinosaurs, etc. (Kanner 1943; Lewis and Bodfish 1998; South et al. 2005). The circumscribed interests are evident in children and adolescents (Sasson et al. 2008), as well as in 2–5 year-olds (Sasson et al. 2011), as shown by fewer fixations onto faces or people compared to controls when they freely view arrays containing both faces and non-face objects. Moreover, 2 year-olds with autism orient to non-social contingencies rather than biological motion (Klin et al. 2009). Taken together, people with autism show a different saliency representation of social stimuli versus non-social stimuli compared to normals.

When the stimuli are restricted to faces, people with autism show impaired face discrimination and recognition and use atypical strategies for processing faces characterized by reduced attention to the eyes and piecemeal rather than configural strategies (Dawson et al. 2005). In particular, when viewing naturalistic social situations, people with autism demonstrate abnormal patterns of social visual pursuit (Klin et al. 2002). They viewed core feature areas of the faces (i.e., eyes, nose, and mouth) significantly less compared to neurotypical controls (Pelphrey et al. 2002). They showed a greater tendency to saccade away from the eyes when information was present in those regions (Spezio et al. 2007b), but showed increased preference to the location of the mouth (Neumann et al. 2006) and relied primarily on information from the mouth (Spezio et al. 2007a). Eye-tracking data revealed a pronounced influence of active avoidance of direct eye contact on atypical gaze in people with autism (Kliemann et al. 2010). These results again shows a different saliency representation of faces in autism compared to normals.

In tasks with top-down instructions such as visual search, attention is guided toward likely targets by a limited set of stimulus attributes such as color and size (Wolfe and Horowitz 2004; Wolfe 2012). Several studies have shown superior visual search skills by individuals with autism (Plaisted et al. 1998; O'Riordan and Plaisted 2001; O'Riordan et al. 2001; O'Riordan 2004; Kemner et al. 2008), particularly in relatively difficult tasks. This superiority has been attributed to enhanced memory for distractor locations already inspected, and enhanced ability to discriminate between target and distractor stimulus features (O'Riordan and Plaisted 2001), while it is also arguable that the superiority is due to the anomalously enhanced perception of stimulus features (Joseph et al. 2009). However, studies investigating the role of top-down excitation and inhibition of stimulus representations in children with autism showed that children with autism did not differ from controls in excitatory or inhibitory top-down control of stimulus representations (O'Riordan 2000), leaving open the possibility that the autism advantage in visual search (O'Riordan et al. 2001) derived from enhanced bottom-up perception of stimulus attributes (Joseph et al. 2009). It is important to note that the stimuli in the above-mentioned studies are low-level features and inanimate stimuli (e.g. letters and shapes) but not complex images or social stimuli. Using both social stimuli of faces and people and non-social autism special-interest stimuli as search objects, we have demonstrated that people with autism have reduced attention to target-congruent objects in the search array, especially social attention (Wang et al. 2014b). Furthermore, some studies employed visual search to investigate recognition abilities of facial expressions in children with autism and found that faces with certain emotions are detected faster than others (Farran et al. 2011; Rosset et al. 2011). However, when compared with age-matched controls, no significant differences were found anymore.

Social rewards are salient cues (see above) and they share common neural basis as monetary rewards (Izuma et al. 2008; Lin et al. 2012b). However, people with autism show specific impairment in learning to choose social rewards, compared to monetary rewards (Lin et al. 2012a). Furthermore, people with autism are not influenced by the presence of an observer in a charity donation task as compared with healthy controls who donate significantly more in the observer's presence than absence, showing insensitivity in people with autism to social reputation (Izuma et al. 2011). People with autism also have reduced preference and sensitivity to donations to people charities compared with donations to the other charities (Lin et al. 2012c). In conclusion, people with autism also show altered saliency representation of more abstract social cues like social rewards.

5 Amygdala Theory of Autism

The amygdala is proposed to be part of a neural network comprising the "social brain" (Brothers 1990), while autism is a neuropsychiatric condition that disrupts the development of social intelligence. It is thus plausible that autism may be caused by an amygdala abnormality (Baron-Cohen et al. 2000). This hypothesis is supported by the following evidence. The abnormal facial scanning patterns in people with autism (Adolphs et al. 2001; Klin et al. 2002; Pelphrey et al. 2002; Neumann et al. 2006; Spezio et al. 2007a, b; Kliemann et al. 2010) are rather similar as seen in patients with amygdala damage, who fail to fixate on the eyes in faces (Adolphs et al. 2005), while neuroimaging studies in healthy individuals have shown that amygdala activation is specifically enhanced for fearful faces when saccading from the mouth to the eye regions (Gamer and Büchel 2009). Besides abnormal eye fixations onto faces, several studies have found reliable, but weak, deficits in the ability to recognize emotions from facial expressions in autism (Smith et al. 2010; Philip et al. 2010; Wallace et al. 2011; Kennedy and Adolphs 2012) (for review, see Harms et al. 2010), while on the other hands, patients with amygdala lesions also show abnormal recognition of emotion from facial expressions (Adolphs et al. 1999), and abnormal recognition of mental states from the eye region of faces (Adolphs et al. 2002), providing further support for the amygdala's involvement in autism.

When directly testing the amygdala function in people with autism, the amygdala-mediated orientation towards eyes seen in BOLD-fMRI is reported to be dysfunctional in autism (Kliemann et al. 2012). Activation in the amygdala is strongly correlated with the time spent fixating the eyes in the autistic group (Dalton et al. 2005), but compared to neurotypically developed controls, the amygdala activation was significantly weaker in the people with autism (Kleinhans et al. 2011), consistent with behavioral findings of reduced fixations onto the eyes. Recent studies with single neuron recordings in the human amygdala have even found weaker response to the eyes but stronger response to the mouth in patients with autism compared to control patients (Rutishauser et al. 2013). Despite considerable variability in reports of abnormal face processing in autism, these evidence largely support a link between abnormal processing of faces in autism and amygdala function.

Although there is evidence for global dysfunction at the level of the whole brain in autism (Piven et al. 1995; Geschwind and Levitt 2007; Amaral et al. 2008; Anderson et al. 2010), several studies emphasize abnormalities in the amygdala both morphometrically (Ecker et al. 2012) and in terms of functional connectivity (Gotts et al. 2012). The aberrant gaze patterns in individuals with autism has also been associated with an anatomical link supported by findings of similar gaze fixations, brain activation patterns and amygdala volume in their genetic but unaffected siblings, who demonstrate robust differences compared with typically developing control individuals (Dalton et al. 2007). Amygdala volume can predict gaze patterns in humans (Nacewicz et al. 2006), and even in monkeys (Zhang et al. 2012), consistent with a substantial literature showing structural abnormalities (Bauman and Kemper 1985; Schumann et al. 2004; Schumann and Amaral 2006; Amaral et al. 2008; Ecker et al. 2012) and atypical activation (Gotts et al. 2012; Philip et al. 2012) in the amygdala in autism.

Finally, it is important to note that autism is well known to be highly heterogeneous at the biological and behavioral levels and it is arguable that there will be no single genetic or cognitive cause for the diverse symptoms defining autism (Happe et al. 2006). No unanimously endorsed hypothesis for a primary deficit has emerged that can plausibly account for the full triad of social, communicative and rigid/repetitive difficulties (Happe 2003). It is also worth noting that a bona fide lesion of the amygdala shows no autistic symptoms by clinical examination and autism diagnosis (Paul et al. 2010).

References

Adams RB, Kleck RE (2003) Perceived gaze direction and the processing of facial displays of emotion. Psychol Sci 14:644–647

Adolphs R (2008) Fear, faces, and the human amygdala. Curr Opin Neurobiol 18:166–172

Adolphs R (2010) What does the amygdala contribute to social cognition? Ann NY Acad Sci 1191:42–61

Adolphs R, Baron-Cohen S, Tranel D (2002) impaired recognition of social emotions following amygdala damage. J Cogn Neurosci 14:1264–1274
Adolphs R, Gosselin F, Buchanan TW, Tranel D, Schyns P, Damasio AR (2005) A mechanism for impaired fear recognition after amygdala damage. Nature 433:68–72
Adolphs R, Sears L, Piven J (2001) Abnormal processing of social information from faces in autism. J Cogn Neurosci 13:232–240
Adolphs R, Tranel D, Damasio AR (1998) The human amygdala in social judgment. Nature 393:470–474
Adolphs R, Tranel D, Damasio H, Damasio A (1994) Impaired recognition of emotion in facial expressions following bilateral damage to the human amygdala. Nature 372:669–672
Adolphs R, Tranel D, Hamann S, Young AW, Calder AJ, Phelps EA, Anderson A, Lee GP, Damasio AR (1999) Recognition of facial emotion in nine individuals with bilateral amygdala damage. Neuropsychologia 37:1111–1117
Akiyama T, Kato M, Muramatsu T, Saito F, Umeda S, Kashima H (2006) Gaze but not arrows: a dissociative impairment after right superior temporal gyrus damage. Neuropsychologia 44:1804–1810
Amaral DG, Behniea H, Kelly JL (2003) Topographic organization of projections from the amygdala to the visual cortex in the macaque monkey. Neuroscience 118:1099–1120
Amaral DG, Schumann CM, Nordahl CW (2008) Neuroanatomy of autism. Trends Neurosci 31:137–145
Anderson BA, Laurent PA, Yantis S (2011) Value-driven attentional capture. Proc Natl Acad Sci
Anderson JR, Sallaberry P, Barbier H (1995) Use of experimenter-given cues during object-choice tasks by capuchin monkeys. Anim Behav 49:201–208
Anderson JS, Druzgal TJ, Froehlich A, DuBray MB, Lange N, Alexander AL, Abildskov T, Nielsen JA, Cariello AN, Cooperrider JR, Bigler ED, Lainhart JE (2010) Decreased Interhemispheric Functional Connectivity in Autism. Cerebral Cortex
Arcizet F, Mirpour K, Bisley JW (2011) A pure salience response in posterior parietal cortex. Cereb Cortex 21:2498–2506
Argyle M, Ingham R, Alkema F, McCallin M (1973) The different functions of gaze. Semiotica, 7:19
Aviezer H, Trope Y, Todorov A (2012) Body cues, not facial expressions, discriminate between intense positive and negative emotions. Science 338:1225–1229
Bagshaw MH, Mackworth NH, Pribram KH (1972) The effect of resections of the inferotemporal cortex or the amygdala on visual orienting and habituation. Neuropsychologia 10:153–162
Baron-Cohen S, Ring HA, Bullmore ET, Wheelwright S, Ashwin C, Williams SCR (2000) The amygdala theory of autism. Neurosci Biobehav Rev 24:355–364
Basso MA, Wurtz RH (2002) Neuronal activity in substantia nigra pars reticulata during target selection. J Neurosci 22:1883–1894
Batki A, Baron-Cohen S, Wheelwright S, Connellan J, Ahluwalia J (2000) Is there an innate gaze module? Evidence from human neonates. Infant Behav Dev 23:223–229
Bauman M, Kemper TL (1985) Histoanatomic observations of the brain in early infantile autism. Neurology 35:866–874
Baxter MG, Murray EA (2002) The amygdala and reward. Nat Rev Neurosci 3:563–573
Baxter MG, Parker A, Lindner CCC, Izquierdo AD, Murray EA (2000) Control of response selection by reinforcer value requires interaction of amygdala and orbital prefrontal cortex. J Neurosci 20:4311–4319
Bechara A, Tranel D, Damasio H, Adolphs R, Rockland C, Damasio A (1995) Double dissociation of conditioning and declarative knowledge relative to the amygdala and hippocampus in humans. Science 269:1115–1118
Becker B, Mihov Y, Scheele D, Kendrick KM, Feinstein JS, Matusch A, Aydin M, Reich H, Urbach H, Oros-Peusquens A-M, Shah NJ, Kunz WS, Schlaepfer TE, Zilles K, Maier W, Hurlemann R (2012) Fear processing and social networking in the absence of a functional amygdala. Biol Psychiatry 72:70–77

Bermudez MA, Gobel C, Schultz W (2012) Sensitivity to temporal reward structure in amygdala neurons. Curr Biol CB 22:1839–1844

Berridge KC, Robinson TE (1998) What is the role of dopamine in reward: hedonic impact, reward learning, or incentive salience? Brain Res Rev 28:309–369

Bichot NP, Rossi AF, Desimone R (2005) Parallel and serial neural mechanisms for visual search in macaque area V4. Science 308:529–534

Blair IV, Judd CM, Chapleau KM (2004) The influence of afrocentric facial features in criminal sentencing. Psychol Sci 15:674–679

Blakemore S-J, Decety J (2001) From the perception of action to the understanding of intention. Nat Rev Neurosci 2:561–567

Brothers L (1990) The social brain: a project for integrating primate behavior and neurophysiology in a new domain. Concepts Neurosci 1:27–51

Burrows BE, Moore T (2009) Influence and limitations of popout in the selection of salient visual stimuli by area V4 neurons. J Neurosci 29:15169–15177

Byrnit J (2009) Gorillas' (Gorilla gorilla) use of experimenter-given manual and facial cues in an object-choice task. Anim Cogn 12:401–404

Campbell R, Heywood CA, Cowey A, Regard M, Landis T (1990) Sensitivity to eye gaze in prosopagnosic patients and monkeys with superior temporal sulcus ablation. Neuropsychologia 28:1123–1142

Carlin JD, Calder AJ, Kriegeskorte N, Nili H, Rowe JB (2011) A head view-invariant representation of gaze direction in anterior superior temporal sulcus. Curr Biol CB 21:1817–1821

Chelazzi L, Miller EK, Duncan J, Desimone R (1993) A neural basis for visual search in inferior temporal cortex. Nature 363:345–347

Constantinidis C, Steinmetz MA (2001) Neuronal responses in area 7a to multiple-stimulus displays: i. neurons encode the location of the salient stimulus. Cereb Cortex 11:581–591

Dalton KM, Nacewicz BM, Alexander AL, Davidson RJ (2007) Gaze-fixation, brain activation, and amygdala volume in unaffected siblings of individuals with autism. Biol Psychiatry 61:512–520

Dalton KM, Nacewicz BM, Johnstone T, Schaefer HS, Gernsbacher MA, Goldsmith HH, Alexander AL, Davidson RJ (2005) Gaze fixation and the neural circuitry of face processing in autism. Nat Neurosci 8:519–526

Dawson G, Webb SJ, McPartland J (2005) Understanding the nature of face processing impairment in autism: insights from behavioral and electrophysiological studies. Dev Neuropsychol 27:403–424

Desimone R, Gross CG (1979) Visual areas in the temporal cortex of the macaque. Brain Res 178:363–380

DSM-5 (2013) Diagnostic and statistical manual of mental disorders: DSM-5. American Psychiatric Association

Dunbar RIM (2010) The social role of touch in humans and primates: behavioural function and neurobiological mechanisms. Neurosci Biobehav Rev 34:260–268

Ecker C, Suckling J, Deoni SC, Lombardo MV, Bullmore ET, Baron-Cohen S, Catani M, Jezzard P, Barnes A, Bailey AJ, Williams SC, Murphy DGM (2012) Brain anatomy and its relationship to behavior in adults with autism spectrum disorder: a multicenter magnetic resonance imaging study. Arch Gen Psychiatry 69:195–209

Emery NJ, Capitanio JP, Mason WA, Machado CJ, Mendoza SP, Amaral DG (2001) The effects of bilateral lesions of the amygdala on dyadic social interactions in rhesus monkeys (Macaca mulatta). Behav Neurosci 115:515–544

Emery NJ, Lorincz EN, Perrett DI, Oram MW, Baker CI (1997) Gaze following and joint attention in Rhesus monkeys (Macaca mulatta). J Comp Psychol 111:286–293

Farran EK, Branson A, King BJ (2011) Visual search for basic emotional expressions in autism; impaired processing of anger, fear and sadness, but a typical happy face advantage. Res Autism Spectrum Disord 5:455–462

Field T (2003) Touch. MIT Press, Cambridge

Fitzgerald DA, Angstadt M, Jelsone LM, Nathan PJ, Phan KL (2006) Beyond threat: Amygdala reactivity across multiple expressions of facial affect. NeuroImage 30:1441–1448
Freese JL, Amaral DG (2006) Synaptic organization of projections from the amygdala to visual cortical areas TE and V1 in the macaque monkey. J Comp Neurol 496:655–667
Friesen CK, Kingstone A (1998) The eyes have it! Reflexive orienting is triggered by nonpredictive gaze. Psychon Bull Rev 5:490–495
Gallace A, Spence C (2010) The science of interpersonal touch: an overview. Neurosci Biobehav Rev 34:246–259
Gamer M, Büchel C (2009) Amygdala activation predicts gaze toward fearful eyes. J Neurosci 29:9123–9126
Gazzola V, Spezio ML, Etzel JA, Castelli F, Adolphs R, Keysers C (2012) Primary somatosensory cortex discriminates affective significance in social touch. Proc Natl Acad Sci
Geschwind DH, Levitt P (2007) Autism spectrum disorders: developmental disconnection syndromes. Curr Opin Neurobiol 17:103–111
Ghashghaei HT, Barbas H (2002) Pathways for emotion: interactions of prefrontal and anterior temporal pathways in the amygdala of the rhesus monkey. Neuroscience 115:1261–1279
Gothard KM, Battaglia FP, Erickson CA, Spitler KM, Amaral DG (2007) Neural responses to facial expression and face identity in the monkey amygdala. J Neurophysiol 97:1671–1683
Gottlieb JP, Kusunoki M, Goldberg ME (1998) The representation of visual salience in monkey parietal cortex. Nature 391:481–484
Gotts SJ, Simmons WK, Milbury LA, Wallace GL, Cox RW, Martin A (2012) Fractionation of social brain circuits in autism spectrum disorders. Brain 135:2711–2725
Grabenhorst F, Hernadi I, Schultz W (2012) Prediction of economic choice by primate amygdala neurons. Proc Natl Acad Sci
Grant EC, Mackintosh JH (1963) A comparison of the social postures of some common laboratory rodents. Behaviour 21:246–259
Gross CG (1994) How inferior temporal cortex became a visual area. Cereb Cortex 4:455–469
Hadj-Bouziane F, Liu N, Bell AH, Gothard KM, Luh W-M, Tootell RBH, Murray EA, Ungerleider LG (2012) Amygdala lesions disrupt modulation of functional MRI activity evoked by facial expression in the monkey inferior temporal cortex. Proc Natl Acad Sci 109: E3640–E3648
Hall ET (1966) The hidden dimension. Doubleday, Garden City, N.Y
Happe F (2003) Cognition in autism: one deficit or many? Novartis Found Symp 251:198–207
Happe F, Ronald A, Plomin R (2006) Time to give up on a single explanation for autism. Nat Neurosci 9:1218–1220
Hariri AR, Mattay VS, Tessitore A, Kolachana B, Fera F, Goldman D, Egan MF, Weinberger DR (2002) Serotonin transporter genetic variation and the response of the human amygdala. Science 297:400–403
Harms M, Martin A, Wallace G (2010) Facial emotion recognition in autism spectrum disorders: a review of behavioral and neuroimaging studies. Neuropsychol Rev 20:290–322
Herry C, Bach DR, Esposito F, Di Salle F, Perrig WJ, Scheffler K, Luthi A, Seifritz E (2007) Processing of temporal unpredictability in human and animal amygdala. J Neurosci 27:5958–5966
Hickey C, Chelazzi L, Theeuwes J (2010) Reward changes salience in human vision via the anterior cingulate. J Neurosci 30:11096–11103
Hickey C, van Zoest W (2012) Reward creates oculomotor salience. Curr Biol 22:R219–R220
Hietanen J (2002) Social attention orienting integrates visual information from head and body orientation. Psychol Res 66:174–179
Hietanen JK (1999) Does your gaze direction and head orientation shift my visual attention? NeuroRep Rapid Commun Neurosci Res 10:3443–3447
Hoffman EA, Haxby JV (2000) Distinct representations of eye gaze and identity in the distributed human neural system for face perception. Nat Neurosci 3:80–84
Hoffman KL, Gothard KM, Schmid MC, Logothetis NK (2007) Facial-expression and gaze-selective responses in the monkey amygdala. Curr Biol CB 17:766–772

Hood BM, Willen JD, Driver J (1998) Adult's eyes trigger shifts of visual attention in human infants. Psychol Sci 9:131–134

Ikemoto S, Panksepp J (1999) The role of nucleus accumbens dopamine in motivated behavior: a unifying interpretation with special reference to reward-seeking. Brain Res Rev 31:6–41

Ipata AE, Gee AL, Goldberg ME, Bisley JW (2006) Activity in the lateral intraparietal area predicts the goal and latency of saccades in a free-viewing visual search task. J Neurosci 26:3656–3661

Izuma K, Matsumoto K, Camerer CF, Adolphs R (2011) Insensitivity to social reputation in autism. Proc Natl Acad Sci 108:17302–17307

Izuma K, Saito DN, Sadato N (2008) Processing of social and monetary rewards in the human striatum. Neuron 58:284–294

Joseph RM, Keehn B, Connolly C, Wolfe JM, Horowitz TS (2009) Why is visual search superior in autism spectrum disorder? Dev Sci 12:1083–1096

Kanner L (1943) Autistic disturbances of affective contact. Nerv Child 2:217–250

Kemner C, van Ewijk L, van Engeland H, Hooge I (2008) Brief report: eye movements during visual search tasks indicate enhanced stimulus discriminability in subjects with PDD. J Autism Dev Disord 38:553–557

Kennedy DP, Adolphs R (2012) Perception of emotions from facial expressions in high-functioning adults with autism. Neuropsychologia 50:3313–3319

Kennedy DP, Glascher J, Tyszka JM, Adolphs R (2009) Personal space regulation by the human amygdala. Nat Neurosci 12:1226–1227

Kingstone A, Friesen CK, Gazzaniga MS (2000) Reflexive joint attention depends on lateralized cortical connections. Psychol Sci 11:159–166

Kingstone A, Tipper C, Ristic J, Ngan E (2004) The eyes have it!: An fMRI investigation. Brain Cogn 55:269–271

Kleinhans NM, Richards T, Johnson LC, Weaver KE, Greenson J, Dawson G, Aylward E (2011) fMRI evidence of neural abnormalities in the subcortical face processing system in ASD. NeuroImage 54:697–704

Kliemann D, Dziobek I, Hatri A, Jr Baudewig, Heekeren HR (2012) The role of the amygdala in atypical gaze on emotional faces in autism spectrum disorders. J Neurosci 32:9469–9476

Kliemann D, Dziobek I, Hatri A, Steimke R, Heekeren HR (2010) Atypical reflexive gaze patterns on emotional faces in autism spectrum disorders. J Neurosci 30:12281–12287

Klin A, Jones W, Schultz R, Volkmar F, Cohen D (2002) Visual fixation patterns during viewing of naturalistic social situations as predictors of social competence in individuals with autism. Arch Gen Psychiatry 59:809–816

Klin A, Lin DJ, Gorrindo P, Ramsay G, Jones W (2009) Two-year-olds with autism orient to non-social contingencies rather than biological motion. Nature 459:257–261

Kling AS, Brothers LA (1992) The amygdala: neurobiological aspects of emotion, memory and mental dysfunction

Kreiman G, Koch C, Fried I (2000) Category-specific visual responses of single neurons in the human medial temporal lobe. Nat Neurosci 3:946–953

LaBar K, LeDoux J, Spencer D, Phelps E (1995) Impaired fear conditioning following unilateral temporal lobectomy in humans. J Neurosci 15:6846–6855

Langton SRH, O'Malley C, Bruce V (1996) Actions speak no louder than words: symmetrical cross-modal interference effects in the processing of verbal and gestural information. J Exp Psychol Hum Percept Perform 22:1357–1375

Langton SRH, Watt RJ, Bruce V (2000) Do the eyes have it? Cues to the direction of social attention. Trends Cogn Sci 4:50–59

Law Smith MJ, Montagne B, Perrett DI, Gill M, Gallagher L (2010) Detecting subtle facial emotion recognition deficits in high-functioning Autism using dynamic stimuli of varying intensities. Neuropsychologia 48:2777–2781

LeDoux JE (1993) Emotional memory systems in the brain. Behav Brain Res 58:69–79

Leonard CM, Rolls ET, Wilson FA, Baylis GC (1985) Neurons in the amygdala of the monkey with responses selective for faces. Behav Brain Res 15:159–176

Lewis MH, Bodfish JW (1998) Repetitive behavior disorders in autism. Ment Retard Dev Disabil Res Rev 4:80–89
Lin A, Adolphs R, Rangel A (2012a) Impaired learning of social compared to monetary rewards in autism. Front Neurosci 6
Lin A, Adolphs R, Rangel A (2012b) Social and monetary reward learning engage overlapping neural substrates. Soc Cogn Affect Neurosci 7:274–281
Lin A, Tsai K, Rangel A, Adolphs R (2012c) Reduced social preferences in autism: evidence from charitable donations. J Neurodev Disord 4:8
Logothetis NK, Sheinberg DL (1996) Visual object recognition. Annu Rev Neurosci 19:577–621
Loken LS, Wessberg J, Morrison I, McGlone F, Olausson H (2009) Coding of pleasant touch by unmyelinated afferents in humans. Nat Neurosci 12:547–548
Malkova L, Gaffan D, Murray EA (1997) Excitotoxic lesions of the amygdala fail to produce impairments in visual learning for auditory secondary reinforcement but interfere with reinforcer devaluation effects in Rhesus monkeys. J Neurosci 17:6011–6020
Mason WA, Capitanio JP, Machado CJ, Mendoza SP, Amaral DG (2006) Amygdalectomy and responsiveness to novelty in rhesus monkeys (Macaca mulatta): generality and individual consistency of effects. Emotion 6:73–81
Mazer JA, Gallant JL (2003) Goal-related activity in V4 during free viewing visual search: evidence for a ventral stream visual salience map. Neuron 40:1241–1250
McGaugh JL (2000) Memory–a century of consolidation. Science 287:248–251
McGaugh JL (2004) The amygdala modulates the consolidation of memories of emotionally arousing experiences. Annu Rev Neurosci 27:1–28
McPeek RM, Keller EL (2002) Superior colliculus activity related to concurrent processing of saccade goals in a visual search task. J Neurophysiol 87:1805–1815
McPeek RM, Keller EL (2004) Deficits in saccade target selection after inactivation of superior colliculus. Nat Neurosci 7:757–763
Miklösi A, Polgárdi R, Topál J, Csányi V (1998) Use of experimenter-given cues in dogs. Anim Cogn 1:113–121
Moeller S, Freiwald WA, Tsao DY (2008) Patches with links: a unified system for processing faces in the macaque temporal lobe. Science 320:1355–1359
Mormann F, Dubois J, Kornblith S, Milosavljevic M, Cerf M, Ison M, Tsuchiya N, Kraskov A, Quiroga RQ, Adolphs R, Fried I, Koch C (2011) A category-specific response to animals in the right human amygdala. Nat Neurosci 14:1247–1249
Morris JS, Frith CD, Perrett DI, Rowland D, Young AW, Calder AJ, Dolan RJ (1996) A differential neural response in the human amygdala to fearful and happy facial expressions. Nature 383:812–815
Murthy A, Ray S, Shorter SM, Schall JD, Thompson KG (2009) Neural control of visual search by frontal eye field: effects of unexpected target displacement on visual selection and saccade preparation. J Neurophysiol 101:2485–2506
Nacewicz BM, Dalton KM, Johnstone T, Long M, McAuliff E, Oakes T, Alexander AL, Davidson RJ (2006) Amygdala volume and nonverbal social impairment in adolescent and adult males with autism. Arch Gen Psychiatry 63:1417–1428
Neumann D, Spezio ML, Piven J, Adolphs R (2006) Looking you in the mouth: abnormal gaze in autism resulting from impaired top-down modulation of visual attention. Soc Cogn Affect Neurosci 1:194–202
Nummenmaa L, Hyönä J, Hietanen JK (2009) I'll walk this way: eyes reveal the direction of locomotion and make passersby look and go the other way. Psychol Sci 20:1454–1458
O'Riordan M (2000) Superior modulation of activation levels of stimulus representations does not underlie superior discrimination in autism. Cognition 77:81–96
O'Riordan M, Plaisted K (2001) Enhanced discrimination in autism. Q J Exp Psychol A 54:961–979
O'Riordan M, Plaisted K, Driver J, Baron-Cohen S (2001) Superior visual search in autism. J Exp Psychol Hum Percept Perform 27:719–730
O'Riordan MA (2004) Superior visual search in adults with autism. Autism 8:229–248

Ogawa T, Komatsu H (2004) Target selection in area V4 during a multidimensional visual search task. J Neurosci 24:6371–6382

Olausson H, Lamarre Y, Backlund H, Morin C, Wallin BG, Starck G, Ekholm S, Strigo I, Worsley K, Vallbo AB, Bushnell MC (2002) Unmyelinated tactile afferents signal touch and project to insular cortex. Nat Neurosci 5:900–904

Oosterhof NN, Todorov A (2008) The functional basis of face evaluation. Proc Natl Acad Sci 105:11087–11092

Osterling J, Dawson G (1994) Early recognition of children with autism: A study of first birthday home videotapes. J Autism Dev Disord 24:247–257

Paton JJ, Belova MA, Morrison SE, Salzman CD (2006) The primate amygdala represents the positive and negative value of visual stimuli during learning. Nature 439:865–870

Paul L, Corsello C, Tranel D, Adolphs R (2010) Does bilateral damage to the human amygdala produce autistic symptoms? J Neurodev Disord 2:165–173

Peignot P, Anderson JR (1999) Use of experimenter-given manual and facial cues by gorillas (Gorilla gorilla) in an object-choice task. J Comp Psychol 113:253–260

Pelphrey K, Sasson N, Reznick JS, Paul G, Goldman B, Piven J (2002) Visual scanning of faces in autism. J Autism Dev Disord 32:249–261

Pelphrey KA, Singerman JD, Allison T, McCarthy G (2003) Brain activation evoked by perception of gaze shifts: the influence of context. Neuropsychologia 41:156–170

Pelphrey KA, Viola RJ, McCarthy G (2004) When strangers pass: processing of mutual and averted social gaze in the superior temporal sulcus. Psychol Sci 15:598–603

Perrett DI, Smith PAJ, Potter DD, Mistlin AJ, Head AS, Milner AD, Jeeves MA (1985) Visual cells in the temporal cortex sensitive to face view and gaze direction. Proc R Soc Lond B Biol Sci 223:293–317

Phelps EA, LeDoux JE (2005) Contributions of the amygdala to emotion processing: from animal models to human behavior. Neuron 48:175–187

Philip RCM, Dauvermann MR, Whalley HC, Baynham K, Lawrie SM, Stanfield AC (2012) A systematic review and meta-analysis of the fMRI investigation of autism spectrum disorders. Neurosci Biobehav Rev 36:901–942

Philip RCM, Whalley HC, Stanfield AC, Sprengelmeyer R, Santos IM, Young AW, Atkinson AP, Calder AJ, Johnstone EC, Lawrie SM, Hall J (2010) Deficits in facial, body movement and vocal emotional processing in autism spectrum disorders. Psychol Med 40:1919–1929

Piech RM, McHugo M, Smith SD, Dukic MS, Van Der Meer J, Abou-Khalil B, Most SB, Zald DH (2011) Attentional capture by emotional stimuli is preserved in patients with amygdala lesions. Neuropsychologia 49:3314–3319

Piech RM, McHugo M, Smith SD, Dukic MS, Van Der Meer J, Abou-Khalil B, Zald DH (2010) Fear-enhanced visual search persists after amygdala lesions. Neuropsychologia 48:3430–3435

Piven J, Arndt S, Bailey J, Havercam S, Andreasen N, Palmer P (1995) An MRI study of brain size in autism. Am J Psychiatry 152:1145–1149

Plaisted K, O'Riordan M, Baron-Cohen S (1998) Enhanced visual search for a conjunctive target in autism: a research note. J Child Psychol Psychiatry 39:777–783

Povinelli DJ, Eddy TJ (1996) Chimpanzees: joint visual attention. Psychol Sci 7:129–135

Puce A, Allison T, Bentin S, Gore JC, McCarthy G (1998) Temporal cortex activation in humans viewing eye and mouth movements. J Neurosci 18:2188–2199

Purcell BA, Weigand PK, Schall JD (2012) Supplementary eye field during visual search: salience, cognitive control, and performance monitoring. J Neurosci 32:10273–10285

Ristic J, Friesen C, Kingstone A (2002) Are eyes special? It depends on how you look at it. Psychon Bull Rev 9:507–513

Rolls E (1984) Neurons in the cortex of the temporal lobe and in the amygdala of the monkey with responses selective for faces. Hum Neurobiol 3:209–222

Rolls ET (2010) The affective and cognitive processing of touch, oral texture, and temperature in the brain. Neurosci Biobehav Rev 34:237–245

Rolls ET, O'Doherty J, Kringelbach ML, Francis S, Bowtell R, McGlone F (2003) Representations of pleasant and painful touch in the human orbitofrontal and cingulate cortices. Cereb Cortex 13:308–317
Roozendaal B, Castello NA, Vedana G, Barsegyan A, McGaugh JL (2008) Noradrenergic activation of the basolateral amygdala modulates consolidation of object recognition memory. Neurobiol Learn Mem 90:576–579
Rosset D, Santos A, Da Fonseca D, Rondan C, Poinson F, Deruelle C (2011) More than just another face in the crowd: Evidence for an angry superiority effect in children with and without autism. Res Autism Spectrum Disord 5:949–956
Rutishauser U, Tudusciuc O, Neumann D, Mamelak AN, Heller AC, Ross IB, Philpott L, Sutherling WW, Adolphs R (2011) Single-unit responses selective for whole faces in the human amygdala. Curr Biol CB 21:1654–1660
Rutishauser U, Tudusciuc O, Wang S, Mamelak AN, Ross IB, Adolphs R (2013) Single-neuron correlates of atypical face processing in autism. Neuron 80:887–899
Sander D, Grandjean D, Pourtois G, Schwartz S, Seghier ML, Scherer KR, Vuilleumier P (2005) Emotion and attention interactions in social cognition: brain regions involved in processing anger prosody. NeuroImage 28:848–858
Sasson N (2006) The development of face processing in autism. J Autism Dev Disord 36:381–394
Sasson NJ, Elison JT, Turner-Brown LM, Dichter GS, Bodfish JW (2011) Brief report: circumscribed attention in young children with autism. J Autism Dev Disord 41:242–247
Sasson NJ, Turner-Brown LM, Holtzclaw TN, Lam KSL, Bodfish JW (2008) Children with autism demonstrate circumscribed attention during passive viewing of complex social and nonsocial picture arrays. Autism Res 1:31–42
Sato W, Okada T, Toichi M (2007) Attentional shift by gaze is triggered without awareness. Exp Brain Res 183:87–94
Sato W, Yoshikawa S, Kochiyama T, Matsumura M (2004) The amygdala processes the emotional significance of facial expressions: an fMRI investigation using the interaction between expression and face direction. NeuroImage 22:1006–1013
Scaife M, Bruner JS (1975) The capacity for joint visual attention in the infant. Nature 253:265–266
Scheumann M, Call J (2004) The use of experimenter-given cues by South African fur seals (Arctocephalus pusillus). Anim Cogn 7:224–230
Schirmer A, Teh KS, Wang S, Vijayakumar R, Ching A, Nithianantham D, Escoffier N, Cheok AD (2011) Squeeze me, but don't tease me: Human and mechanical touch enhance visual attention and emotion discrimination. Soc Neurosci 6:219–230
Schumann CM, Amaral DG (2006) Stereological analysis of amygdala neuron number in autism. J Neurosci 26:7674–7679
Schumann CM, Hamstra J, Goodlin-Jones BL, Lotspeich LJ, Kwon H, Buonocore MH, Lammers CR, Reiss AL, Amaral DG (2004) The amygdala is enlarged in children but not adolescents with autism; the hippocampus is enlarged at all ages. J Neurosci 24:6392–6401
Sheinberg DL, Logothetis NK (2001) Noticing familiar objects in real world scenes: the role of temporal cortical neurons in natural vision. J Neurosci 21:1340–1350
Shen K, Paré M (2007) Neuronal activity in superior colliculus signals both stimulus identity and saccade goals during visual conjunction search. J Vis 7
Shen K, Paré M (2014) Predictive saccade target selection in superior colliculus during visual search. J Neurosci 34:5640–5648
South M, Ozonoff S, McMahon W (2005) Repetitive behavior profiles in asperger syndrome and high-functioning autism. J Autism Dev Disord 35:145–158
Spezio ML, Adolphs R, Hurley RSE, Piven J (2007a) Abnormal use of facial information in high-functioning autism. J Autism Dev Disord 37:929–939
Spezio ML, Adolphs R, Hurley RSE, Piven J (2007b) Analysis of face gaze in autism using "Bubbles". Neuropsychologia 45:144–151
Tanaka K (1997) Mechanisms of visual object recognition: monkey and human studies. Curr Opin Neurobiol 7:523–529

Thomas NWD, Paré M (2007) Temporal processing of saccade targets in parietal cortex area lip during visual search. J Neurophysiol 97:942–947
Thompson KG, Bichot NP (2005) A visual salience map in the primate frontal eye field. Prog Brain Res 147:249–262
Thompson KG, Hanes DP, Bichot NP, Schall JD (1996) Perceptual and motor processing stages identified in the activity of macaque frontal eye field neurons during visual search. J Neurophysiol 76:4040–4055
Todd RM, Talmi D, Schmitz TW, Susskind J, Anderson AK (2012) Psychophysical and neural evidence for emotion-enhanced perceptual vividness. J Neurosci 32:11201–11212
Todorov A, Mandisodza AN, Goren A, Hall CC (2005) Inferences of competence from faces predict election outcomes. Science 308:1623–1626
Tolias AS, Moore T, Smirnakis SM, Tehovnik EJ, Siapas AG, Schiller PH (2001) eye movements modulate visual receptive fields of v4 neurons. Neuron 29:757–767
Tsao DY, Freiwald WA, Tootell RBH, Livingstone MS (2006) A cortical region consisting entirely of face-selective cells. Science 311:670–674
Tye KM, Stuber GD, de Ridder B, Bonci A, Janak PH (2008) Rapid strengthening of thalamo-amygdala synapses mediates cue-reward learning. Nature 453:1253–1257
van Boxtel JJA, Lu H (2011) Visual search by action category. J Vis 11
van Boxtel JJA, Lu H (2012) Signature movements lead to efficient search for threatening actions. PLoS ONE 7:e37085
Vuilleumier P (2002) Perceived gaze direction in faces and spatial attention: a study in patients with parietal damage and unilateral neglect. Neuropsychologia 40:1013–1026
Vuilleumier P (2005) How brains beware: neural mechanisms of emotional attention. Trends Cogn Sci 9:585–594
Vuilleumier P, Richardson MP, Armony JL, Driver J, Dolan RJ (2004) Distant influences of amygdala lesion on visual cortical activation during emotional face processing. Nat Neurosci 7:1271–1278
Wallace G, Case L, Harms M, Silvers J, Kenworthy L, Martin A (2011) Diminished sensitivity to sad facial expressions in high functioning autism spectrum disorders is associated with symptomatology and adaptive functioning. J Autism Dev Disord 41:1475–1486
Wang S, Fukuchi M, Koch C, Tsuchiya N (2012a) Spatial attention is attracted in a sustained fashion toward singular points in the optic flow. PLoS ONE 7:e41040
Wang S, Krajbich I, Adolphs R, Tsuchiya N (2012b) The role of risk aversion in non-conscious decision-making. Front Psychol 3
Wang S, Tudusciuc O, Mamelak AN, Ross IB, Adolphs R, Rutishauser U (2014a) Neurons in the human amygdala selective for perceived emotion. Proc Natl Acad Sci USA 111:E3110–E3119
Wang S, Xu J, Jiang M, Zhao Q, Hurlemann R, Adolphs R (2014b) Autism spectrum disorder, but not amygdala lesions, impairs social attention in visual search. Neuropsychologia 63:259–274
Wang S, Tsuchiya N, New J, Hurlemann R, Adolphs R (2015) Preferential attention to animals and people is independent of the amygdala. Soc Cogn Affect Neurosci 10:371–380
Watson KK, Platt ML (2012) Social signals in primate orbitofrontal cortex. Curr Biol CB 22:2268–2273
Willis J, Todorov A (2006) First impressions: making up your mind after a 100-ms exposure to a face. Psychol Sci 17:592–598
Willis ML, Palermo R, Burke D, McGrillen K, Miller L (2010) Orbitofrontal cortex lesions result in abnormal social judgements to emotional faces. Neuropsychologia 48:2182–2187
Wise RA (2004) Dopamine, learning and motivation. Nat Rev Neurosci 5:483–494
Wolfe JM (2012) The rules of guidance in visual search. In: Kundu M et al. (ed) Perception and machine intelligence, vol 7143, pp 1–10. Springer, Heidelberg
Wolfe JM, Horowitz TS (2004) What attributes guide the deployment of visual attention and how do they do it? Nat Rev Neurosci 5:495–501

Yang J, Bellgowan PSF, Martin A (2012) Threat, domain-specificity and the human amygdala. Neuropsychologia 50:2566–2572

Zhang B, Noble PL, Winslow JT, Pine DS, Nelson EE (2012) Amygdala volume predicts patterns of eye fixation in rhesus monkeys. Behav Brain Res 229:433–437

Vision and Memory: Looking Beyond Immediate Visual Perception

Cheston Tan, Stephane Lallee and Bappaditya Mandal

1 Introduction

The topic of vision is "often studied as if our conscious experience were the ultimate end-product of visual processing" (Hayhoe 2009), with research sometimes overly focused on immediate visual perception—processes aimed at understanding the current visual scene or episode. However, in many situations, visual perception alone is insufficient, and what is also needed is visual memory, defined as "the preservation of visual information after the optical source of that information is no longer available to the visual system" (Palmer 1999).

For example, you may well recognize the identity of a person, such as your mother. But where and when did you last meet her? Does she look the same as before, or does it look like her health might have deteriorated (or gotten better)? Is that purplish blotch on her forearm something she's had for a while? Has her hair greyed a lot recently? Does she look happier than usual? Is she wearing a new outfit? Is her usual necklace missing, or does she usually not wear jewelry in the first place? All of these questions require some form of visual memory to be answered. Beyond the above example, Table 1 lists more such questions which require visual memory, illustrating the many ways in which visual memory is used at some point for most everyday tasks.

The need for visual memory in the examples found in Table 1 are due to various reasons. The human visual system has high resolution only at the fovea, thus requiring attentional shifts and eye movements. Visual memory is needed during integration of information across these shifts. More fundamentally, any practical visual system has some limitation on its field of view, and thus requires visual memory to integrate information across shifts in the field of view. The structure of

C. Tan (✉) · S. Lallee · B. Mandal
Institute for Infocomm Research, 1 Fusionopolis Way,
#21-01 Connexis South, Singapore 138632, Singapore
e-mail: cheston-tan@i2r.a-star.edu.sg

Q. Zhao (ed.), *Computational and Cognitive Neuroscience of Vision*,
Cognitive Science and Technology, DOI 10.1007/978-981-10-0213-7_9

Table 1 Visual memory is used in many everyday situations; visual perception from the "snapshot" of one single eye-fixation alone is insufficient

Task	Examples of questions that involve visual memory
Shopping	Which shirt/dress/skirt looks nicer on me: this one or the one I just tried on? Have I browsed through this grocery store aisle yet?
Cooking and baking	Where did I keep the flour, sugar and cream? Does what I made look like what I saw on the cooking show (or in the cookbook)?
Eating and drinking	What have I eaten this week? (How much salt/sugar/oil/fiber/etc. have I taken?) How many teaspoons of sugar did I just add to my coffee?
Driving	Is this the way home? Where should I look to find out what road I'm on? What did I see in my rear- or side-view mirror when I glanced at it just now?
Taking public transport	Where are empty seats in the bus? (Which is closest or looks most comfortable?) Does it look like the bus is nearing my destination?
Navigating	Given what I saw when I was looking around just now, where am I on this map? On the map, where is my destination, and what route do I take to get there?
Walking	When I looked down just now, were there obstacles I should avoid? When I looked just now, was there one more step, or have I reached the top of the stairs?
Returning home	When did I last check my mailbox for letters? Does my apartment/house look the same as I left it, or has someone been here?
Watching TV or movies	What happened before the camera cut to the current shot? How does that previous shot relate to the current shot?
Using a tablet or computer	Based on past experience, what will happen when I click/tap on this icon? Based on past experience, how do I navigate to a particular program/app/menu?

the world also plays an important part. As things move around, they occlude each other temporarily, and visual memory helps to overcome the temporary loss of visual information. Also, things in the world are separated by time and/or space, so visual memory is needed to retain information about one thing in order to relate it to another thing (e.g. for visual comparison). In Sect. 5.1, we will describe in more detail how visual memory plays a number of different functional roles in overcoming these constraints.

Natural visual behavior in the real world requires much more than just understanding the present scene, and often involves the deliberate choice and planning of a sequence of routines or operations involving both vision and memory (Hayhoe 2009). These routines are under the active control of the system (whether human, animal, or robot), and dynamically affect what is seen: the choice of routine affects what is seen, and conversely, what is seen affects subsequent choice of routine. We will describe some of these routines in more detail in Sect. 5.2. Thus, in real-world tasks, visual perception and memory are intrinsically linked.

1.1 An Under-Investigated Research Topic

If interactions between vision and memory are ubiquitous, then why has the study of such interactions not been more prominent (Palmeri and Tarr 2008)? We examine the reasons by area of study. In computer vision, images and videos are processed at uniform resolution, so there is no notion of saccades to shift the center of high-resolution processing. Thus, there is no need for visual memory as a buffer for integration of information over saccades. For robots that can make eye, head or body movements, thus requiring some sort of information integration over time, digital computer memory is highly precise, can easily be maintained, and capacity is not a major issue. The same holds true of databases of computer vision features, used for example in image retrieval (Chandrasekhar et al. 2014a). Hence, there is seemingly little to be researched regarding visual memory in computer vision and robotics.

In neuroscience, the fields of vision and memory do not intersect often. Memory researchers tend to prefer to study spatial memory in rodents for a number of reasons, including ease of training and recording, as well as the availability of experimental techniques. Visual perception and the transformation of information along the visual pathway is still far from fully understood, so the study of downstream visual memory may be perceived by some as still too preliminary. Practical difficulties also contribute. Studying visual memory, rather than visual perception, implies recording from the same neurons for extended periods of time, which is harder to achieve in practice. Longer trials for memory experiments means fewer trials can be done, fewer conditions can be tested, or experiments have to be longer.

In cognitive psychology, memory was traditionally studied using words (Luck and Vogel 2013), because such stimuli were easier to present, manipulate and test. In recent years, however, the study of visual memory has taken off, and associated topics such as visual search and perceptual learning are mainstream topics in their own right. Importantly, however, the study of visual memory is still relatively young, and fundamental issues are still debated (Luck and Vogel 2013; Suchow, Fougnie et al. 2014). Only in recent years has the field begun to mature towards studying realistic stimuli and modeling underlying neural mechanisms (Brady et al. 2011; Franconeri et al. 2013; Orhan and Jacobs 2014). Crucially, "relatively little work has been conducted to understand how component operations of vision and memory are coordinated to support real-world perception, memory and behavior" (Hollingworth 2009). For instance, "very little empirical or theoretical work has been done on the role of visual short-term memory in cognition" (Luck 2008), and "models [of scene-viewing] have rarely been tested in the context of natural behavior" (Tatler 2014).

1.2 Chapter Overview

The main message of this chapter is that for real-world visual tasks, immediate visual perception is not sufficient, and it interacts with visual memory (or other forms of memory) virtually all the time, as discussed in Sect. 1.

Next, in Sect. 2, we will provide some background on the standard neuroscience and psychology perspectives on visual memory.

In Sect. 3, we go from vision to memory, describing the visual attributes that influence memorability. In Sect. 4, we go in the reverse direction, from memory to vision, and describe how past memories can influence visual perception.

In Sect. 5, we will describe how visual memory is used during real-world visual tasks, illustrating the variety of ways in which vision and memory interact seamlessly in our everyday lives. We will describe the different functional roles that visual memory can play during real-world visual tasks, i.e. what visual memory is used for. We then propose a hierarchical framework that organizes a set of routines that utilize visual memory in its various functional roles to accomplish various real-world tasks.

Finally, in Sect. 6, we will summarize the chapter, and finish off by discussing some open research questions and possible applications.

2 Background: Vision and Memory

In this section, we briefly but critically review the standard textbook treatments of the relationship between vision and memory from the perspectives of neuroscience and psychology. Unlike this chapter, which puts forth the view that the interaction between vision and memory is important—and not as prominently studied as it should be—these standard perspectives either view vision and memory as mostly separate, or that vision is just another input modality for memory (as commonly viewed by neuroscience and psychology respectively).

2.1 Neuroscience Perspective

From the neuroscience perspective, there is actually no clear separation between vision and memory at the neuronal level. However, some accounts of the neuroscience of vision may give the impression that there is a distinct "visual system" that is not involved in memory. In this sub-section, we will first explain why this is the case, and then show that there is evidence for a continuum—rather than distinct separation—between vision and memory in the brain.

Standard accounts of the neuroscience of vision often talk of a "ventral visual pathway" (Felleman and Essen 1991), starting from primary visual cortex (V1), and ending in inferotemporal cortex (area IT) in macaque monkeys. IT projects to brain areas that are polysensory (or polymodal), i.e. they respond to stimuli of various sensory modalities, and thus are no longer purely visual. From this perspective, it makes sense to call the V1-to-IT pathway a visual pathway.

Simplified accounts of this visual pathway traditionally explain the pathway as a series of processing stages that produce representations that are increasingly complex and also increasingly invariant to changes in position and scale, suitable to be used by downstream brain areas for invariant recognition of complex objects (Serre et al. 2007).

However, from the memory perspective, this account is less compelling. Many accounts of this visual pathway focus on neural responses to static images presented briefly to passively fixating monkeys. These accounts downplay studies that investigate the changes in neural responses over repeated presentations, and studies that involve monkeys performing memory tasks.

When the properties of IT neurons are studied explicitly with regard to memory, studies have shown that these neurons can clearly be both visual and mnemonic simultaneously. Fuster and Jervey (1982) trained monkeys to perform a color match-to-sample task: a sample color was first shown, and after a brief delay, the monkey had to choose that same color from among several choices, i.e. this task required visual memory. One of their results was the existence of neurons that were selective for the sample color, but only responded strongly during the delay period, i.e. after the sample color had disappeared and before the color choices appeared. In other words, such neurons were clearly involved in visual short-term memory for this task.

Over longer time scales, it is also clear that IT neurons are involved in memory. Messinger et al. (2001) trained monkeys to associate pairs of objects. They found that over the course of training, the neural responses to the paired objects became more similar. These changes depended on successful behavioral learning of the associations, and were long-lasting. Crucially, these changes could be observed within the course of a single training session. In other words, this kind of visual associative memory rapidly caused neuronal changes in IT.

In summary, there is evidence that neurons in the visual pathway can be considered visual in the sense of responding only to visual stimuli, but they can also be considered mnemonic in the sense of being involved in short-term and long-term memory.

2.2 Psychology Perspective

From the psychological perspective, the research at the intersection of vision and memory primarily focuses on memory for visual stimuli, rather than other interactions, such as the co-ordination of vision and memory for action. Compared to other topics within memory research, however, visual memory is a relatively new area, as explained in Sect. 1. As such, the taxonomy for visual memory and the resulting sub-types of visual memory are similar to that for other forms of memory (e.g. word memory).

Visual memory can be classified according to five basic characteristics: duration, content, loss, capacity and maintenance (Palmer 1999). Respectively, these refer to how long the visual memory lasts, the kind of information contained, how that information is lost from memory, how much information the memory can hold, and whether the information can be maintained or refreshed. Using these characteristics, three kinds of visual memory have been distinguished, and these types are similar to memory for other modalities.

Iconic memory is a low-level type of visual memory of high capacity but lasts only for a brief duration on the order of less than a second. The information in iconic memory is highly spatiotopic, i.e. not tolerant to changes in retinal position. There does not appear to be any way (e.g. repeated rehearsal or visualization) to maintain the information in iconic memory. The existence of iconic memory and its characteristics are relatively well studied, but its functional role is much less clear (Palmer 1999). Due to its spatiotopic nature and short duration, iconic memory could be related to some aspect of saccade planning, or could simply be a temporary buffer for the construction of short-term visual memory. However, there is little evidence for these (and other) hypotheses (Palmer 1999).

Visual short-term memory (VSTM), compared to iconic memory, is of longer duration, represented in more abstract, non-spatiotopic form, more limited in capacity, and can be maintained through active visualization (Palmer 1999). The duration of information in VSTM is at least 10 s, possibly longer if maintained through active rehearsal (visualization). The information is VSTM is non-spatiotopic in that it is not tied to exact retinal position, but spatial information and relationships are maintained. The contents are post-categorical, i.e. semantic meaning has been extracted, but this semantic information is presumably contained elsewhere (and linked from VSTM).

Visual long-term memory (VLTM) is usually thought of as simply the visual aspects of a general long-term memory system that contains information from multiple senses, along with more abstract information (Palmer 1999). As such, similar to LTM in general, VLTM can be sub-categorized into semantic, procedural and episodic memory. Visual semantic memory is information about the visual appearance of objects, such as dogs in general, a specific breed of dog and individual dogs. Visual procedural memory includes things like the visual knowledge of how to walk from one familiar location to another. Visual episodic memory is simply the visual component of the layman conception of "memory" (specific experiences from a person's past). The detailed relationship between VLTM and VSTM representation of objects is unclear (Luck 2008).

3 From Vision to Memory: Image Memorability

As described above in Sect. 2.2, the study of visual memory in psychology has historically been mostly limited to characterizing the different types of visual memory, in terms of capacity, duration etc. In a way, such research has focused on the memory aspects of visual memory, rather than the visual aspects. For instance, some papers draw conclusions about the capacity of visual short-term memory by using a relatively small number of artificial stimuli, with little consideration given to how the results might vary with the type of stimuli (or even each individual stimulus).

As such, the question of how the visual attributes of images affect memory has been relatively neglected. It is only in recent years that this question has been addressed in a systematic way, particularly in conjunction with computational modelling. Below, we describe these recent studies on image memorability, first for visual scenes in general, and then for the special domain of faces.

3.1 Visual Attributes and Scene Memorability

Humans have a surprisingly large capacity for remembering images (Standing 1973). We remember not only the gist of the scene, but also recognize which specific scene we saw and other details (Rock and Englestein 1959). We remember some pictures for a very long time, but forget others quickly. Even when an image does not contain any particularly memorable objects or people, the image as a whole may still be remembered (Brady et al. 2008; Konkle et al. 2010). What visual attributes contribute towards memorability?

Studies conducted by Isola et al. (2011), Konkle et al. (2010) showed that simple image features, including mean saturation values and pixel intensity statistics, do not play an important role in memorability. They have also shown that general, non-semantic object statistics (like total object count, object size, etc.) do not play a large role in memorability.

On the other hand, semantics play an important role in scene memorability (Konkle et al. 2010). Experiments conducted by Isola et al. (2011) show that attributes related to semantics (such as presence, count and size of labeled objects) boosted memorability. They concluded that the objects with semantics are a primary substrate of scene memorability. Additionally, their analysis suggest that people, interiors, foregrounds and human-scale objects tend to contribute positively to memorability. However, exteriors, wide angle vistas, backgrounds and natural scenes tend to contribute negatively to memorability.

When global features such as SIFT (Lowe 2004), HOG (Dalal and Triggs 2005), Gist (Oliva and Torralba 2001) and SSIM (Shechtman and Irani 2007) are used together in combination, a fair amount of correlation with human scene memorability can be achieved. However, object semantics was shown to have the highest contribution towards memorability scores.

3.2 Visual Attributes and Face Memorability

It is well-known that human memory for familiar faces is very impressive even under poor viewing conditions, including bad lighting, changes in viewpoint, viewing from a distance, seeing the face in motion, and large variability in age and facial expression (Lander and Chuang 2005; Sinha et al. 2006). However, human memory for unfamiliar faces is comparatively poor. Lander and Chuang (2005) showed that recognition of identity was significantly better for familiar faces than unfamiliar (or less familiar) faces, shown either in motion (e.g. while talking) or as single static images. For unfamiliar faces, what factors determine whether they are remembered or not?

Numerous studies have shown that face distinctiveness (or uniqueness) influences memorability, defined as the degree to which the faces are remembered or forgotten (Bruce et al. 1994; Busey 2001). Apart from distinctiveness, many other facial attributes have been investigated for their role in face memory, including race, age, dominance, gender, emotional expressions, trustworthiness, attractiveness, etc.

Recently, Bainbridge et al. (2013) performed a large-scale experiment investigating the relationship between face memorability and facial attributes (e.g. attractive, unhappy, sociable, emotionally stable, mean, boring, aggressive, weird, intelligent, confident, caring, egotistic, responsible, trustworthy), memory-related attributes (typicality, familiarity) and attributes previously reported to be significantly correlated with memorability (commonness, emotional magnitude, friendliness).

This large-scale visual memory experiment involved 877 online participants on Amazon Mechanical Turk, and followed the same experimental protocols as Isola et al. (2011). From the 10 k US Adult Faces Database, they randomly selected 2,222 photos as target images and 6,468 photos as filler images. Memorability measures were hit rates (HRs) and false alarm rates (FARs) for repetition detection on target photographs, where repetitions were spaced 91 to 109 photographs apart. Each target photo was seen by an average of 81.7 participants. On average, target faces were correctly recognized in 51.6 % of trials (SD 12.6 %). The average rate of false alarms was 14.4 % (SD 8.7 %). In this experiment and others, high consistency in memorability scores across participants was found.

Analysis of the results indicate that six significant contributing factors affecting face memorability are interestingness, irresponsibility, kindness, unhappiness, emotionally instability and unattractiveness. These findings suggest that unlike previous findings (Vokey and Read 1992), face memorability may not solely be determined by memory-related attributes but is also influenced by other social and personality traits, even when the face images are presented briefly (Bainbridge et al. 2013).

3.3 Discussion

Quantitative, large-scale research into visual attributes that influence memory for images is still in its infancy. Beyond still images, research could be extended to videos. The current memory timeframe being studied is on the order of minutes, and both shorter and longer timeframes can be investigated. It remains to be seen if memorability of an image is a static property, or if it can vary with timeframe.

Predicting memorability—and possibly even manipulating it (Khosla et al. 2013)—has applications in computer vision and graphics, such as choosing a book cover, designing a logo or advertisement, organizing images in a photo album and selecting an image for a website. More specifically for faces, algorithms for predicting memorability could be used for choosing highly memorable faces for advertisements, magazine covers and articles, or company websites.

4 From Memory to Vision: Past Experience Influences Perception

In the previous section, we went from vision to memory, describing how visual attributes influence memorability. In this section, we go in the reverse direction, and describe how past memories can influence visual perception. Throughout life, the visual system is exposed to a world which obeys certain physical laws and contains strong regularities, and the brain picks up those regularities from the sensory signals it receives. As a result, the perceptions generated by the bottom up sensory streams are shaped by past experience, either through lateral or top-down connections (Gregory 1970).

4.1 Memory Influencing Vision Through Top-Down Processing

The influence of past experience as a force that shapes our perceptual processes can be illustrated using visual illusions. An image is a collection of light intensities that a bottom-up process interprets, and the image is ultimately labelled through semantic description (e.g. using words). The perceptual process is shaped by expectations about the image content, and through semantic description, one can modify in a top-down manner how an image is perceived. An example of this phenomenon is shown in Fig. 1, an image which can be difficult to interpret meaningfully when seen for the first time. However, when the semantic hint of "cow" is given, various parts of the image (such as the cow's ears, nose and eyes) become more easily interpretable. Furthermore, this effect is often irreversible; after seeing the cow in the image, many people reportedly cannot "unsee" the cow ever

Fig. 1 Illustration of the effect of top-down knowledge on bottom-up perception. The image can appear uninterpretable or meaningless until the viewer has been primed with the concept of "cow"

again on subsequent viewings. This well-known "cow illusion" demonstrates that a semantic hint about image content can be sufficient to strongly shape perception.

The effect of semantic concepts on visual perception has also been investigated in the case of bi-stable illusions (Balcetis and Dale 2003). Bi-stable illusions refer to images that can be interpreted in two different ways, although only one perception can be active at any given time. Balcetis and Dale (2003) showed that the perceived figure can depend on priming with semantic knowledge either in the form of a simple sentence or embedded in the context of a discussion.

Although the top-down influence of knowledge can be shown through explicit semantic priming, it can also be shown solely through perceptual stimulation. Indeed, the way we perceive things is deeply biased by how we usually perceive them: an example of this effect is the famous "hollow mask illusion" (Fig. 2). When looking at a hollow mask of a face from the back, the concave surface is often perceived to be a convex one. This is likely to be due to our overwhelming experience of faces as convex objects; this experience overrides the real visual cues present in the image (e.g. shadows) that provide evidence of a concave surface.

4.2 Memory Influencing Vision Through Multimodal Contingencies

Vision can also be influenced by past experiences (i.e. memory) indirectly, through the shaping of associations across different sensory modalities. Physical entities and events are often perceptible through several different senses. For example, a lighting

Fig. 2 The "hollow mask illusion". A concave hollow face is often perceived as convex. The world we experience daily contains convex faces, therefore inducing a strong bias to perceive face-like surfaces as convex

strike (visual) will usually be accompanied by corresponding thunder (auditory), as well as the smell of ozone (olfactory) preceding the storm. Over time, the brain is exposed to many such multi-modal contingencies, and a specific pattern of activity in one modality can become tightly coupled with specific activity in other modalities. For events that are frequent and have high cross-modal correlation, our perception becomes shaped so that our senses affect each other. In short, based on associations in previous experiences, each modality can bias perception in other modalities.

The most investigated multimodal interaction involves audition and vision, as this type of interaction is easy to conceptualize and reproduce. The interactions between vision and audition can be demonstrated through a galore of experiments (for a review, see Spence 2011), especially in the context of speech. The typical example of this interaction is the ventriloquist effect, where the origin of a sound gets falsely attributed to the visual location that appears likely to produce it. An example of direct modulation of visual perception from sound is the double flash illusion. A single flash of light is accompanied by two auditory beeps, which induces the perception of two flashes of light (Shamset al. 2000). So far, this is one of the rare cases of auditory dominance over vision.

Perhaps the most well-known audio-visual contingency is the McGurk effect (McGurk and MacDonald 1976), in which a subject presented with a video of someone uttering a specific syllable ("bah") dubbed with the auditory track of another syllable ("gah") induces the illusory perception of an in-between syllable ("dah"). This study illuminates the interaction between the visual and auditory modalities by showing how visual information can interfere with auditory percepts. More interestingly for our current topic, the reverse effect has been found: Sweeny et al. demonstrated that specific sounds affects visual judgments about the size of oval shapes in a way that is congruent with the mouth shape used to produce those sounds (Sweeny et al. 2012).

Another famous illusion demonstrates the interplay between touch and vision: the rubber hand illusion (Botvinick and Cohen 1998). A subject's hand is hidden under the table and a visible rubber hand is roughly superposed. During a conditioning phase, both the real and the rubber hand are touched synchronously, so that the subject assimilates the rubber hand as his or her own limb. During the test phase, the subject will report feeling touch when observing a contact with the rubber hand, even in the absence of contact with the real hand. Conversely, the estimated position of the hand has been shown to be dependent on both modalities (for a review and model see Van Beers et al. 1999). Moreover, the congruence between hand posture and the picture of a hand biases reaction time to the visual stimulus (Craighero et al. 2002) and ability to mentally rotate the hand (Shenton et al. 2004).

4.3 Discussion

All the above-mentioned effects and illusions have one thing in common: they are all based on top-down influence from past experiences, illustrating how memory can shape visual perception. The exact mechanisms involved are still unknown, but one influential framework is Damasio's Convergence Divergence Zone concept, which proposes that modal sensory pathways are fused in a hierarchical manner with some amodal conceptual areas at the top of the hierarchy (Damasio 1989; Meyer and Damasio 2009). In his framework, the experience of a percept would come from the partial reactivation, in a top-down manner, of the sensory hierarchy. When a concept gets activated by a given modality, all other modal representations of this concept get activated as well. As a result, when you perceive a lip motion you hear the associated sound and vice versa. This framework is appealing as it accounts for some cross modal effects that originate from high-level semantic representation, such as reading of certain words that generates activity in auditory (Kiefer et al. 2008) or olfactory (González et al. 2006) areas. It also explains how semantic priming could bias perception of ambiguous figures through shaping of the top-down reconstruction.

5 Visual Memory in Real-World Tasks

In the previous two sections, we described how vision influenced memory and vice versa. In this section, we will describe how visual memory is used during real-world visual tasks, illustrating the variety of ways in which vision and memory interact seamlessly in our everyday lives. We first describe the different functional roles (i.e. what visual memory is used for) that visual memory can play during real-world visual tasks. We then propose a hierarchical framework that organizes a collection of visual routines (Ullman 1984) that utilize visual memory in its various functional roles to accomplish these real-world tasks.

5.1 Functional Roles of Visual Memory

The interactions listed previously in Table 1 point to a number of functional roles played by visual memory (Henderson 2008). We can use the simple example of making a cup of coffee to illustrate these different functional roles, summarized in Table 2. As I enter the kitchen to make coffee, I search for my cup. After a little bit of searching, I know I've found my cup because the currently attended object matches the visual representation of my cup held in a *visual buffer*. After I take my cup, place it on the table, and turn to get the coffee powder, the *active scene representation* in visual memory allows me to know that the cup is still on the table (and should still be there later) even though it is currently out of sight. I get the container of coffee powder from the lower-left cabinet, because *episodic scene memory* tells me that is where I last put the coffee powder (and also where I usually

Table 2 Functional roles of visual memory

Functional role	Description of functional role	Time scale	Memory type	Usage example
Visual buffer	Temporary store of visual representations for active manipulation or comparison	Current perceptual episode	VSTM	During search, an object matches the representation of my cup held in memory
Active scene representation	Knowledge about a specific scene in the current perceptual episode	Current perceptual episode	VSTM and/or VLTM	My cup is still on the table, even though I turned away
Episodic scene memory	Knowledge about a particular scene or scenes from the past	Beyond current episode	VLTM	I last left the container of coffee powder in the lower-left cabinet
Scene schema knowledge	Knowledge about a particular scene category	Beyond current episode	VLTM	The container of coffee powder is probably in one of the cabinets

put it). If it's not there, perhaps because someone moved it, *scene schema knowledge* guides me to start my search at the most likely locations, such as the other cabinets. Below, we describe these functional roles of visual memory in more detail.

5.1.1 Visual Buffer

One functional role of visual memory is to serve as a buffer for visual representations (Luck 2008), which is filled either bottom-up through perception, or top-down through memory recollection. This buffer is needed to hold visual representations during manipulation or comparison. One example is mental rotation (Shepard and Metzler 1971), in which subjects are shown a target, and have to select the 3D-rotated version of the target from among several choices. In order to accomplish the task, subjects have to represent the target in visual memory, and mentally rotate that representation in 3D while comparing it to the choices. In this case, visual memory serves as a buffer for both manipulation and comparison.

Comparison is a common visual routine, used by other routines. For example, during visual search, the target must be compared to possible matches, and therefore must be held in some sort of visual memory buffer that is distinct from the current visual perception. Comparison is described in more detail in Sect. 5.2.2. The memory construct corresponding to this role as a buffer is VSTM, rather than iconic memory or VLTM (Luck 2008).

Another use of visual memory as a buffer is during drawing or copying. After viewing a given object or scene, the subject (or artist) needs to momentarily hold part of that image in visual memory while saccading to the canvas or paper, and then reproduce as best as possible the contents of visual memory.

5.1.2 Active Scene Representation

When our eyes saccade, vision is suppressed (Matin 1974). Nonetheless, a stable internal representation of the visual environment is constructed over multiple fixations (Castelhano and Henderson 2005). The constructed scene representation is not simply stitched together or superimposed from low-level sensory information in iconic memory, in the way that one pieces together a jigsaw puzzle (Irwin et al. 1983; Rayner and Pollatsek 1983). Rather, higher-level memory systems (but not sensory memory systems) accumulate visual information to construct a representation of a natural scene (Hollingworth and Henderson 2002). There is evidence that this functional role is supported by both VSTM (Hollingworth 2009; Palmer 1999) and VLTM (Hollingworth 2009). However, there is evidence for a larger VLTM component to this online scene representation (Hollingworth 2005). This is consistent with the fact that episodic scene viewing often occurs on the order of minutes or longer (i.e. similar to the timescale of VLTM).

What are the characteristics of such scene representations? These visual memory scene representations are specific to the perceived scene, rather than derived from general scene schemas. The representations contain spatial information such as scene layout and spatial configuration of objects relative to this layout. The visual representations are abstract in the sense that they do not simply contain low-level sensory information, but represent visual properties such as shape and viewpoint. Attention is not required to maintain these scene representations, they are robust across fixations and shifts of attention, and therefore these representations are not limited to the currently attended object (Henderson 2008; Hollingworth and Henderson 2002).

5.1.3 Episodic Scene Memory

In the example in Table 2, episodic memory of where I last left the coffee powder enables me to rapidly and accurately find it. This form of visual long-term memory retains schematic and semantic information from previously-viewed scenes (Henderson 2008). Such memories are relatively stable over time (Hollingworth 2004), and are created incidentally during normal scene viewing, i.e. do not require active effort to learn or maintain (Castelhano and Henderson 2005; Williams et al. 2005).

Related to the above, the same scene can be experienced multiple times (possibly with minor variations across episodes), such as a person's living room or work desk. For such familiar scenes, the effect of visual memory can be very strong. Take your work desk, for example. Unlike an unfamiliar desk, you can probably locate with minimal error your coffee mug, stapler, telephone without a second thought or without any searching. During the course of work, you probably sometimes reach for these desk-top items before they are even in view, and only use vision for fine-grained localization when your hand is close to grasping the item.

5.1.4 Scene Schema Knowledge

Beyond episodic memory for specific scenes, another functional role of visual long-term memory is to provide scene schemas (Mandler and Ritchey 1977; Potter et al. 2004): abstract representations for a particular scene type (e.g. "desk") specifying the categories of objects that are statistically common in such scenes, as well as the typical spatial relationships of these objects (Alba and Hasher 1983; Mandler and Parker 1976). At an unfamiliar desk, you probably need to perform some visual search to locate the stapler, but you don't search randomly. Rather, scene schemas for desks (combined with common sense) guide you to search the desk surface first, rather than behind the computer or inside a file folder.

5.2 Routines for Visual Memory

In this section, our main message is that natural, everyday tasks that involve vision (e.g. driving, cooking, etc.) can be hierarchically decomposed into lower-level sub-tasks or routines that are shared across multiple tasks, as shown in Fig. 3 (found at this end of this chapter, before the list of references). Many of these routines involve visual memory, since information is needed regarding a visual object or scene that is no longer present or that has changed.

To illustrate, let's revisit the example of making a cup of coffee. First, I may need to find my cup among several cups on the tabletop. This requires the *visual search* routine. Before that, I need to know what to search for, thus requiring the *recollection* of what my cup looks like. After finding my cup and adding a packet of instant coffee powder, I fill my cup with hot water. I may need to *monitor* the fluid level and compare it to my *recollection* of how much water I usually add, so that the coffee is at my preferred level of thickness. As I walk back to my desk, I carefully hold the cup of hot coffee, constantly *scanning* for any events that may cause the coffee to spill (e.g. bumping into someone around a corner, a door suddenly opening, etc.)

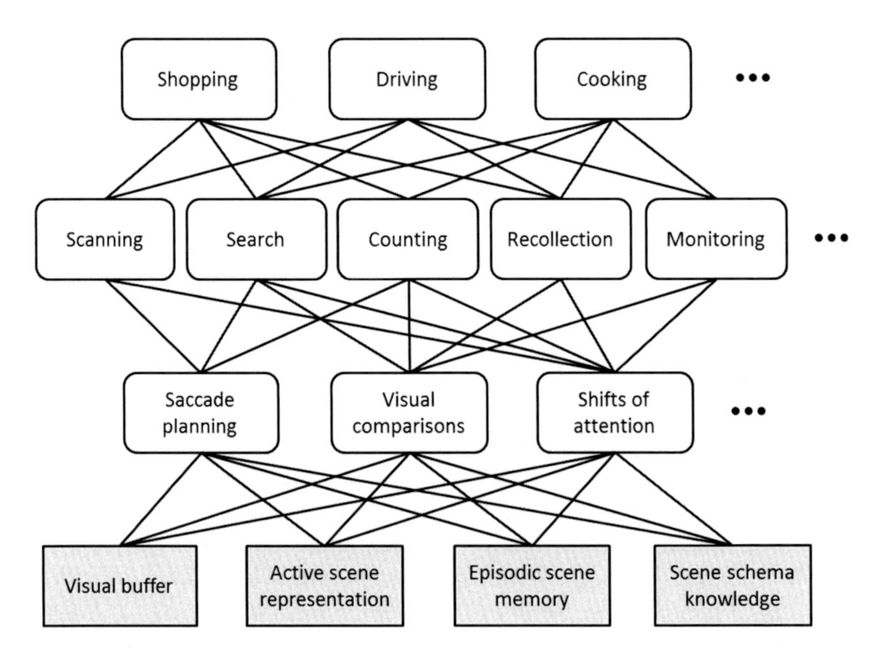

Fig. 3 Hierarchical organization of routines. Filled rectangles indicate functional roles of visual memory. Lines connect lower-level routines (or functional roles) to higher-level routines that utilize them

The above visual memory routines can themselves be composed of other lower-level routines that are shared. For example, *visual search* and *monitoring* first require recollection of some target visual object or visual condition. They then also require a *visual comparison* routine to compare the current visual percept with the target. Both *visual search* and *scanning* involve *shifts of attention*, which have been shown to precede saccades. Before these shifts of attention, *saccade planning* is also needed to determine where attention and saccades should be directed next.

Many of these routines can be sub-conscious and implicit, in the sense that we do not need to think about which routines to perform as part of a higher-level task, and in many cases the routines themselves can be performed automatically, without conscious awareness.

In the rest of this section, we will discuss in more detail a subset of the visual memory routines shown in Fig. 3. We will proceed up the (loose) hierarchy, where higher-level routines may invoke multiple lower-level routines. It is important to note that Fig. 3 is simply for illustrative purposes, meant to indicate one possible way to organize such tasks and routines. We do not claim that the organization depicted is unique, objective, or even accurate.

5.2.1 Saccade Planning

One role of visual memory is to represent the current visual scene, to be used as a basis for planning the next action (or sequence of actions), including saccades and other motor movements. This visual routine of saccade planning is also used by other, higher-level routines such as visual search.

There is evidence that people maintain relative accurate 3D spatial representations of the environment around them, and use these representations as a basis for planning saccades and other actions (Loomis and Beall 2004). Karn and Hayhoe (2000) showed that visual memory can be spatially very precise, such that it can be used to support planning of saccades accurately.

Experiments show that particularly for locations outside the current field of view, information in visual memory is important for saccade planning (Land et al. 1999). The visual system can use this information, rather than making a head or eye movement to get the location back into the field of view. In fact, even when there is a conflict between the information in memory and the current visual information, subjects planned saccades based on the information in memory (Aivar et al. 2005). Additionally, in patients with blind fields due to damage to early visual cortex can make accurate saccade into their blind fields based on visual memory (Martin et al. 2007).

Overall, there is evidence that saccade planning can use visual memory (together with peripheral visual information) for locations both inside and outside the current field of view (Edelman et al. 2002).

5.2.2 Visual Comparisons

Possibly one of the most common visual memory routines is making perceptual comparisons, i.e. comparing part of the current visual percept to part of visual memory. Such comparison operations are required almost constantly for real-world tasks, because many other visual (and general cognitive) routines require the ability to make comparisons (Markman and Gentner 2000).

Just as a visual system alone without any form of memory would be practically useless, a memory system that cannot be compared to current perception would be equally useless. Visual comparison operations can be used to compare information separated by time, space, or perceptual disruptions (such as eye-blinks, occlusion, saccades) (Hollingworth 2008).

For example, during visual search, a representation of the target must be maintained in memory, and as the subject attends sequentially to different objects, the target representation is compared with the representation of the current object (Duncan and Humphreys 1989). In addition, the next location to be search is biased towards objects that seem to match the target (Chelazzi et al. 1993). This means that comparisons are performed not only at the currently attended location, but also at other locations, in order to pick the best location to attend next.

5.2.3 Establishing Saccade Correspondence

Related to the above visual memory routine of making perceptual comparisons is establishing saccade correspondence (Luck 2008). This refers to the comparison of the intended saccade target with the postsaccade visual input, in order to verify that the saccade was made correctly. While saccades can be relatively accurate, the average saccade error is 10 % of the saccade distance, even in the simplest laboratory conditions (Kapoula 1985). Saccade correspondence is required to verify the saccade and possibly perform a correction. There is evidence that VSTM is used for this purpose (Hollingworth et al. 2008), i.e. to remember visual properties of the target, so that a corrective saccade can be made to the correct target among a number of possible distractors.

Even for saccade targets that were originally in the periphery just before the saccade, subjects were able to locate and saccade to such targets, provided that these targets were fixated previously (Edelman et al. 2002). This suggests that saccade correspondence can use visual information from not just prior to the saccade, but also from more distant visual memory.

5.2.4 Search

Visual search is one of the most common visual tasks performed by humans (Hollingworth 2012). How does visual memory influence visual search (Wolfe 1998)? First and foremost, prior knowledge of either the general layout of the scene, or the specific location of the target, efficiently guides visual search (Zelinsky et al. 1997). More generally, repeatedly searching through the same scene facilitates subsequent searches, with facilitation even after a single search (Brockmole and Henderson 2006). Familiarity with the general scene category can also guide visual search, even without specific knowledge of the particular scene (Torralba et al. 2006). Overall, while in certain cases memory may not be able to facilitate visual search, it is clear from many studies that visual memory generally does aid visual search (Hollingworth 2008).

Other than reducing search times, another aspect of visual search influenced by memory is the phenomenon of “inhibition of return”, whereby subjects tend to avoid revisiting previously search locations (Klein 2000). In a sense, this can also be viewed as a strategy that utilizes memory to speed up search times, because if a location has already been searched, searching it again simply slows down the search process without increasing the probability of finding the target.

5.2.5 Counting

A visual routine that has similarities to visual target search is counting, a routine that seems simple but is actually a complex, high-level routine that builds on other routines. One form of counting can be thought of as repeated visual search. For example, counting the number of chairs in a room can be performed by repeatedly searching for chairs. However, one additional constraint is that the already-counted items must be kept track of, and the next saccade must be planned such that it is to an item that has not been counted yet. One common strategy is to proceed systematically in some rough spatial order, such as left-to-right. This strategy is not perfect, however, especially when the items to be counted are numerous and arranged close to each other in a haphazard manner.

A different form of counting ignores specific visual appearance, and does not involve searching for a target (e.g. when counting the number of objects in a box). Visual memory is nonetheless required to keep track of what’s been counted, and to plan the next saccade.

6 Conclusion

Vision and memory are among two of the largest topics of research in neuroscience and psychology. However, research on their intersection and interaction is relatively underwhelming in scope and scale. In vision research, the sub-area of visual memory is viewed as just another sub-area like object recognition or visual motion. The converse case in memory research is also true. In computer vision, visual memory research is still a niche topic in its infancy, and it remains to be seen if it will gain mainstream acceptance.

In this chapter, we have tried to advocate that the interaction between vision and memory should be more than just another sub-area of study within each research community. During the performance of real-world tasks, which is perhaps the most important goal (and arguably the only ultimate goal) of vision and of memory, it is striking that the interaction between vision and memory is pervasive.

To summarize the chapter, we first illustrated how practically all everyday tasks involve the interaction of vision and memory at some point. We then proceeded to discuss how visual attributes influence what is remembered, and also conversely, how memory can influence visual perception. We examined in detail the various functional roles played by visual memory, and how processes that utilize visual memory can be organized as a loose hierarchy of routines or tasks that build upon each other to achieve "visual intelligence" for real-world tasks (Tan et al. 2013).

We will conclude the chapter by looking to the future: discussing open research questions and possible technology applications related to visual memory.

6.1 Open Research Questions

In psychology, as mentioned earlier, little research has been done in term of the functional aspects of visual memory, and this remains an area that has many opportunities. For example, when scene schemas are constructed from individual episodes, what goes into the schema and what is left out, and why? Is it possible to predict what an individual subject's active scene representation contains, if given the precise history of previous fixations? Among the contents of VSTM, which become part of VLTM, and why? Given a particular cue for recollection, which relevant episodes are retrieved from VTLM, and why?

In the neuroscience of vision, it is widely acknowledged that top-down feedback plays an important—but poorly understood—role. Among the various notions of top-down feedback, memory-related top-down processing could play a very important role. For example, even for the task of object recognition, which does not have any apparent links to memory, it may be the case that visually similar objects are automatically retrieved from memory, and information from prior episodes is used to help object recognition.

For the types of memory mentioned in this chapter, how are they formed, what are their exact representations, and how are these computed? How are the contents of VSTM "transferred" to VLTM during long-term memory formation, and vice versa for recollection? How exactly is the online scene representation from multiple fixations represented, and how is that representation constructed? Neurally, what are the brain correlates, or is the neural representation so distributed as to make that question meaningless? If neural correlates exist, do they form a distinct brain area or network, or do they simply re-use many of the same neurons as for immediate visual perception?

A hierarchical taxonomy of visual memory routines, such as that proposed in Fig. 3, can be laid out in more detail. Figure 3 is not meant to be comprehensive, but simply an illustrative proposal. There are potentially more routines and sub-routines that remain to be included, and their functional relationships remain to be elaborated in more detail (more than just arrows in a diagram).

There is also the question of how such routines come to exist. Are they innate, are they acquired/learnt through experience, or does the answer depend on the specific routine? For the routines that are learnt, how does learning occur, particularly in the absence of explicit teaching? Furthermore, how are sub-routines created? For a new task, how does the brain figure out which existing sub-routines to use, in what order, and which new sub-routines may be needed?

These are just some of the many general questions. For each routine, there are more specific open questions and opportunities. For example, in visual search, much research has been performed to produce reaction-time curves in order to understand the efficiency of visual search for different types of stimuli. However, much of this research uses artificial stimuli, and results are aggregated over many subjects and many trials. How does visual search function for real-world scenes? Are we able to produce models that can predict fixations and search times for individual trials for specific subjects?

6.2 *Possible Applications*

Much of this chapter has dealt with the scientific understanding of visual memory. In these last few paragraphs, we switch gears to discuss related engineering-oriented applications. Several trends bode well for applications involving visual memory, i.e. applications that go beyond the understanding of each image, video frame or video snippet, and instead integrate visual information over longer timescales.

Recent years have seen tremendous increases in the number of surveillance and in-vehicle cameras. Currently, these are primarily being used as recording devices, the contents of which are only analyzed when need arises, e.g. after traffic accidents or intrusions occur. However, there is immense potential in going beyond this, for example to automatically piece together a series of short events to make sense of the longer episode, e.g. to detect would-be robbers "casing" (checking out) a bank on several different occasions prior to a planned robbery.

Furthermore, people's lives are increasingly being recorded in images and videos, which are then archived or shared online. Much of this data is never analyzed or used in any meaningful way by computers. These images and videos could potentially be mined for information related to changes in eating habits, activities, and fashion choices over the years. They could also be used a form of external autobiographical memory, e.g. how did I celebrate my 21st birthday? The trend of "life-logging", i.e. extensively documenting one's life through images and videos (Tan et al. 2014), may increase significantly due to the recent announcement of a variety of "smartglasses" by large consumer companies, such as Google, Sony, Baidu and Epson. Such immense life-logging data naturally lends itself to applications for rapidly and accurately retrieving visual memories (Chandrasekhar et al. 2014b).

One more application trend that is just starting is that of humanoid robots that interact constantly with human co-workers or companions. In order to interact effectively with humans, such robots must be able to keep track of previous interactions, so that individual human habits and preferences can be learnt, for example. Going back to the example in the introduction, being able to identify a person is just the first step. Being able to rapidly retrieve meaningful information about previous interactions is also crucial for human interactions in virtually all kinds of situations.

6.3 Final Remarks

The future will see more research into the interaction of vision and memory. As computer vision systems reach commercially-viable levels of performance, such systems will inevitably be used to process the vast amount of visual information already currently available. Consequently, methods for storing, indexing and retrieving this vast information—in other words, memory—will become increasingly important. In neuroscience, one set of important-but-unanswered questions deals with top-down processing. The medial temporal lobe system, involved in memory, receives inputs from the visual system and feeds back into it. The interaction between memory and vision is thus a natural candidate for studying top-down processing. In psychology, as we understand more and more about vision and memory using simple artificial stimuli in laboratory conditions, attention will inevitably turn towards more realistic stimuli in more realistic conditions, under which vision and memory are naturally intertwined.

References

Aivar MP, Hayhoe MM, Chizk CL, Mruczek REB (2005) Spatial memory and saccadic targeting in a natural task. J Vis 5(3):3

Alba JW, Hasher L (1983) Is memory schematic? Psychol Bull 93(2):203

Bainbridge WA, Isola P, Oliva A (2013) The intrinsic memorability of face photographs. J Exp Psychol Gen 4(142):1323–1334

Balcetis E, Dale R (2003) There is no naked eye: higher-order social concepts clothe visual perception. In: Proceedings of the twenty-fifth annual meeting of the cognitive science society. pp 109–114

Botvinick M, Cohen J (1998) Rubber hands feel touch that eyes see. Nature 391:756

Brady TF, Konkle T, Alvarez GA (2011) A review of visual memory capacity: beyond individual items and toward structured representations. J Vis 11(5)

Brady TF, Konkle T, Alvarez GA, Oliva A (2008) Visual long-term memory has a massive storage capacity for object details. Proc Natl Acad Sci USA 105:14325–14329
Brockmole JR, Henderson JM (2006) Using real-world scenes as contextual cues for search. Vis Cogn 13(1):99–108
Bruce V, Burton AM, Dench N (1994) What's distinctive about a distinctive face? Q J Exp Psychol A 47(1):119–141
Busey TA (2001). Formal models of familiarity and memorability in face recognition. In: Wenger MJ, Townsend JT (eds) Computational, geometric, and process perspectives on facial cognition: contexts and challenges, Lawrence Erlbaum Associates, pp 147–191
Castelhano M, Henderson J (2005) Incidental visual memory for objects in scenes. Vis Cogn 12 (6):1017–1040
Chandrasekhar V, Wu M, Li X, Tan C, Mandal B, Li L, Lim JH (2014) Efficient retrieval from large-scale egocentric visual data using a sparse graph representation. In: IEEE conference on computer vision and pattern recognition workshops. IEEE, pp 541–548
Chandrasekhar V, Wu M, Li X, Tan C, Li L, Lim JH (2014) Incremental graph clustering for efficient retrieval from streaming egocentric video data. In: international conference on pattern recognition. IEEE, pp 2631–2636
Chelazzi L, Miller EK, Duncan J, Desimone R (1993) A neural basis for visual search in inferior temporal cortex. Nature 363:345–347
Craighero L, Bello A, Fadiga L, Rizzolatti G (2002) Hand action preparation influences the responses to hand pictures. Neuropsychologia 40(5):492–502
Dalal N, Triggs W (2005) Histograms of oriented gradients for human detection. In: ieee conference on computer vision and pattern recognition. IEEE, pp 886–893
Damasio AR (1989) Time-locked multiregional retroactivation: a systems-level proposal for the neural substrates of recall and recognition. Cognition 33(1–2):25–62
Duncan J, Humphreys GW (1989) Visual search and stimulus similarity. Psychol Rev 96(3):433
Edelman JA, Cherkasova MV, Nakayama K (2002) A spatial memory system for the guidance of eye movements in crowded visual scenes. J Vis 2(7):572
Felleman DJ, Van Essen DC (1991) Distributed hierarchical processing in the primate cerebral cortex. Cereb Cortex 1(1):1–47
Franconeri SL, Alvarez GA, Cavanagh P (2013) Flexible cognitive resources: competitive content maps for attention and memory. Trends Cogn Sci 17(3):134–141
Fuster J, Jervey J (1982) Neuronal firing in the inferotemporal cortex of the monkey in a visual memory task. J Neurosci 2(3):361–375
González J, Barros-Loscertales A, Pulvermüller F, Meseguer V, Sanjuán A, Belloch V, Avila C (2006) Reading cinnamon activates olfactory brain regions. NeuroImage 32(2):906–912
Gregory R (1970) The intelligent eye. McGraw-Hill
Hayhoe MM (2009) Visual memory in motor planning and action. In: Brockmole JR (ed) The visual world in memory. Psychology Press, pp 117–139
Henderson JM (2008) Eye movements and scene memory. In: Luck S, Hollingworth A (eds) Visual memory. Oxford University Press, pp 87–121
Hollingworth A (2004) Constructing visual representations of natural scenes: the roles of short-and long-term visual memory. J Exp Psychol Hum Percept Perform 30(3):519
Hollingworth A (2005) The relationship between online visual representation of a scene and long-term scene memory. J Exp Psychol Learn Mem Cogn 31(3):396
Hollingworth A (2008) Visual memory for natural scenes. In: Luck S, Hollingworth A (eds) Visual memory. Oxford University Press, pp 123–161
Hollingworth A (2009) Memory for real-world scenes. In: Brockmole JR (ed) The visual world in memory. Psychology Press, pp 89–116
Hollingworth A (2012) Guidance of visual search by memory and knowledge. In Dodd MD, Flowers JH (eds) The Influence of Attention, Learning, and Motivation on Visual Search. Springer, New York, pp 63–89
Hollingworth A, Henderson JM (2002) Accurate visual memory for previously attended objects in natural scenes. J Exp Psychol Hum Percept Perform 28(1):113

Hollingworth A, Richard AM, Luck SJ (2008) Understanding the function of visual short-term memory: transsaccadic memory, object correspondence, and gaze correction. J Exp Psychol Gen 137(1):163
Irwin DE, Yantis S, Jonides J (1983) Evidence against visual integration across saccadic eye movements. Percept Psychophys 34(1):49–57
Isola P, Xiao JX, Torralba A, Oliva A (2011) What makes an image memorable? In: IEEE CVPR. pp 145–152
Kapoula Z (1985) Evidence for a range effect in the saccadic system. Vis Res 25(8):1155–1157
Karn KS, Hayhoe MM (2000) Memory representations guide targeting eye movements in a natural task. Vis Cogn 7(6):673–703
Khosla A, Bainbridge WA, Torralba A, Oliva A (2013) Modifying the memorability of face photographs. In: International conference on computer vision. pp 145–152
Kiefer M, Sim E-J, Herrnberger B, Grothe J, Hoenig K (2008) The sound of concepts: four markers for a link between auditory and conceptual brain systems. J Neurosci 28:12224–12230
Klein RM (2000) Inhibition of return. Trends Cogn Sci 4(4):138–147
Konkle T, Brady TF, Alvarez GA, Oliva A (2010) Conceptual distinctiveness supports detailed visual long-term memory for real-world objects. J Exp Psychol-Gen 139(3):558–578
Land M, Mennie N, Rusted J, others (1999) The roles of vision and eye movements in the control of activities of daily living. Perception 28(11):1311–1328
Lander K, Chuang L (2005) Why are moving faces easier to recognize? Vis Cogn 12:429–442
Loomis JM, Beall AC (2004) Model-based control of perception/action. In: Optic flow and beyond. Springer, pp 421–441
Lowe DG (2004) Distinctive image features from scale-invariant keypoints. Int J Comput Vis 60 (2):91–110
Luck S (2008) Visual short-term memory. In Luck S Hollingworth A (ed) Visual memory. Oxford University Press, pp 43–85
Luck SJ, Vogel EK (2013) Visual working memory capacity: from psychophysics and neurobiology to individual differences. Trends Cogn Sci 17(8):391–400
Mandler JM, Parker RE (1976) Memory for descriptive and spatial information in complex pictures. J Exp Psychol: Hum Learn Mem 2(1):38
Mandler JM, Ritchey GH (1977) Long-term memory for pictures. J Exp Psychol: Hum Learn Mem 3(4):386
Markman AB, Gentner D (2000) Structure mapping in the comparison process. Am J Psychol Psychol 113(4):501
Martin T, Riley ME, Kelly KN, Hayhoe M, Huxlin KR (2007) Visually-guided behavior of homonymous hemianopes in a naturalistic task. Vis Res 47(28):3434–3446
Matin E (1974) Saccadic suppression: a review and an analysis. Psychol Bull 81(12):899
McGurk H, MacDonald J (1976) Hearing lips and seeing voices. Nature 264(5588):746–748
Messinger A, Squire LR, Zola SM, Albright TD (2001) Neuronal representations of stimulus associations develop in the temporal lobe during learning. Proc Natl Acad Sci USA 98 (21):12239–12244
Meyer K, Damasio A (2009) Convergence and divergence in a neural architecture for recognition and memory. Trends Neurosci 32(7):376–382
Oliva A, Torralba A (2001) Modeling the shape of the scene: A holistic representation of the spatial envelope. Int J Comput Vis 42:145–175
Orhan AE, Jacobs R (2014) Toward ecologically realistic theories in visual short-term memory research. Atten Percept Psychophys 76(7):2158–2170
Palmer SE (1999) Vision science: photons to phenomenology. MIT Press, Cambridge, MA
Palmeri TJ, Tarr MJ (2008) Visual object perception and long-term memory. In: Luck S, Hollingworth A (eds), Visual memory. Oxford University Press, pp163–207
Potter MC, Staub A, O'Connor DH (2004) Pictorial and conceptual representation of glimpsed pictures. J Exp Psychol Hum Percept Perform 30(3):478
Rayner K, Pollatsek A (1983) Is visual information integrated across saccades? Percept Psychophys 34(1):39–48

Rock I, Englestein P (1959) A study of memory for visual form. Am J Psychol
Serre T, Oliva A, Poggio T (2007) A feedforward architecture accounts for rapid categorization. Proc Natl Acad Sci USA 104(15):6424–6429
Shams L, Kamitani Y, Shimojo S (2000) Illusions: what you see is what you hear. Nature 408 (6814):788
Shechtman E, Irani M (2007) Matching local self-similarities across images and videos. In: Proceedings of the IEEE computer society conference on computer vision and pattern recognition. pp 1–8
Shenton JT, Schwoebel J, Coslett HB (2004) Mental motor imagery and the body schema: evidence for proprioceptive dominance. Neurosci Lett 370(1):19–24
Shepard RN, Metzler J (1971) Mental rotation of three-dimensional objects. Science 171 (3972):701–703
Sinha P, Balas B, Ostrovsky Y, Russell R (2006) Face recognition by humans: Nineteen results all computer vision researchers should know about. Proc IEEE 94(11):1948–1962
Spence C (2011) Crossmodal correspondences: A tutorial review. Atten Percept Psychophys 73 (4):971–995
Standing L (1973) Learning 10000 pictures. Q J Exp Psychol 25(2):207–222
Suchow JW, Fougnie D, Brady TF, Alvarez GA (2014) Terms of the debate on the format and structure of visual memory. Atten Percept Psychophys 76(7):2071–2079
Sweeny TD, Guzman-Martinez E, Ortega L, Grabowecky M, Suzuki S (2012) Sounds exaggerate visual shape. Cognition 124(2):194–200
Tan C, Goh H, Chandrasekhar V, Li L, Lim J-H (2014) Understanding the nature of first-person videos: characterization and classification using low-level features. In: 2014 IEEE conference on computer vision and pattern recognition workshops. IEEE, pp 549–556
Tan C, Leibo JZ, Poggio T (2013) Throwing down the visual intelligence gauntlet. In: Cipolla R, Battiato S, Farinella GM (eds) Machine learning for computer vision. Springer, pp 1–15
Tatler BW (2014) Eye movements from laboratory to life. In Horsley M, Eliot M, Knight BA, Reilly R (eds) Current trends in eye tracking research. Springer International Publishing, pp 17–35
Torralba A, Oliva A, Castelhano MS, Henderson JM (2006) Contextual guidance of eye movements and attention in real-world scenes: the role of global features in object search. Psychol Rev 113(4):766
Ullman S (1984) Visual routines. Cognition 18(1):97–159
Van Beers RJ, Sittig AC, van der Gon JJD (1999) Integration of proprioceptive and visual position-information : An experimentally supported model. J Neurophysiol 81(3):1355–1364
Vokey JR, Read JD (1992) Familiarity, memorability, and the effect of typicality on the recognition of faces. Mem Cogn 20(3):291–302
Williams CC, Henderson JM, Zacks F (2005) Incidental visual memory for targets and distractors in visual search. Percept Psychophys 67(5):816–827
Wolfe JM (1998) Vis search Attention 1:13–73
Zelinsky GJ, Rao RPN, Hayhoe MM, Ballard DH (1997) Eye movements reveal the spatiotemporal dynamics of visual search. Psycholo Sci 448–453

Approaches to Understanding Visual Illusions

Chun Siong Soon, Rachit Dubey, Egor Ananyev and Po-Jang Hsieh

1 Introduction

Visual illusions occur when sensory inputs give rise to distorted perception of the physical sources. Based on the potential underlying mechanisms, we categorize visual illusion into three types: physiological/pathological illusions, perceptual illusions, and ambiguous (bistable/multistable) illusions.

Physiological/pathological illusions arise from excessive or abnormal stimulation. An example of a physiological illusion is an afterimage that occurs after viewing a bright light source. Afterimages can also occur due to pathological causes. For example, illusory palinopsia (Bender et al. 1968; Gersztenkorn and Lee 2014) is caused by migraines or head trauma, and results from abnormal alterations in neuronal activities, during which a visual percept may persist or reoccur even after the original stimulus disappears.

Perceptual illusions refer to phenomena whereby the same physical properties are perceived differently in distinctive visual contexts or environments. For example, in the vertical-horizontal illusion (Shipley et al. 1949), people tend to overestimate the length of a vertical line in comparison to a horizontal line of identical length. In the simultaneous contrast illusion (Wallach 1948), an object is perceived as brighter on a less luminant background than the same object on a more luminant background. These perceptual illusions may arise from differential responses of the brain interpreting the stimuli according to the current contexts and/or past experiences.

C.S. Soon · R. Dubey · E. Ananyev · P.-J. Hsieh (✉)
Neuroscience and Behavioral Disorders Program,
Duke-NUS Medical School, Singapore, Singapore
e-mail: pojang.hsieh@duke-nus.edu.sg

Q. Zhao (ed.), *Computational and Cognitive Neuroscience of Vision*,
Cognitive Science and Technology, DOI 10.1007/978-981-10-0213-7_10

Ambiguous illusions are characterized by a change in the visual percept due to the ambiguity of the stimulus. Examples include the Necker cube (Necker 1832) and binocular rivalry (Wade 1996), whereby the same physical stimulus may elicit involuntary perceptual switches between alternative percepts. These ambiguous illusions may result from the brain making inferences and decisions under uncertainty.

In this chapter, we will briefly describe three major theoretical frameworks of perception that can potentially explain why each *type* of illusions—rather than specific illusions—arise.

2 Conventional Approach and Physiological/Pathological Visual Illusions

A conventional approach towards understanding visual perception is that neurons in the primary and higher order visual cortices detect and encode specific informative features within their receptive fields (Atick and Redlich 1992; Field 1994; Bell and Sejnowski 1997; Simoncelli and Olshausen 2001) from retinal inputs. Essentially, the hierarchy of detected features result in percepts that correspond to the physical properties of the stimuli sources in the real world. For example, the activity of visual cortical neurons has been shown to be associated with specific stimulus properties, such as orientation (Hubel and Wiesel 2005), motion (Zeki 1974; Maunsell and Van Essen 1983), color (Zeki 1983), object size (Dobbins et al. 1998) and even higher level features like faces (Desimone et al. 1984) and personal identity (Quiroga et al. 2005). Based on this framework, illusions or misperceptions are essentially discrepancies between the responses of feature detectors and the visual stimuli. This could arise from neurobiological constraints on the function of these feature detection neurons, or imperfect operation of feature detection processes. For instance, neuronal adaptations of feature detectors (as well as photoreceptors), and lateral competition between them may result in illusory phenomena such as afterimages (Barlow and Hill 1963) and perceptual rivalries (Marr 1982). Other perceptual distortions and pathological illusions may also be observed when neurons are lesioned or artificially stimulated. For example, when electrically stimulating neurons in extrastriate area MT that are selective to a particular direction of motion, the monkeys would move their eyes to indicate a modified percept of motion (Sugrue et al. 2005). Similarly, electrical stimulation of human fusiform face-selective regions may distort face perception (Parvizi et al. 2012). Some illusions may also arise due to malfunctions of a physiological nature, e.g., distorted perceptions due to the effects of psychoactive substances (Freedman 1969; Nichols 2004).

However, this conventional feature detection framework does not seem to provide a clear explanation for why certain perceptual and ambiguous illusions occur. For example, in the vertical-horizontal illusion, the length of a vertical line is perceived as longer in comparison to a horizontal line of identical length. To explain this phenomenon with the conventional feature detection approach, one would have to assume that there are inherent differences between vertical and horizontal feature detectors in terms of responsivity or detector distribution that result in biased perception. However, there is no known evidence supporting this idea, and it is not immediately evident why and how such differences arise if the modus operandi of the visual system is the direct detection of visual features in objective reality.

3 The Inverse Problem and Alternative Approaches to Visual Illusions

Due to the transformation of sources in 3D space onto 2D projections, the real world properties of a stimulus, such as its speed, position, and trajectory cannot be uniquely specified for any given projection on the retina (Purves and Lotto 2003). This is also known as the inverse problem: source information is underdetermined because different objects or physical parameters may give rise to similar retinal images (Fig. 1). In other words, features detected from 2D retinal images could have arisen from different possible 3D sources.

Contrary to the view that vision involves solely feature detections, and visual illusions arise from abnormal activity in the feature detection processes, a large corpus of work point out that the human visual system uses past experiences in some way or other in order to generate useful percepts to a world that cannot be known directly (Knill and Richards 1996; Rao et al. 2002; Weiss et al. 2002; Purves and Lotto 2003; Howe and Purves 2005a; Purves et al. 2008). According to this view, visual illusions are not necessarily due to defects or artifacts of visual

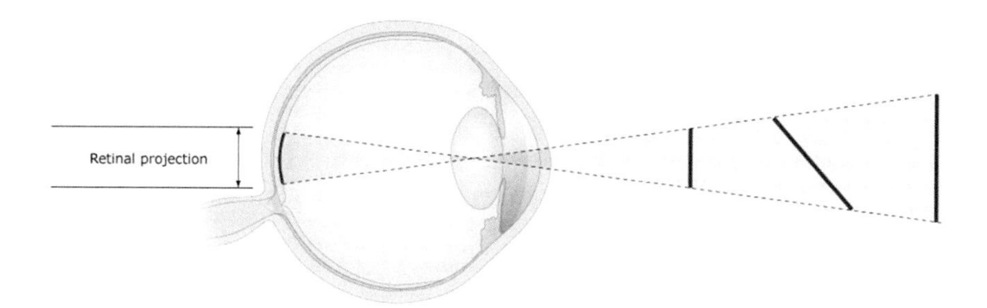

Fig. 1 The uncertain relationship between real world stimuli and retinal images. The same retinal projection can be generated by objects of different sizes at different distances from the observer, and in different orientations. (Figure adapted from Purves et al. 2008)

processing, but are the inevitable and most probable estimations about objective reality.

Many theorists have tried to explain how the visual system gives rise to successful behavior when the physical properties of physical sources cannot be specified by the retinal images. Based on Helmholtz's view on 'unconsciousness inferences', these approaches suggest that visual percepts are not just the result of detecting physical properties of the environment, but are actively influenced by the empirical significance of the stimuli. Two different frameworks have been proposed to address how visual perception and illusions can be explained in these terms: empirical ranking theory and Bayesian decision theory.

4 Empirical Ranking Theory and Perceptual Illusions

The empirical ranking theory abandons altogether the notion that what we see is an accurate reflection of the properties of the physical world, conceding that the inverse problem of vision is inherently not solvable (Purves et al. 2008). Instead, this theory postulates that the perceptual qualities of any specific aspect of a retinal stimulus is a function of the relative frequency of occurrence of that parameter, in relation to all other instances of the same stimulus parameter in accumulated past experiences of a given visual system. The perceptual experience is defined by the trial and error history of the visual system's interaction with the world through its sensory apparatus, rather than just the physical signals from the source stimuli. In other words, our percepts cannot be mapped directly onto physical reality, and visual perception cannot be said to be a veridical representation of the external world—nor does it try to achieve such a goal. Rather, the goal of the visual system is to support behavior that is evolutionarily advantageous, and it achieves this through statistically informed perceptual responses to sensory inputs based on previous behavioral outcomes.

According to the empirical ranking approach, illusions would not be considered the result of an imperfect visual system, but are rather signatures of its core strategy. Validation of this hypothesis usually involves the collection of statistics based on natural image databases which serves as a proxy for our accumulated past experiences of the visual world. If our visual system has indeed evolved to generate percepts empirically, then the visual percepts of a given stimulus, including illusions, should be predictable on the basis of such data.

For example, Howe and Purves (2002) used the empirical ranking theory to explain the variable perception of length as a function of stimulus orientation—a well-known visual illusion that the apparent length of a linear stimulus in the retinal image changes with respect to the orientation (Pollock and Chapanis 1952). As shown in Fig. 2, the perceived length of a line of fixed length is dependent on its orientation; a line appears to be shortest when it is horizontal, and longest when it is oriented 20–30° from the vertical axis. Howe and Purves (2002) used a large range scanner to acquire a database of natural images, which included the location in 3D

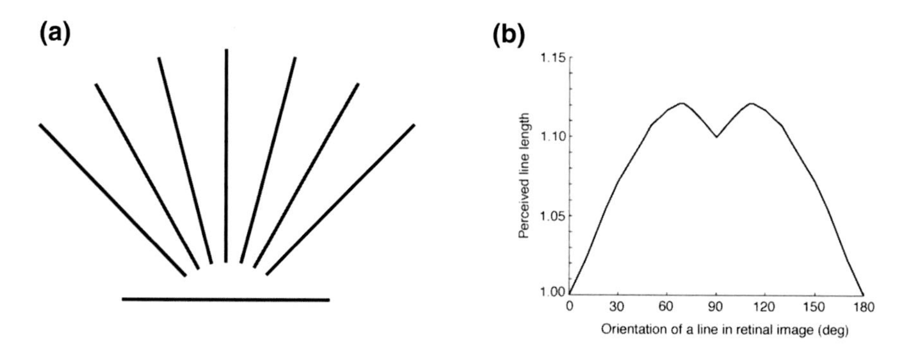

Fig. 2 Variable perception of length based on stimulus orientation. **a** A demonstration of the illusion. **b** The average of the psychophysical data, which shows the variation in the apparent length of a linear stimulus as a function of its orientation in the retinal image. (Figure adapted from Howe and Purves 2002)

space of every pixel in the scenes. This database served as a proxy for accumulated past experiences. These range images were then analyzed to establish whether the probabilistic relationship between the length of intervals in images and length of the physical sources could explain the perceptual bias.

Figure 3a shows the frequency distributions of the physical length to projected length ratio (λ, metres/pixel) at different orientation values (θ) generated from the range image database. This distribution varies systematically for lines projected in different orientations. The frequency distribution of the ratio λ was further analyzed as a function of the projected interval orientation θ. Figure 3b shows the normalized mean of λ averaged across all values of projected interval length with respect to θ. Notice that this curve is similar to the psychophysical function in Fig. 2b. Thus,

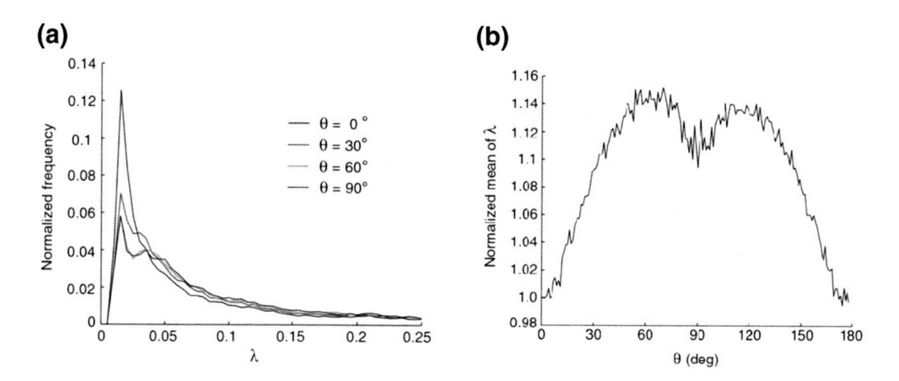

Fig. 3 Accumulated human experience with line angles and length. **a** Frequency distribution of the physical length to projected length ratio (λ) at different θ values. **b** Normalized mean of λ averaged across all values of projected interval length plotted as a function of θ. Notice the similarity of this curve with the psychophysical function in Fig. 2b. (Figure adapted from Howe and Purves 2002)

these results demonstrate how statistics from natural scenes can predict the variable perception of line length with respect to changes in orientation.

The empirical approach to visual perception has been similarly used to explain a variety of well-known perceptual illusions such as the Cornsweet illusion (Purves et al. 1999), the Chubb illusion (Lotoo et al. 2001), the Mueller-Lyer illusion (Howes and Purves 2005b), and the 'flash-lag' effect (Wojtach et al. 2008). The empirical approach has also been applied to explain various motion illusions (Wojtach et al. 2009) and the perception of light intensity, angles, and line orientation in general (Yang and Purves 2004, Howe and Purves 2005c).

Although the empirical ranking theory is very powerful in explaining perceptual illusions and how statistical regularities shape our perception, it does not say much about how perception is affected by attention, emotion and expectations (Nour and Nour, 2015). Particularly, it is less clear how empirical ranking theory can explain the phenomenon of percept-switching in ambiguous illusions (e.g., binocular rivalry, Necker cube, young-old woman and rabbit-duck illusions) since statistics accumulated from years of visual experiences would presumably not have changed within the short time frame of perceptual switches to trigger such a switch. The Bayesian decision theory, on the other hand, seems to be able to provide the requisite tools to account for how the visual system makes inferences about the world based on different contexts and mental states, which leads to the phenomenon of percept-switching in ambiguous illusions.

5 Bayesian Decision Theory and Cognitive Illusions

Like the empirical ranking approach, the Bayesian decision theory approach to perception does not accept that the inverse problem can be solved with simple feature detection. However, it postulates that statistical inferences can be made about external reality based on sensory inputs. Bayesian decision theory is based on Bayes' theorem, a framework for making inferences and decisions under uncertainty. Here we present the central tenets of this framework in the context of visual perception. The problem of perception can be framed thus: given sensory evidence (E) detected by the visual system, what inferences can be made about the sources of such evidence, i.e., the state of the world (H)? To illustrate, given that a vertical line of a certain length is detected in the retinal image (E), is this due to a short vertical line very close to the eye (H_1), a slanted line that is slightly longer and further away (H_2), or a vertical line that is even longer and further away (H_3) (Fig. 1)? Bayesian Decision Theory provides a means of evaluating the probability of these hypotheses, providing a best guess as to what in the world gave rise to the sensory evidence.

The probability of both events H and E occurring can be written as $P(H, E)$. This can also be defined as the probability of one of the events occurring, e.g., $P(H)$, multiplied by the probability of the second event happening, given that the first event has already occurred, i.e., $P(E|H)$. Alternatively, $P(H, E)$ can also be defined as $P(E)$ multiplied by $P(H|E)$. This gives us $P(H)P(E|H) = P(E)P(H|E)$. Since the two expressions are identical, dividing both by $P(E)$ gives us the Bayes' rule defined as:

$$P(H|E) = \frac{P(E|H)P(H)}{P(E)}$$

According to Bayes' theorem, the posterior probability $P(H|E)$ is proportional to the prior probability $P(H)$ multiplied by the likelihood $P(E|H)$. Returning to our example, given that a vertical line is detected on the retinal image (E), the posterior probability that there is a short vertical line in the real world very close to the eye (H_1) is proportional to the prior probability of H_1 occurring, and the probability of detecting E given that H_1 occurs. Since we do not have direct access to how frequently H_1 actually occurs, the prior probability is essentially a *belief* based on past experiences. Thus, Bayes' rule allows for the incorporation of information based on our prior beliefs to modify our visual perception in favor of the most probable hypothesis, i.e., the highest $P(H_n|E)$. However, the retinal image is compatible with infinite source stimuli with different combinations of lengths, orientations and distance from the eye, so it would not be feasible that the posterior probability of each candidate line is evaluated as a discrete hypothesis. In this example, continuous probability distributions could be separately evaluated along each dimension, and combined to identify the hypothesis with the highest posterior probability. If this best guess closely approximates the real world object, then there should be minimal error when compared against the sensory evidence.

This perspective of vision as Bayesian inference has been used to successfully explain multiple visual illusions. According to this framework, illusions are a result of an inherent uncertainty associated with the sensory input for an optimal (albeit imperfect) system. For example, Weiss et al. (2002) developed an optimal Bayesian estimator to explain the moving rhombus illusion, which, while seemingly simple, had eluded a clear explanation based on the existing computational frameworks of motion perception. In this illusion, a thin, low-contrast rhombus (Fig. 4b) moving horizontally is perceived to move diagonally, even though both a thick, low-contrast rhombus and a thin, high-contrast rhombus are seen as moving horizontally (Fig. 4a, c). These differences in apparent motion are reflected in the posterior probability modeled by the Bayesian estimator for each rhombus as probability distributions of velocities. The prior is estimated based on the assumption that image velocities are generally expected to be low. The likelihood term consists of local velocity probability distributions for *each* point of the rhombus. These estimates reflect stimulus contrast: low contrast implies greater uncertainty in the sensory evidence, corresponding to a less precise distribution (blurred bands in Fig. 4b, c). Multiplication of the prior and likelihood terms for each of the three scenarios yields the posterior probability distributions, whose peaks predict the perceived direction of motion.

Another classical visual illusion phenomenon, binocular rivalry, has been explained in Bayesian terms by a predictive coding account that conceptualizes brain functions as embodying Bayesian prediction mechanisms (Hohwy et al. 2008). In binocular rivalry, each eye is presented with a different image. Faced with

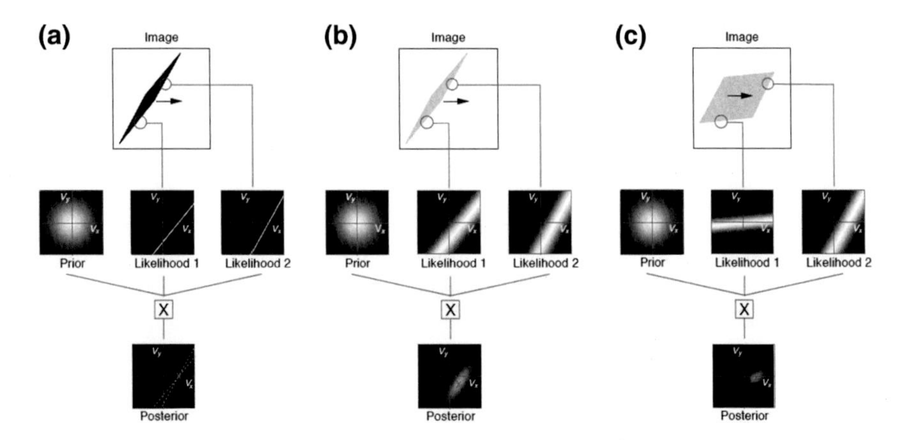

Fig. 4 A Bayesian explanation for moving rhombus illusion (see demo at http://www.cs.huji.ac.il/~yweiss/Rhombus/rhombus.html). The images in (**a**) and (**c**) are perceived to move to the right, with no vertical velocity component. However, a low-contrast thin rhombus moving rightward (**b**) produces an illusion that it moves diagonally downward. The low-velocity prior is the same for all three scenarios. However, the likelihood functions, estimated for *each* point on the rhombus, differ in two respects. Due to low contrast, images (**b**) and (**c**) produce less precise (highly variable) likelihood distributions, shown here as blurred bands for two sample points, each from one of the leading edges (Likelihoods 1 and 2). These two bands have similar slopes for thin rhombuses (**a** and **b**) but differ for the thick rhombus (**c**). Multiplying these likelihood functions by the prior probability yields the posterior probability, whose center falls almost precisely on the x-axis in velocity space for images (**a**) and (**c**). However, the posterior probability for the thin, low-contrast rhombus has a noticeable vertical component, indicating (apparent) downward diagonal motion (**b**). (Figure adapted from Weiss et al. 2002)

these unnatural conditions, what the perceiver sees at a given point in time is *either* image, or a transient composite. The image currently perceived is called the *dominant* image, while the other one is called the *suppressed* image. Eventually, the dominant image percept yields to the previously suppressed one, which then becomes dominant, and the cycle repeats indefinitely.

This effect has typically been described in terms of the *inhibition* (suppression) of the non-dominant image, and the gradual *adaptation* to the dominant image that eventually leads to the loss of dominance (Blake and Wilson 2011). The inhibition of the non-dominant image is essential to ensure the dominance of the other image (while it lasts), while adaptation is meant to explain why the dominance ends after a certain period of time. Although these mechanisms are compatible with the feature detection theory, it is less clear why we do not simply perceive a stable composite image combining features detected from binocular retinal inputs, since we are capable of perceiving such composite stimuli.

In contrast, the explanation of binocular rivalry in Bayesian terms centers around the assumption that the brain's objective in *all* scenarios (inclusive of binocular rivalry) is to minimize the prediction error (Hohwy et al. 2008). In other words, the leading hypothesis constructed by the brain needs to explain most of the sensory signal, while still being ecologically plausible.

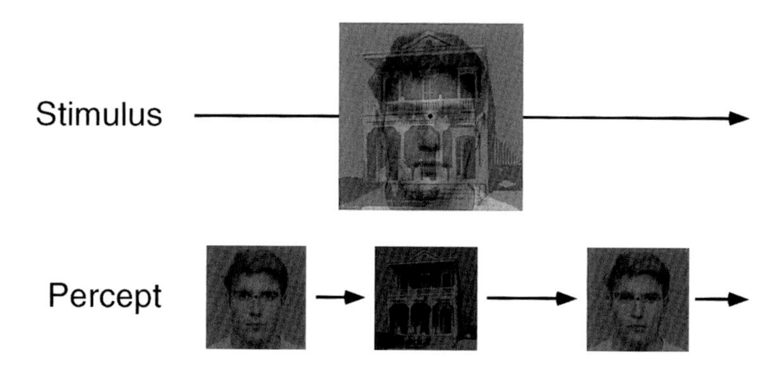

Fig. 5 Binocular rivalry of face/house stimuli. When the image of a house is presented to one eye and a face to the other, one's percept alternates between a house and a face. (Figure adapted from Tong et al. 1998)

Hohwy et al. (2008) illustrated this with the example of a house being presented to one eye and a face to the other (Fig. 5). The *prior* probabilities (the probabilities of the hypotheses *regardless* of the available evidence) of the two rivalling hypotheses, house, $p(H_h)$, and face, $p(H_f)$, are assumed to be approximately the same. At the same time, either hypothesis has a higher prior probability than the *combined* hypothesis, i.e., some sort of house-o-face, $p(H_{h+f})$, which the authors point out is very improbable (has a low prior probability). On the other hand, the combined hypothesis, H_{h+f}, can explain more of the available evidence – it can account for twice the portion of the incoming sensory signal as either hypothesis. Therefore, it is said that it has a higher likelihood, i.e., probability that the sensory evidence can be explained given the hypothesis, $p(E|H_{h+f})$, than either of the two likelihoods of the individual hypotheses, $p(E|H_h)$ and $p(E|H_f)$. Nevertheless, $p(H_{h+f}|E)$ remains low due to the low prior probability, $p(H_{h+f})$. The imbalance between the likelihood and prior probabilities for the three competing hypotheses means that there is no clear winner that is ecologically plausible and can account for all or most of the sensory evidence. As a result, while a single hypothesis may be dominant at any given time, prediction errors persist either due to low likelihood or low prior probability, and the corresponding percept is not stable.

However, it is worth noting that we have no problems perceiving composites of house and face images that are superimposed (see Fig. 5, Stimulus), if the same composite image were presented to either or both eyes. Even if there might be a bias towards either seeing house or face at any point in time, our perception certainly does not switch drastically between house only and face only. Thus, an alternative explanation would be that all three hypotheses, $p(H_h)$, $p(H_f)$ and $p(H_{h+f})$, leave a large error when compared against the retinal image input of each eye. As the visual system continually adjusts its hypothesis in order to minimize prediction errors, but fails to reach a stable error minima for the sensory inputs of both eyes, the dominant percept switches periodically. This also appears to be a more parsimonious explanation for binocular rivalry in general. Indeed, while there is some evidence for such eye-specific neuronal signals in early visual cortex (Schwarzkopf et al. 2010a, b), it is not clear whether they can be considered *error* signals rather than simple feedforward sensory signals.

To sum up, according to the predictive coding explanation of binocular rivalry, faced with the unnatural conditions of having non-matching stimuli in the two eyes that cannot be fused into a coherent 3D percept, our system copes by selecting an 'optimal' hypothesis with the lowest prediction error at any given time, and switching whenever it becomes less 'plausible' than the competing one.

Even when the two eyes receive the same stimulus under normal viewing conditions, perceptual switching can also occur, for example, in ambiguous figures like the Necker cube, young-old woman and rabbit-duck illusions. Traditional explanations for this phenomenon also involve adaptation. In predictive coding terms, adaptation can be conceived of in terms of the decay in the prior probability term, because "a static hypothesis will quickly lose its clout in a changing world" (Hohwy et al. 2008, p. 692). Adaptation may be considered a built-in mechanism to avoid committing to a hypothesis that may no longer be true. This decay in the prior probability of the dominant percept is presumed to cause the switch to the other hypothesis.

Crucial to the predictive coding implementation of Bayesian brain is the notion of perceptual hierarchy, which maintains, modifies, and switches hypotheses to accommodate the incoming sensory data in a dynamic manner. As such, this approach has the strength of offering a single, unified framework for the lateral competition occurring at each level, as well as a concrete computational account for the cross-level interactions in the form of hypothesis (top-down) and error (bottom-up) propagation. This approach therefore makes it easier to accommodate the high-level influences on lower-level perception, which is often necessary to shed light on complex phenomena such as psychopathological illusions (Nour and Nour 2015).

While Bayesian decision theory has shown huge explanatory potential across a wide range of visual phenomena, and brain function in general, its flexibility is also a point of concern (Bowers and Davis 2012), as illustrated by the alternative explanation for binocular rivalry presented above. Thus, it is important to constrain Bayesian decision theory accounts of visual phenomena with neuroscientific evidence. The search for the neuronal implementation of the Bayesian framework is underway, and plausible neurophysiological accounts have been offered (Friston 2005). Further studies of Bayesian inference (Yuille and Kersten 2006; Stocker and Simoncelli, 2006; Chater et al. 2006) support the idea that visual illusions arise as a result of the normal functioning of an optimal—albeit imperfect—system, which is designed to make inferences under uncertainty. This is in stark contrast to viewing illusions as arising from faulty processing.

6 Conclusion

A long-standing challenge in vision is that visual systems cannot directly access or measure the physical properties of the real world. Recent research on Bayesian decision theory and empirical ranking theory aims to explain how our visual system addresses this issue and points to the direction that visual percepts do not map directly to reality, but are generated on the basis of the statistical appearance of the real-world sources. These frameworks rationalize visual perception by taking into account the empirical nature of vision. The fundamental difference between these general frameworks is their philosophical conception of the nature of visual perception. However, it remains to be seen if they are really mutually exclusive, or may in fact be empirically compatible to some extent; after all, both are based on Helmholtz's unconscious inferences view, and postulate the neural embodiment of statistical methods for solving the problems of perception. For example, empirical ranking theory emphasizes that the goal of perception is to support evolutionarily advantageous behavior rather than to represent external reality. However, it is quite plausible that a visual system which best approximates external reality under *most* circumstances—constrained by the sensorial accessibility of physical signals—will, in the long run, be most evolutionarily advantageous. Similarly, from the perspective of Bayesian decision theory, frequency-based perceptual biases could be seen as nonlinear frequency-based weighting of error signals. Further research is needed to verify whether this may in fact be a highly efficient strategy for minimizing prediction errors across the visual hierarchy, or perhaps the entire brain. Extending the Bayesian error-reduction mechanisms across the interconnected hierarchies of the whole neural system also appears to be not incompatible with the perception-for-action view of the empirical ranking theory. While it may never be possible—or desirable—to completely eliminate errors from the whole system, differential weighting of errors (or selective error tolerance) may facilitate the flexible adaptation to diverse contexts, thus supporting evolutionarily advantageous behavior.

Illusion, in all its myriad forms, deviates from the veridical perception of physical sources. Whether it is seen as a malfunction or a natural consequence of the normal operations of the visual system, this deviation provides a rich test bed for the evaluation of theories of perception. Here we have merely provided a brief overview of different types of illusions in relation to two important alternatives to the conventional feature detection theory, namely the empirical ranking theory and the Bayesian decision theory. Ultimately, the relative ability of these two approaches in explaining a range of unresolved visual illusions, with the support of neurophysiological evidence, could help evaluate which of these theoretical frameworks is closer to explaining visual perception in general.

References

Atick JJ, Redlich AN (1992) What does the retina know about natural scenes? Neural Comput 4 (2):196–210

Barlow HB, Hill RM (1963) Evidence for a physiological explanation of the waterfall phenomenon and figural after effects. Nature 200:1434–1435

Bell AJ, Sejnowski TJ (1997) The 'independent components' of natural scenes are edge filters. Vision Res 37:3327–3338

Bender MB, Feldman M, Sobin AJ (1968) Palinopsia. Brain: J Neurol 91(2): 321–338

Blake R, Wilson H (2011) Binocular vision. Vision Res 51(7):754–770

Bowers JS, Davis CJ (2012) Bayesian just-so stories in psychology and neuroscience. Psycho Bull 138(3):389–414. http://doi.org/10.1037/a0026450

Chater N, Tenenbaum JB, Yuille A (2006) Probabilistic models of cognition: Conceptual foundations. Trends Cogn Sci 10(7):287–291

Desimone R, Albright TD, Gross CG, Bruce C (1984) Stimulus-selective properties of inferior temporal neurons in the macaque. J Neurosci 4:2051–2062

Dobbins AC, Jeo RM, Fiser J, Allman JM (1998) Distance modulation of neural activity in the visual cortex. Science 281:552–555

Field DJ (1994) What is the goal of sensory coding? Neural Comput 6:559–601

Freedman DX (1969) The psychopharmacology of hallucinogenic agents. Annu Rev Med 20:409–418

Friston K (2005) A theory of cortical responses. Philos Trans R Soc Lond B Biol Sci 360 (1456):815–836. doi:10.1098/rstb.2005.1622

Friston K, Mattout J, Kilner J (2011) Action understanding and active inference. Biol Cybern 104 (1–2):137–160

Gersztenkorn D, Lee AG (2014) Palinopsia revamped: a systematic review of the literature. Surv Ophthalmol 60:1–35

Hohwy J, Roepstorff A, Friston K (2008) Predictive coding explains binocular rivalry: an epistemological review. Cognition 108(3):687–701

Howe CQ, Purves D (2002) Range image statistics can explain the anomalous perception of length. Proc Natl Acad Sci 99(20):13184–13188

Howe CQ, Purves D (2005a) Natural-scene geometry predicts the perception of angles and line orientation. Proc Natl Acad Sci USA 102:1228–1233

Howe CQ, Purves D (2005b) Perceiving geometry: geometrical illusions explained by natural scene statistics. Springer Press, New York

Howe CQ, Purves D (2005c) The Müller-Lyer illusion explained by the statistics of image–source relationships. Proc Natl Acad Sci USA 102(4):1234–1239

Hubel DH, Wiesel TN (2005) Brain and visual perception. Oxford Press, New York

Knill DC, Richards W (Eds.) (1996) Perception as Bayesian inference. Cambridge University Press, Cambridge

Lotto RB, Purves D (2001) An empirical explanation of the Chubb illusion. J Cogn Neurosci 13:547–555

Marr D (1982) Vision: a computational investigation into the human representation and processing of visual information. W.H. Freeman, San Francisco, Reprint: The MIT Press, Massachusetts

Maunsell JHR, Van Essen DC (1983) Functional properties of neurons in the middle temporal visual area (MT) of the macaque monkey: I. Selectivity for stimulus direction, speed and orientation. J Neurophysiol 49:1127–1147

Necker LA (1832) Observations on some remarkable optical phenomena seen in Switzerland; and on an optical phenomenon which occurs on viewing a figure of a crystal or geometrical solid. Lond Edinb Philos Mag J Sci 1(5):329–337

Nichols D (2004) Hallucinogens. Pharmacol Ther 101(2):131–181

Nour MM, Nour JM (2015) Perception, illusions and Bayesian inference. Psychopathology 48:217–221

Parvizi J et al (2012) Electrical stimulation of human fusiform face-selective regions distorts face perception. J Neurosci 32(43):14915–14920
Pollock WT, Chapanis A (1952) The apparent length of a line as a function of its inclination. Q J Exp Psychol 4(4):170–178
Purves D, Lotto B (2003) Why we see what we do: an empirical theory of vision. Sinauer, Massachusetts
Purves D, Shimpi A, Lotto RB (1999) An empirical explanation of the Cornsweet effect. J Neuro sci 19:8542–8551
Purves D, Wojtach WT, Howe C (2008) Visual illusions: an empirical explanation. Scholarpedia 3 (6):3706
QuianQuiroga R, Reddy L, Kreiman G, Koch C, Fried I (2005) Invariant visual representation by single neurons in the human brain. Nature 435:1102–1107
Rao RP, Olshausen BA, Lewicki MS (2002) Probabilistic models of the brain: perception and neural function. MIT Press, Massachusetts
Schwarzkopf DS, Schindler A, Rees G (2010a) Knowing with which eye we see: utrocular discrimination and eye-specific signals in human visual cortex. PLoS ONE 5(10):e13775. doi:10.1371/journal.pone.0013775
Schwarzkopf DS, Schindler A, Rees G (2010b) Knowing with which eye we see: utrocular discrimination and eye-specific signals in human visual cortex. PLoS ONE 5(10):e13775. doi:10.1371/journal.pone.0013775
Shipley WC, Nann BM, Penfield MJ (1949) The apparent length of tilted lines. J Exp Psychol 39:548–551
Simoncelli EP, Olshausen BA (2001) Natural image statistics and neural representation. Annu Rev Neurosci 24:1193–1216
Stocker AA, Simoncelli EP (2006) Noise characteristics and prior expectations in human visual speed perception. Nat Neurosci 9(4):578–585
Sugrue LP, Corrado GS, Newsome WT (2005) Choosing the greater of two goods: neural currencies for valuation and decision making. Nat Rev Neurosci 6:363–375
Tong F, Nakayama K, Vaughan JT, Kanwisher N (1998) Binocular rivalry and visual awareness in human extrastriate cortex. Neuron 21:753–759
Wade NJ (1996) Descriptions of visual phenomena from Aristotle to Wheatstone. Perception 25 (10):1137–1175
Wallach H (1948) Brightness constancy and the nature of achromatic colors. J Exp Psychol 38 (3):310–324
Weiss Y, Simoncelli EP, Adelson EH (2002) Motion illusions as optimal percepts. Nat Neurosci 5 (6):598–604
Wojtach WT, Sung K, Truong S, Purves D (2008) An empirical explanation of the flash-lag effect. Proc Natl Acad Sci 105(42):16338–16343
Wojtach WT, Sung K, Purves D (2009) An empirical explanation of the speed-distance effect. PLoS ONE 4(8):e6771
Yang Z, Purves D (2004) The statistical structure of natural light patterns determines perceived light intensity. Proc Natl Acad Sci USA 101:8745–8750
Yuille A, Kersten D (2006) Vision as Bayesian inference: analysis by synthesis? Trends Cogn Sci 10(7):301–308
Zeki SM (1974) Functional organization of a visual area in the posterior bank of the superior temporal sulcus of the rhesus monkey. J Physiol (Lond) 236:549–573
Zeki S (1983) Colour coding in the cerebral cortex: the reaction of cells in monkey visual cortex to wavelengths and colors. Neuroscience 9:741–765

Impact of Neuroscience in Robotic Vision Localization and Navigation

Christian Siagian and Laurent Itti

1 Introduction

One important reason to examine the mechanisms of how we see is for the advancement of technology. Robotics as a field has a distinct appeal because its goal is to build fully functioning intelligent systems that can operate robustly in the real world. Because the world is filled with many unexpected challenges, the systems need to have a deep understanding of their surrounding. Vision becomes an attractive mode of perception because of its low cost, low power, and versatility in deciphering the multitude of cues in the environment. These cues can be used to solve many problems, such as robot navigation and object recognition. By implementing many subtasks within a single perceptual framework, a system can contextualize these components into a single cohesive explanation. In addition, robotics, which is comprised of many fields such as computer science, electrical, and mechanical engineering, offers a unique interdisciplinary and practical avenue to apply knowledge discovered in research of biological systems, because capabilities that robotic systems try to emulate are commonly the ones neuroscience tries to explain.

Robotics has recently made strides in emulating complex human abilities in the form of autonomous self-driving cars Montemerlo et al. (2008), Thrun (2011), home and service robots Marder-Eppstein et al. (2010), Willow (2009), and bipedal humanoid robots American Honda Motor Co Inc. 2009. Foundational among these systems is the ability to move about one's environment (also known as navigation) and to recognize one's location (also known as localization). However, despite advancements in these fields, such as on the road for autonomous cars and indoors for assistant robots, ample opportunities remain for substantial progress in less constrained pedestrian environments such as a university campus or an outdoor shopping center. This type of setting encompasses a major portion of working environments

C. Siagian (✉) · L. Itti
Department of Computer Science, University of Southern California,
Los Angeles, CA 90089, USA
e-mail: christian.g.siagian@gmail.com

Q. Zhao (ed.), *Computational and Cognitive Neuroscience of Vision*,
Cognitive Science and Technology, DOI 10.1007/978-981-10-0213-7_11

Fig. 1 Examples of tasks for service robots. From *left* to *right* patrolling alongside law enforcement personnel, searching for target objects, providing information to visitors, and assessing situations and conditions of the area of responsibility

for human-sized service robots whose tasks are exemplified in Fig. 1. Such robots can be utilized in sectors such as health care, public safety, hospitality, and manufacturing.

Now, more than ever, robots need to understand their surroundings to have greater autonomy. In turn, we can trust them with more difficult responsibilities in more complicated situations; to perform tasks that are dangerous, mundane, and should not be done by humans.

To do so, researchers have looked toward a perceptual system that not only has survived the world we live in but also has thrived in it: the human visual system. Human's ability to break down, compartmentalize, prioritize, and ultimately understand the world full of complex and seemingly contradicting stimuli is the gold standard that the robotic field strives for. Today, with many available studies in human vision, there is a unique opportunity to develop systems that take inspiration from neuroscience and bring a different perspective to robotics research. Also, even though the human brain and modern CPU are very different hardware, they face the same challenges and operate in the same world.

Biologically inspired concepts have been used extensively in robot navigation and localization. In this chapter we will explore the architecture that links localization and navigation, a hierarchical approach to localization which utilizes different ways to analyze a scene, down to lower level biologically-inspired features, such as computing the gist of a scene and recognizing salient landmarks. The presented Beobot 2.0 project Siagian et al. (2011) applies these mechanisms to create a fully working robot that is able to travel for long distances on a college campus, among a crowd of pedestrians.

The project's main contributions are four-fold:

- Using biologically-inspired vision algorithms (visual attention, salient landmark recognition, gist classification) along with computer and robotic vision techniques (localization, road finding) to compute localization and navigation perceptual modules that are robust in busy outdoor pedestrian areas;
- Developing a framework for the integration of these modules using a proposed biologically inspired hierarchical hybrid topological/grid occupancy map representation.

- Exploiting the hierarchical map, along with applying techniques such as forward projection and tracking, to resolve differences in algorithmic complexity, latency, and throughput, to create an overall real-time robotic system.
- Testing over more than 10 km total traveled distance, with routes 400 m or longer, to demonstrate that the approach indeed is viable and robust even in crowded environments like a university campus. As far as we know, this is the most extensive evaluation of a service robot localization and navigation system to date.

2 Related Works

Before describing our biologically-inspired mobile robot vision localization and navigation system, we first examine the related neuroscience background. We start with the system-level computational architecture of the human vision, before going deeper into the individual low-level mechanisms.

2.1 Biologically Inspired Framework

One important quality of biological systems that would markedly improve the current robotic systems is their ability to produce timely, but sufficiently accurate and robust response. The fundamental reason for this, we believe, is because they can perceive a scene of the environment from multiple perspectives.

In the initial viewing of a scene, the human vision system already guides its attention to visually interesting regions within the field of view. This extensively studied early course of analysis Treisman and Gelade (1980), Wolfe (1994), Itti et al. (1998), Itti and Koch (2001) is commonly regarded as perceptual saliency. Saliency-based or "bottom-up" guidance of attention highlights a limited number of possible points of interest in an image, which is useful Frintrop et al. (2006) in selecting landmarks that are most reliable in a particular environment. By focusing on specific sub-regions and not the whole image, the landmark matching process becomes more flexible and less computationally expensive.

Parallel with attention guidance and mechanisms for saliency computation, humans demonstrate exquisite ability at instantly capturing the "gist" of a scene; for example, following presentation of a photograph for just a fraction of a second, an observer may report that it is an indoor kitchen scene with numerous colorful objects on the counter top Potter (1975), Biederman (1982), Tversky and Hemenway (1983), Oliva and Schyns (1997). Such report at a first glance onto an image is remarkable considering that it summarizes the quintessential characteristics of an image, a process previously expected to require much analysis. With very brief exposures (100 ms or below), reports are typically limited to a few general semantic attributes (e.g., indoors, outdoors, office, kitchen) and a coarse evaluation of distributions of visual features (e.g., highly colorful, gray scale, several large masses, many small objects) Sanocki and Epstein (1997), Rensink (2000). However, answering specific questions such as whether an animal was present or not in the scene can be performed reliably down to exposure times of 28 ms Thorpe et al. (1995), Macé et al. (2005), even when the subject's attention is simultaneously engaged by another concurrent visual discrimination task Li et al. (2002). Gist may be computed in brain areas which have been shown to preferentially respond to "places" or visual scene types with a restricted spatial layout Epstein et al. (2000). Spectral contents and color diagnosticity have been shown to influence gist perception Oliva and Schyns (1997, 2000), leading to the development of the existing computational models that emphasize spectral analysis Torralba (2003), Ackerman and Itti (2005).

In what follows, we use the term gist in a more specific sense than its broad psychological definition (what observers can gather from a scene over a single glance): we formalize gist as a relatively low-dimensional (compared to a raw image pixel array) scene representation which is acquired over very short time frames. Thus we represent gist as a vector in some feature space. Scene classification based on gist then becomes possible if and when the gist vector corresponding to a given image can be reliably classified as belonging to a given scene category.

In spite of how contrasting saliency and gist are, both modules rely on raw features that come from the same area, the early visual cortex. Furthermore, the idea that gist and saliency are computed in parallel is demonstrated in a study in which human subjects are able to simultaneously discriminate rapidly presented natural scenes in the peripheral view while being involved in a visual discrimination task in the foveal view Li et al. (2002). From an engineering perspective it is an effective strategy to analyze a scene from opposite coarseness levels, a high-level, image-global layout (corresponding to gist) and detailed pixel-wise analysis (saliency). Also, note that, while saliency models primarily utilize local features Itti and Koch (2001), gist features are almost exclusively holistic Oliva and Torralba (2001), Torralba et al. (2003), Siagian and Itti (2007). Our presented model (Fig. 2) seeks to employ the two complementary concepts of biological vision, implemented faithfully and efficiently, to produce a critical capability such as localization.

After early preprocessing at both retina and LGN (Fig. 2), the visual stimuli arrive at Visual Cortex (cortical visual areas V1, V2, V4, and MT) for low-level feature extractions which are then shared by and serve both attention and gist. Along the Dorsal Pathway or "where" visual processing stream Ungerleider and Mishkin (1982) (posterior parietal cortex), the saliency module builds a saliency map through the use of spatial competition of low-level feature responses throughout the visual field. This competition silences locations which, at first, may produce strong local feature responses but resemble their neighboring locations. Conversely, the competition strengthens points which are distinct from their surroundings. On the contrary, in the Ventral Pathway or the "what" visual processing stream (Inferior Temporal cortex), the low-level feature-detector responses are combined to yield a gist vector as a concise global synopsis of the scene as a whole. Both pathways end up at the pre-frontal cortex where conscious decisions and motor commands are formed.

In addition to applying neuroscientific inspirations for the front-end visual perception, many robotic systems also utilize biologically-inspired topological maps to

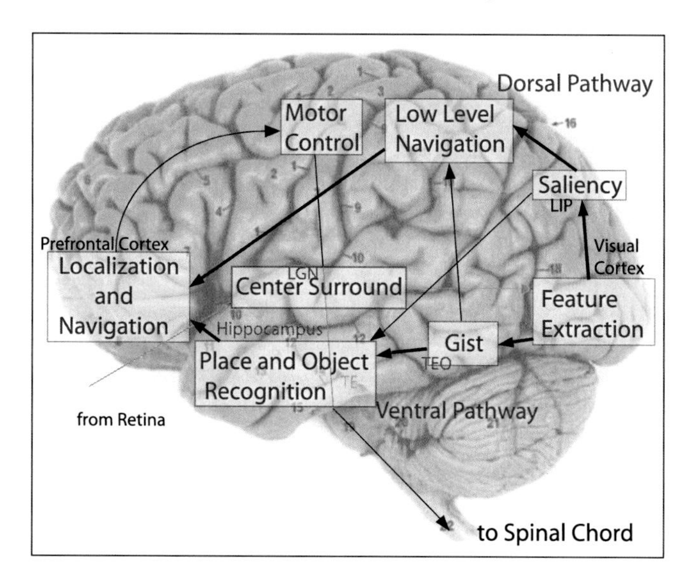

Fig. 2 A sketch of the full system with each sub-system projected onto anatomical locations that may putatively play similar roles in human vision

organize their spatial knowledge. A topological map Thrun (1998), Kuipers (2008), which refers to a graph annotation of an environment, assigns nodes to particular places and edges as paths if direct passage between pairs of places (end-nodes) exist. Often times humans can recall topological information more so than metric. For example, although humans cannot estimate precise distances or directions Tversky (2003), they can draw a detailed and hierarchical topological (or cognitive) map to describe their environments McNamara (1991). Nevertheless, approximate metric information is still deducible within the representation. In addition, the amount of added information is not a heavy burden to update and recall for a robotic system because of the concise nature of a graph. This is in sharp contrast to more traditional metric grid map in robotics localization literature Fox et al. (1999), Thrun et al. (1999), where all parts of the environment must be specified for occupancy, as opposed to assumed untraversable if not specified as places or paths.

In addition, humans can not only localize from a provided map, but they can also automatically create the map, while trying to localize. In robotics, we call this capability SLAM for Simultaneous Localization and Mapping. RatSLAM Milford and Wyeth (2008, 2010), which takes inspiration from research done in how rat brains perform in this task, is an example of such a system. It uses a column average intensity profile taken at specific windows for three different purposes: scene recognition, rotation detection, and speed estimation. The presented system is not a SLAM system. However, we plan to add this extension.

2.2 Gist and Place Recognition

One characteristic of global features such as gist is that they can robustly average out isolated noise. However, the same mechanism limits these methods to recognizing places (e.g. rooms in a building), as opposed to producing exact metric geographical locations. This is because, it is difficult to deduce a change in position with the less precise global features, even after the robot moves considerably.

There are a few place recognition systems that utilize gist features for classification. Renniger and Malik (2004) build a texture descriptor histogram to create an image feature vector, while Ulrich and Nourbakhsh (2000) create color histograms. Oliva and Torralba (2001) encode rough spatial information in their features by applying 2D Fourier Transform within individual image sub-regions on regularly-spaced grid. The resulting set of values, one per grid region, is then reduced using principal component analysis (PCA) to yield a low-dimensional vector. Interestingly, the authors report that the entries in the vector correlate with semantically-relevant dimensions, such as city versus nature, or beach versus forest. In more recent implementation, Torralba et al. (2003) uses steerable wavelet pyramids instead of Fourier transform.

The discrimination power of gist can benefit from a form of coarse scene layout segmentation, which are thought to occur in the V2 and V4 area. The layout can provide explicit region shape and geometry information, which should help detect robot movement more accurately. However, apart from the above mentioned predefined gridding, obtaining robust shape information is quite difficult because of over or under segmentation. Often times there are no clear demarcations in an unconstrained environment, for example in a park full of trees or a bustling street intersection. Furthermore, many of the more complex techniques Liu et al. (2011), Rahtu et al. (2010) are time consuming, making them inappropriate for real-time robotic application. Nevertheless it is also possible that, although the initial gist should be available instantaneously, more detailed information can arrive later, evolving the system knowledge overtime. What is currently being done in utilizing segmentation is applying it to particular environments Katsura et al. (2003), Matsumoto et al. (2000), Murrieta-Cid et al. (2002), where specific visual patterns can be exploited for simplification.

2.3 Salient Landmark Recognition and Vision Localization

Vision localization recognizes a geographical location by identifying a set of visual landmarks known to be present at that location. Salient landmarks is used because they readily stand out in their native environments. However, like any object recognition problems, landmark recognition is also plagued with issues such as occlusion, dynamic background, lighting, and viewpoint changes.

The salient landmark recognition process starts with finding the most salient points in the scene. The system then segments out the regions that these points belong to. The regions can then be matched against a landmark database that associates individual landmarks with metric coordinate locations. The database is constructed during a training run, where the robot is manually driven.

For the initial bottom-up saliency detection, most saliency models Itti et al. (2003), Bruce and Tsotsos (2006), Marat et al. (2009), Zhang et al. (2008), Gao et al. (2008) are appropriate. The general approach is to detect conspicuity using a form of center-surround mechanisms to identify portions of the scene that are significantly different than their respective surroundings. It would be interesting to see how additional localization specific top-down prior knowledge can help produce better salient landmarks. An example of a top-down prior is to exclude moving regions as they could be people walking. Another would be to pick landmarks that are above the horizon line because they are usually parts of a building and less likely to be occluded by pedestrians.

The next step after obtaining the salient points is segmenting the corresponding regions. There are a number of sufficiently fast and robust salient region segmentation techniques Achanta et al. (2009), Walther and Koch (2006), Rosin (2009). For the most part, producing a clean segmentation, i.e. a tight bounding box around a single object, is not necessary. What is most important is that the region is distinct and can be recognized reliably. A serviceable bounding box may highlight a portion of an object or combine multiple objects into one region. In fact, most of the value of saliency is in the reduction in size from the whole image to a few regions of interest, which results in more efficient recognition process compared to systems that bypass this step Se et al. (2005), Goncalves et al. (2005).

A standard way to recognize landmarks is by using local visual features to represent locally dominant image orientation. Among the widely used models, HMAX Serre et al. (2007) is one that is biologically plausible. There are many others that are generally considered non-biological Dalal and Triggs (2005), Bay et al. (2006), with SIFT (Scale-invariant Feature Transform) Lowe (2004) being the most popular. However, Muralidharan and Vasconcelos (2006, 2010) argue that there is a substantial connection, in terms of computations used, between SIFT and biological vision, which explains its success. The main reason for SIFT's effectiveness is robustness in the presence of a small amount of viewpoint change and occlusion, which allows for partial recognition. Because of that, SIFT are utilized by a number of robotic localization systems Se et al. (2005), Goncalves et al. (2005). In addition, there are also systems that use other local features, such as SURF (Speeded Up Robust Features) Valgren and Lilienthal (2008) and GLOH (Gradient Location and Orientation Histogram) Ramisa et al. (2008).

3 System Overview

The presented biologically-inspired localization and navigation system is designed for a human-sized service robot operating in unconstrained pedestrian environments. It primarily uses monocular vision, complemented by a Laser Range Finder (LRF) for obstacle avoidance, and odometry for robot motion tracking. These sensors are necessary to create a fully autonomous system. The author acknowledges that, because the LRF in particular is non-biological, the work is only biologically inspired, not plausible. It is important to note that the author is committed to both studying biological systems (human vision in particular), as well as advancing robot mobility through better understanding of its environment. In the future, he would like to remove the LRF and reliably in detecting surrounding obstacles using vision.

The flow of the overall system is illustrated in Fig. 3. The left side of the image depicts the mobile platform, Beobot 2.0 Siagian et al. (2011), with the highlighted sensors—a forward-pointing camera, an LRF, wheel encoders, an IMU (Inertial Measurement Unit)—and motors. The IMU, which is also equipped with a compass that makes magnetic measurements, is used to improve robot motion estimate. The sensor is analogous to the human vestibular system, which maintains balance and measures motion direction and speed. The IMU is placed high above the computers and motors to minimize magnetic distortion induced by the robot. Also note that ferrous objects in the environment may affect the IMU.

The two major sub-systems, which are displayed in the middle of the figure, are localization (the top of the column) and navigation (bottom). These sub-systems are connected through a hierarchical representation of the environment, displayed in the rightmost column in the figure. The first level of the representation is a global topological map used for full path planning from the current to the destination location. It is an augmented topological map with directed edges Thrun (1998), Blanco et al. (2006). The map has an origin and a rectangular boundary, with each node has a Cartesian coordinate. In addition, each edge has a cost, which is set to the distance between the corresponding end-nodes. The robot state (position and viewing direction) is represented by a point which can lie on a node or an edge.

The next level of representation is a local grid-occupancy map, which details the robot's immediate surroundings. Mechanically, the local map is connected to the global map through an intermediate goal location that is registered in both maps (observe the green dots Fig. 3). The task of the navigation system is to reach the intermediate goal. The robot-centric local map is then advanced along the current graph edge in the global map. This keeps the robot at a fixed location in the local map. The components of the navigation system are the visual road recognition Siagian et al. (2013) to estimate the road direction, the grid map construction, and path planning to generate motor commands. It will be described in Sect. 5.

One critical issue in integrating complex vision algorithms in a real-time system is latency, or processing delay of a module. The system must be able to account for the latency to use the results, and produce an action that does not ignore everything that

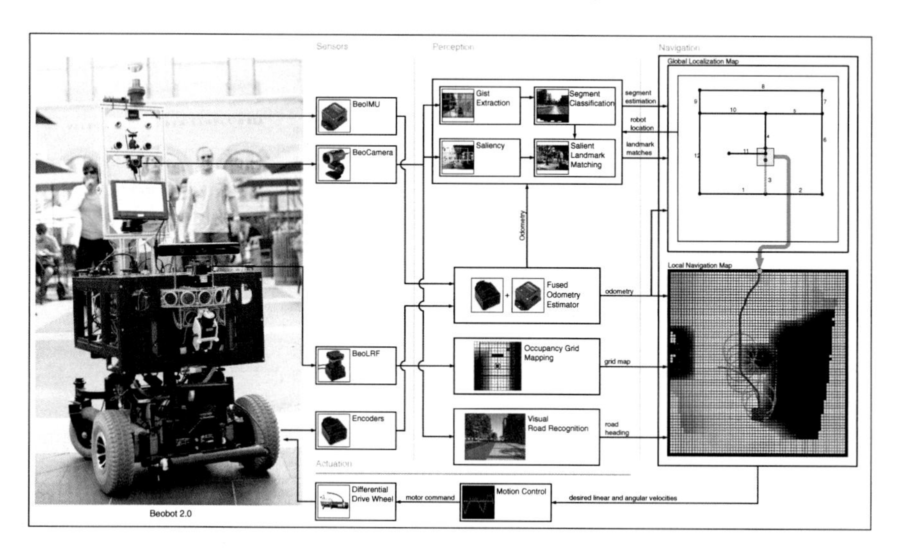

Fig. 3 Diagram for the overall localization and navigation system. We run our system on a wheelchair-based robot called the Beobot 2.0. As shown on the sensor column, the system utilizes an IMU, camera, Laser Range Finder (LRF), and wheel encoders. We use the camera to perform vision localization by integrating results from salient landmark recognition and place recognition. For navigation we use visual road recognition for estimating the road heading, and LRF-based grid-occupancy mapping for obstacle avoidance. The system utilizes a topological map for global location representation and a grid-occupancy map for local surrounding mapping. The grid map, which is represented by the *blue rectangle* on the topological map, is connected with the topological map by the intermediate goal (*green dot*)

occurred during the module processing period. The latency for vision localization is substantial because landmark recognition can take more than seconds.

This is where the hierarchical map construction is advantageous, where navigation is dealt with at the local map level, while localization is at the global level. It effectively separates the localization and navigation tasks, allowing the real-time navigation system to perform independently despite the long latency of the localization system. On the field, localization and navigation interact infrequently, namely, when the robot needs to turn at an intersection or stopping at the goal. Because these events can be anticipated in advance, these reasonable delays can be overcome.

The coming Sect. 4 describe the vision localization system, and Sect. 5 details the navigation system. The system then is thoroughly tested in Sect. 6, with the overall findings summarized in Sect. 7. We end with a discussion and lessons learned from the interdisciplinary collaboration between robotics and neuroscience in Sect. 8.

4 Vision Localization System

The vision localization system, which is illustrated in Fig. 4, is divided into 3 stages: feature extraction, recognition, and localization. The first stage takes a camera image and outputs gist features and salient regions. In the next stage, we associate the outputs with particular locations in the environment. The last stage then integrates these cues to produce the best location estimate.

In the second stage, the system uses the gist features to classify the segment the robot is currently on. A segment is an ordered list of edges in the topological map, with one edge connected to the next to form a continuous path. An example is the selected three-edge segment, highlighted in green, in the map in Fig. 4. A segment is roughly equivalent to the generic term "place" for place recognition systems (mentioned in Sect. 2.2), which refers to a general vicinity of an environment. This grouping is motivated by the fact that the views/layouts within a segment are coarsely similar. Geographically speaking, a segment is usually a portion of a hallway or a road interrupted by a crossing or physical barrier at each end for a natural delineation.

In parallel, the system also tries to match the salient regions in the scene against a landmark database. A match is used to refine the segment classification to a more accurate metric localization. The term salient region refers to a conspicuous area in an image depicting an easily detected part of the environment. An ideal salient region is one that is persistently observed from many points of view and at different times of the day. A salient region does not have to be an isolated object (often times it is part of an object or a jumbled set of objects), it just has to be a consistent pattern of interest in the environment. To this end, a set of salient regions that portray the

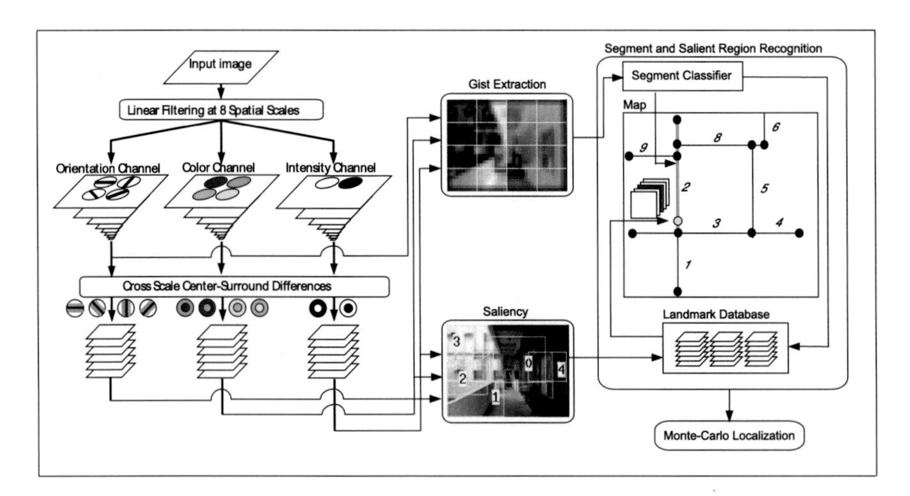

Fig. 4 Diagram of the vision localization system. From an input image the system extracts low-level features consisting of center-surround color, intensity, and orientation channels. They are further processed to produce gist features and salient regions. These cues are then associated with particular locations. The results are used to estimate the robot's location

same pattern of interest are grouped together, and the set is called a landmark. Thus, a salient region can be considered as an evidence of a landmark, and "to match a salient region to a landmark," means to match a region with the landmark's saved regions.

Section 4.1 describes the gist feature computation and segment classification model, while Sect. 4.2 details the salient landmark recognition model. Note that the process of discovering salient regions is done using biological computations, but the process of region matching may not be completely biological.

4.1 Segment Classification

The gist feature generation process starts by computing Visual Cortex features, which is shared by the saliency model Itti et al. (1998) in Sect. 4.2. The system then extracts the gist features from these features using computationally inexpensive mechanisms. The resulting gist features are used to train the segment classifier, which consists of PCA/ICA feature dimension reduction and neural network classification.

At the start, an input image is first filtered through a number of low-level visual "feature channels" at multiple spatial scales in the color, intensity, orientation channels, which are found in Visual Cortex. Some channels, color and orientation, have several sub-channels, color type and orientation, respectively. Each sub-channel runs an image through a nine-scale pyramidal filters, a ratio of 1:1 (level 0) to 1:256 (level 8), both horizontally and vertically. A 5-by-5 Gaussian smoothing is applied in between scales. Within each sub-channel i, the model performs center-surround operations (commonly found in biological-vision which compares image values in center-location to its neighboring surround-locations) between filter output maps $O_i(s)$ at different scales s in the pyramid. This yields feature maps $M_i(c, s)$, given a "center" (finer) c and a "surround" (coarser) scale s. The implementation uses $c = 2, 3, 4$ and $s = c + d$, with $d = 3, 4$ to gather information at multiple scales, with added lighting invariance afforded by the center-surround comparison. Hence, the system computes six feature maps for each domain at scale combinations 2–5, 2–6, 3–6, 3–7, 4–7, and 4–8. Across-scale difference (operator $\ominus$) between two maps is obtained by interpolation to the center (finer) scale and point wise absolute difference (Eq. 1).
For color and intensity channels:

$$M_i(c, s) = |O_i(c) \ominus O_i(s)| = |O_i(c) - \mathrm{Interp}_{s-c}(O_i(s))| \tag{1}$$

The orientation channel applies Gabor filters to the greyscale input image (Eq. 2) at four different angles ($\theta_i = 0, 45, 90, 135°$) and four spatial scales ($c = 0, 1, 2, 3$) for a subtotal of sixteen sub-channels.
For orientation channels:

$$M_i(c) = \mathrm{Gabor}(\theta_i, c) \tag{2}$$

The channel does not apply center-surround on the Gabor filter outputs because these filters are already differential by nature. The color and intensity channels combine to form the three pairs of color opponents derived from Ewald Hering's Color Opponency theories Turner (1994), which identify four primary colors red, green, blue, yellow (denoted as R, G, B, and Y in Eqs. 3, 4, 5, and 6, respectively), and two hue-less dark and bright colors in Eq. 7, computed from the raw camera r, g, b outputs Itti et al. (1998).

$$R = r - (g + b)/2 \tag{3}$$
$$G = g - (r + b)/2 \tag{4}$$
$$B = b - (r + g)/2 \tag{5}$$
$$Y = r + g - 2(|r - g| + b) \tag{6}$$
$$I = (r + g + b)/3 \tag{7}$$

The end results are the red-green and blue-yellow color opponency pairs (Eqs. 8 and 9), and the dark-bright intensity opponency pair (Eq. 10). Each pair is applied at six center-surround scale combinations to make up eighteen color and intensity sub-channels. These, along with the sixteen orientation sub-channels, make up the thirty-four sub-channels used by the gist model. Note that more channels can be added if appropriate.

$$RG(c, s) = |(R(c) - G(c)) \ominus (R(s) - G(s))| \tag{8}$$
$$BY(c, s) = |(B(c) - Y(c)) \ominus (B(s) - Y(s))| \tag{9}$$
$$I(c, s) = |I(c) \ominus I(s)| \tag{10}$$

Figure 5 illustrates the gist model architecture.

To obtain the gist values, the model applies averaging operations, the simplest neurally-plausible computation, in a fixed four-by-four grid over each sub-channel feature map (observe Fig. 5). Equation 11 formalizes the gist feature $G_i^{k,l}(c, s)$ computation for each map $M_i(c, s)$, with indices k and l specifying in the horizontal and vertical coordinate, respectively.
For color, intensity channels:

$$G_i^{k,l}(c, s) = \frac{1}{16WH} \sum_{u=\frac{kW}{4}}^{\frac{(k+1)W}{4}-1} \sum_{v=\frac{lH}{4}}^{\frac{(l+1)H}{4}-1} [M_i(c, s)](u, v) \tag{11}$$

where W and H, the image width and height, are used to normalize the sub-region value sums. The system similarly processes the orientation maps $M_i(c)$ to compute $G_i^{k,l}(c)$. Although more complex statistics such as variance can provide more information, their computational cost is much higher than that of first-order statistics,

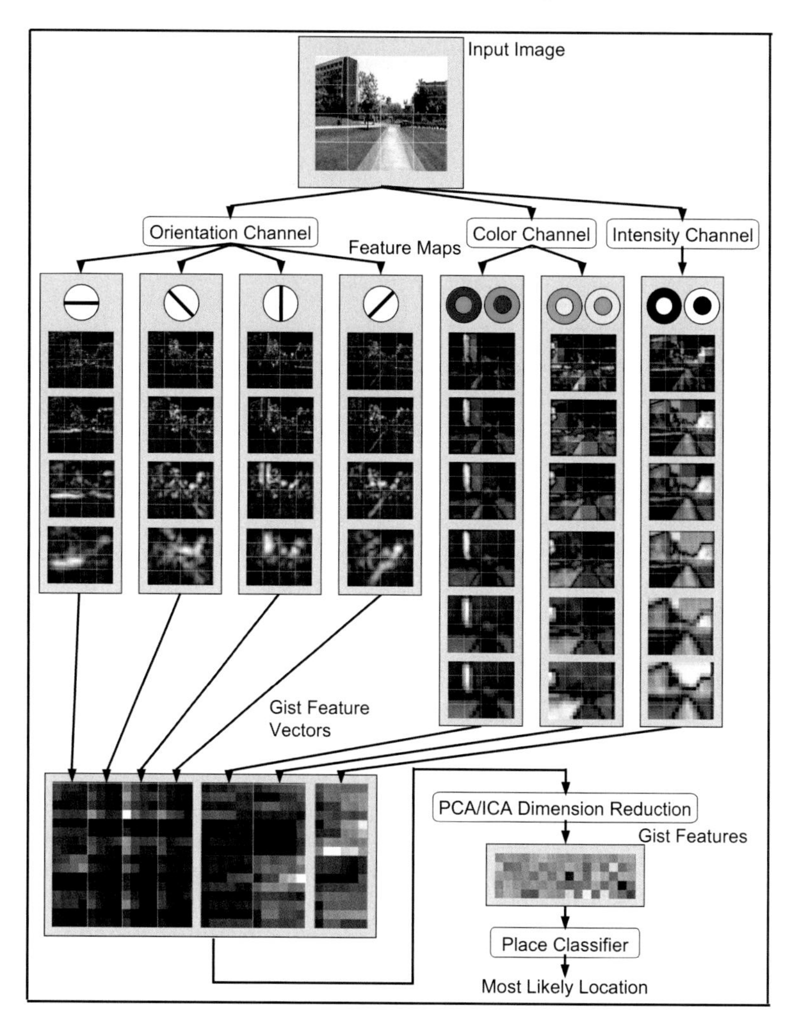

Fig. 5 The Gist Model Architecture, from the visual feature channels, to the gist feature extraction, PCA/ICA dimension reduction, to segment classification

while their biological plausibility remains debated Chubb and Sperling (1988). Thus, here we explore whether first-order statistics over a variety of visual domains is sufficient for satisfactory results.

The total number of the gist feature dimension is 544, which is 34 feature maps times 16 regions per map (Fig. 5). The model then reduces the dimensions using Principal Component Analysis (PCA) and Independent Component Analysis (ICA) using FastICA Hyvrinen (1999) to 80 while preserving up to 97 % of the variance for a set in the upwards of 30,000 campus scenes. For segment classification, the model uses a three-layer neural network (with intermediate layers of 200 and 100 nodes), trained with the back-propagation algorithm. The complete process is illustrated in Fig. 5.

4.2 *Salient Object Recognition*

The saliency model Itti et al. (1998) uses the feature maps produced in Sect. 4.1 to detect conspicuous regions in each channel. It first linearly combines (unweighted pixel-wise addition) the feature maps within each channel to produce a conspicuous map per channel. The model then combines these maps using winner-take-all mechanisms, which emphasize locations that substantially differ from their neighbors, to yield a saliency map. The model further processes the map to produce a set of salient regions (Fig. 6).

The model starts at the pixel location of the saliency map's highest value. To extract a region that contains the point, the model uses a shape estimator algorithm Rutishauser et al. (2004) (region growing with adaptive thresholding) to segment the feature map that gives rise to it. To find the appropriate feature map, the model compares the values of the conspicuity maps at the salient location and select the channel with the highest value (the winning channel). Within the channel, the model then compares values at the same location for all the feature maps. The one with the highest value is the winning center-surround map.

The model then creates a bounding box around the segmented region in the map. Initially, the model fits the smallest-sized rectangle that includes all connected pixels. It then adjusts the size of the box to between 35 and 50 % in both image width and height, if needed. This is because small regions are hard to recognize and overly large ones take too long to match. The model then creates an inhibition-of-return (IOR) mask to suppress that part of the saliency map to move to subsequent regions. This is done by blurring the region with a Gaussian filter to produce a tapering effect at the mask's border. If a new region overlaps any previous regions by more than 66 %, it is discarded but still suppressed.

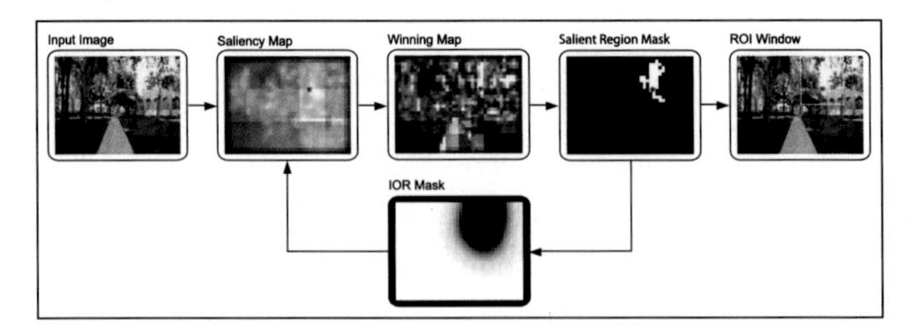

Fig. 6 A salient region is extracted from the center-surround map that gives rise to it. The model uses a shape estimator algorithm to create a region-of-interest (ROI) window and inhibition-of-return (IOR) in the saliency map to find other regions

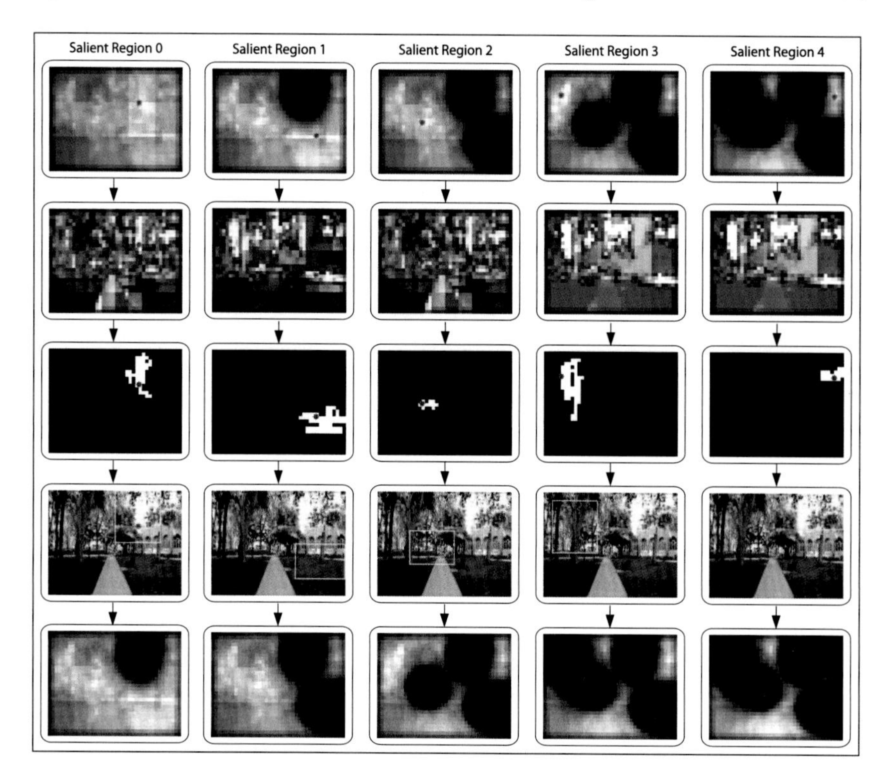

Fig. 7 Process of obtaining multiple salient regions from an image. The IOR mask (last row) shifts the attention of the model to other parts of the image

The model continues until one of three exit conditions occur: unsegmented image area is below 50 %, number of regions processed is 5, and the saliency map value of the next point is lower than 5 % of the first (most salient) point. We limit the regions to 5 because, from experiments, subsequent regions are much less likely of being repeatable. Figure 7 shows extraction of five regions. Extracting multiple regions per image improves selection robustness, given the possibility of occlusion and salient distractions. For example, in the figure, the first region, a ray of sunshine hitting a building, is not an ideal localization cue, while the second is better because it depicts the base of a building.

The model then tries to match the salient regions against a landmark database, which entries are tagged with locations of origins. The procedure to build a landmark database involves a guided robot traversal through all paths in the map, performed several times for lighting coverage. In addition, this also allows the model to select landmarks that are found in multiple traversals, and detected over a long period of time. These requirements take out moving objects (people walking, etc.), which tend to be less reliable for localization purposes. For more complete landmark database construction description, refer to Siagian and Itti (2008). Also note that the model

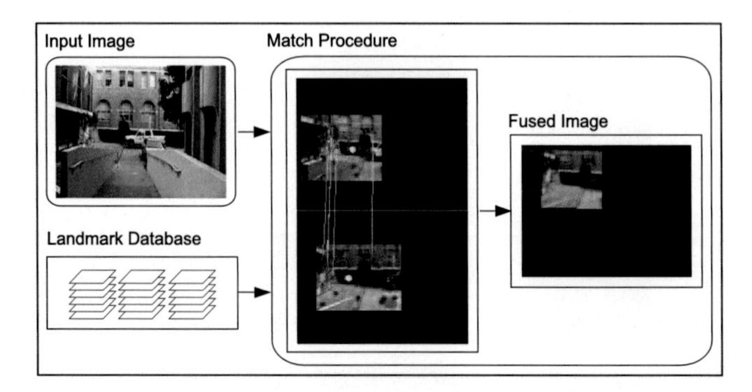

Fig. 8 Matching process of two salient regions using SIFT keypoints (drawn as *red disks*) and salient feature vector, which is a set of feature map values taken at the salient point (drawn as the *yellow disk*). The *lines* indicate the correspondences that are found. The fused image is added to show that the model also estimates the pose change between the pair

simultaneously stores the gist features from each input image for segment classification training.

The model recognizes the salient regions using two sets of signatures: SIFT keypoints Lowe (2004) (using suggested parameters and thresholds) and salient feature vector. For robustness, the model only considers regions that have more than five keypoints. A salient feature vector Siagian and Itti (2007) is a set of values taken from each sub-channel feature map. For each map, the model takes values from a 5-by-5 window centered at the salient point location (yellow disk in Fig. 8) of a region. In total, there are 1050 features (7 sub-channels times 6 feature maps times 5×5 locations). To compare two salient feature vectors, the model accounts for feature similarity (based on the Euclidian-distance in feature space) and location proximity. The model sets a high positive match threshold to virtually eliminate false positives at the expense of slightly higher false negative rate.

To speed up the database matching process, the system first orders the landmarks in the database from the most likely to be matched first. This is done using a number of contextual priors: current robot location estimate, segment classification, and salient feature vector match score. They are linearly combined to make up a single likelihood prediction score Siagian and Itti (2008). This score gives high marks to landmarks that are near the current location estimate, stored in the segment that has high neural network classification output, and whose salient feature vector is most similar to the compared input region.

Because robots are real-time systems, the system does not have time to consider all positive matches to find the best. The database search usually ends once the first match is found. Therefore, the ordering indirectly influences the salient region recognition step. This method of utilization of multiple experts, which is in the spirit of hierarchical recognition, has been shown Zhang and Kosecka (2005), Wang et al. (2006) to speed up the search process.

In addition, the system also employs a search cutoff in the comparison process so that the system can go to the next input region. Thus, instead of going through the whole database, there are times when the matching landmark may not even get to be compared with the input region because of an incorrect prior. However, a more thorough search during uncertain localization should be instituted to allow for recovery for kidnapped robot or during system initialization. At those times, the robot should also be more active in searching for better views, where the landmarks are more likely to be matched.

Siagian and Itti (2008) showed that the contextual search technique cuts down search time by 87 %, a speed up of 8. That is, a search time for a SIFT-only database matching takes much longer than the presented system.

The successful matches are fed to the back-end probabilistic Monte-Carlo Localization (MCL) module, which utilizes Sampling Importance Resampling (SIR) Fox et al. (1999), Thrun et al. (1999), Montemerlo et al. (2002). The details can be read in Siagian and Itti (2009). In summary, the algorithm formulates the location belief as a set of weighted particles. Each particle is a particular location hypothesis, with its weight proportional to the likelihood of observing incoming segment and salient region observations so far. The particles are moved every time the system sends a motor command to the robot by an amount read from the odometry. In addition, the weight is updated whenever an observation is available. The segment observation is applied at every frame, while the salient region observation is only incorporated when positive matches are obtained. Note that, before the resulting region locations are considered, they are first projected forward by the accumulated odometry reading of the robot movements during the matching process. This is done to account for the long latency of the process. The most likely location is taken from the particle with the highest weight. By maintaining a population of particles the system can simultaneously consider many hypotheses, and, thus, accounts for uncertainties in the observations and odometry readings. Because of that, the system keeps a small percentage of random particles to allow for belief diversity in the population.

Figure 9 illustrates how the system works together. The raw feature extraction, gist, and saliency computations, which are run on a 16-core 2.6 GHz machine and operating on 160 × 120 images, takes 25 ms per frame. The segment classification takes less than 1 ms, while the landmark matching takes 3 s on average (0.33 Hz), despite a multi-threaded landmark search implementation. The back-end localization module take 1 ms to compute and outputs results at 30 Hz, the input image frame rate.

The system is naturally distributed and can be run on multiple-core processor architecture. Each perceptual module operates on its own time frame, by accounting for its own internal latency, allowing it to send out results as if it is real time. In addition, because the topological map formulation is one dimensional along the graph edges and needs only 100 particles to localize, processing these perceptual results also takes a small amount of time. Because of this loose coupling, the system can be easily expanded with new observations and modalities just by adding more processors.

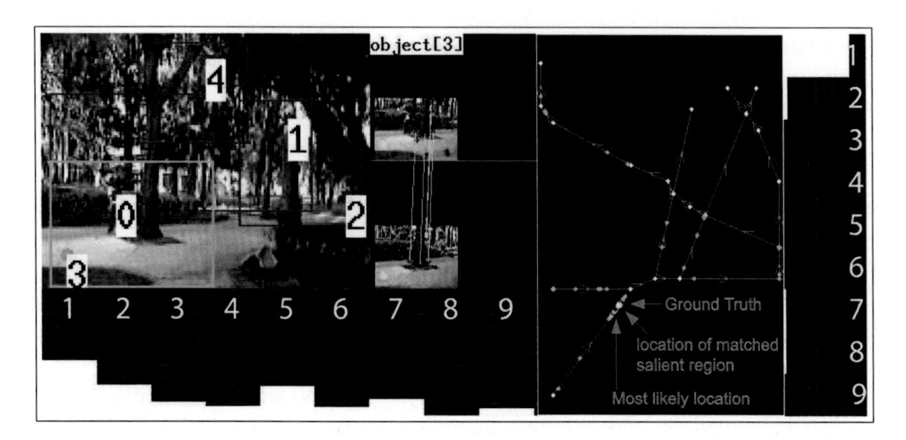

Fig. 9 A snapshot of the localization system test-run in an environment that comprises 9 street segments. *Top-left* (main) image contains the salient region windows. *Green window* means a database match, while *red* is not. A salient region match is displayed next to the main image. Below the main image is the segment estimation vector derived from gist (there are 9 possible segments in the environment corresponding to the *9 vertical bars*; here, from gist alone, the system believes most strongly that it is on segment 1). The *middle* image projects the robot state onto the map: *cyan disks* are the particles, the *yellow disks* are the location of the matched database salient region, and the *blue disk* (the center of the *blue circle*, here partially covered by a *yellow disk*) is the most likely location. The radius of the *blue circle* is equivalent to five feet. The right-most histogram is the number of particles at each of the 9 possible segments, with most of the particles residing in segment 1. The robot believes that it is moving toward the end of the first segment, which is correct within a few feet

5 Navigation System

The components of the navigation system are the visual road recognition Siagian et al. (2013) to estimate the road direction, grid map construction, and path planning to generate motor commands.

The local map represents the robot's surroundings with a 64 by 64 grid occupancy map Moravec and Elfes (1985). Each grid cell spans 15 cm by 15 cm spatial area for a 9.6 m by 9.6 m local map extent. 15 cm or 1/4 the robot size (Beobot 2.0 is 60 cm in length and width) is selected because it is a balance between sufficiently detailed and manageable for real-time update.

As shown in Fig. 3, the robot is displayed as a red rectangle and is placed at the local map horizontal center and three quarters down the vertical length to increase front-of-robot coverage. In addition, there is a layer of grid surrounding the map to represent goal locations that are outside the map. The figure shows the goal to be straight ahead, which, by convention, is the direction of the road. Thus a large portion of motion control is trying to minimize the deviation between the robot and road heading.

Given the robot autonomous maximum speed of 0.89 m/s maximum (0.66 m/s average), the 7.2 m (0.75 $\times$ 64 $\times$ 15 cm) mapping of the robot front area affords at least 8.1 s to avoid static obstacles. For dynamic obstacles such as people, the system latency of as much as 100 ms (or two periods of local map update rate of 20 Hz) translates to only 0.14 m movement or within one grid size using the average human walking speed of 1.4 m/s.

5.1 Visual Road Recognition

The visual road recognition module Siagian et al. (2013), which is illustrated in Fig. 10, utilizes a novel vanishing point (VP) algorithm, and outputs absolute road direction. The algorithm relies on long and robust contour segments. In addition, the module also implements a tracking mechanism to run in real time.

The module first takes an input image and applies Canny edge detection to create an edge map. From here it has two ways to recognize the road. One (the top pipeline in the image) is through a full but slower recognition process, which detects segments in the edge map, votes for the most likely VP, and constructs the road lines that emanate from the VP. On the other hand, the bottom pipeline runs in real-time because it only tracks the available road lines to update the VP. Note that the top pipeline also uses tracking to catch up to the current frame by projecting newly discovered road lines through unprocessed frames accumulated during recognition.

At the end of both pipelines, the module converts the VP pixel location to angle deviation from the road direction. This is done by linear approximation, from 0° at the center of the image to 27° at the edge (or 0.16875° per pixel). The module

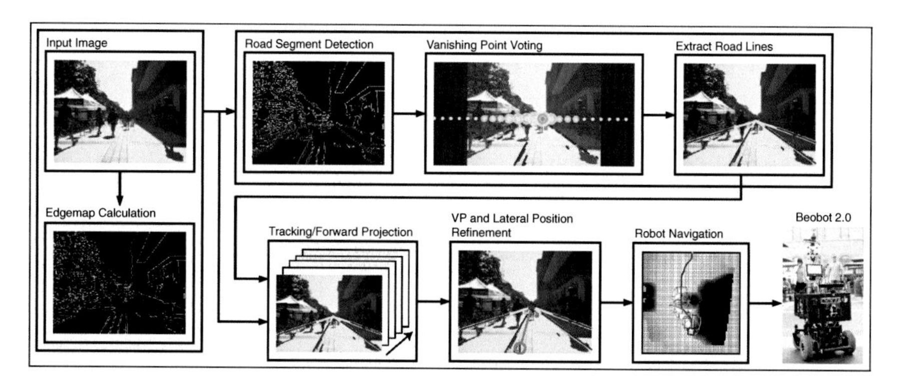

Fig. 10 The visual road recognition module. The algorithm starts by applying a Canny edgemap to the input image. The module has two ways to estimate the road. One is using the slower full recognition step (the top pipeline), where it detects road segments, used them to vote for the vanishing point (VP) and extract road lines. The second (*bottom*) is by tracking the previously discovered road lines to update the VP location as well as the robot lateral position. The tracker is also used to project forward output road lines from the top pipeline to the current frame. The road recognition module outputs the VP-based road heading as well as lateral position to the navigation system to generate motor commands

then calculates the absolute road heading by summing the angular deviation with the current IMU reading. By approximating the road direction in absolute heading, the module does not directly couple the vision-based road recognition with motion generation. The road recognition module is responsible for estimating the road direction, while the IMU is responsible for maintaining the robot heading with respect to that direction. This differentiation is essential when the road is momentarily not visible, e.g., when going under a bridge. Here the robot heading is still maintained, despite the fact that the road direction is not updated. The system applies a temporal filtering to take out erroneous values, and locks the robot to its initial lateral position by maintaining the coordinates where the road lines intersect the bottom of the image.

Siagian et al. (2013) have shown that the module can follow curved roads. The key is extracting as much of the straight image segments on the curved path as possible. In this case, these segments point toward a moving vanishing point, which coincides with the changing direction of the road.

5.2 Grid Occupancy Map Construction

The system updates the grid occupancy values using LRF proximity distances that are already appropriately aligned with respect to the road heading. The grid values are initially set to -1.0 to indicate unknown occupancy. The grid cells are updated only if they are hit by or in line with a LRF ray. For each ray, the system draws a line from the robot location to the endpoint. If the laser line goes through a grid cell, indicating that it is free, the occupancy value is set to 0.0. If the laser line ends on the grid cell, it is set to 1.0. This discrete valued map is stored and updated at each time step. To generate a path, the system first blurs the map to soften the grid occupancy discretization, which helps warn the robot that an obstacle is forthcoming. The blurring is not applied to cells with unknown occupancy to minimize their influence. They are given a conservative value of 0.99 to allow the path planning algorithm to go through them only if other options have been exhausted. The resulting map is called the blurred map B.

As the robot moves longitudinally toward the goal, the grid values are shifted down one row at a time, keeping the robot location as is. This means that the local map is robot-centric in denoting the location, but allocentric in heading direction, aligning with the road. The robot movement estimation, which is computed using the fused odometry of wheel encoders and IMU, is sufficiently accurate for grid shifting purposes. Figure 11 illustrates the grid occupancy process, which includes a path biasing step, described in the next section. The grid construction process, because the map uses a relatively coarse 15 cm by 15 cm resolution, is computed in less than 1 ms, and outputs results at 20 Hz after each LRF reading.

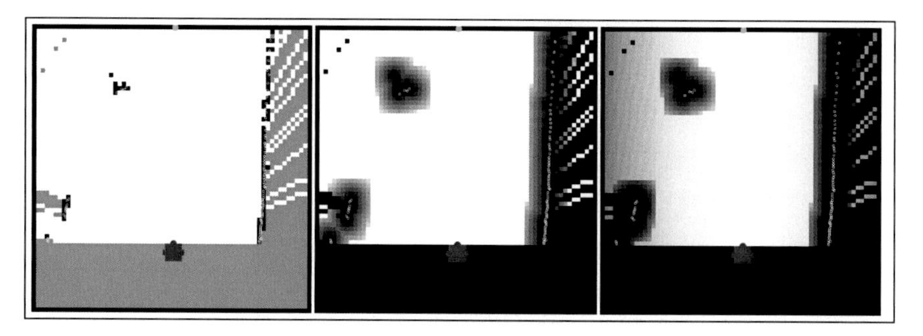

Fig. 11 Three stages of the grid occupancy map. The first, leftmost, image is the discrete map D of all discovered occupancy values, with the *white* color being unoccupied, *black* fully occupied, and *cyan* unknown. The second map is the blurred map B, while the last occupancy map F is biased by most recently produced path to encourage temporal path consistency. Notice that the unknown grid cells, such as the ones below the robot, do not extend their influence out in both the blurred and biased map

5.3 Path Generation

The system first applies A* search to the local map to create a coarse path trajectory to the goal. For each grid cell, the search process creates eight directed edges to the surrounding nodes. The edge weight is computed by adding the edge length and occupancy value of the destination cell, which is the sum of the destination $B(i,j)$ and a bias term to encourage path consistency. Observe Fig. 11 for illustration. The bias term is proportional to the proximity of the destination cell to the corresponding closest point in the previous A* path. We find that adding a bias and generating a path on every frame, which is computed efficiently because of the coarse 15 cm grid dimension, is more practical for encouraging temporal smoothness than correcting a path from the previous frame.

To optimize obstacle avoidance and smooth out the discrete path, the system applies a modified elastic band algorithm Quinlan and Khatib (1993) using the LRF laser scans as individual obstacles. Each path point undergoes an iterative process, where at each time step the point is moved by an attractive force to straighten the path, and a repulsive force to repel it perpendicular to the closest obstacle. Note that, when identifying the closest obstacle points, the system models the robot body more accurately as a square.

To compute motor command using the smoothed path, the system takes the path's first step and applies the Dynamic Window Approach (DWA) Fox et al. (1997). It not only calculates the deviation between the current robot heading and heading of the initial path step, it also takes into account the robot dynamics. This is done by only considering commands within the allowable velocity space based on the estimated velocity calculated from the previous command, its execution delay, and motor and body dynamics. Furthermore, the approach then accurately simulates each resulting arc trajectory of the allowable commands (translational and rotational velocity),

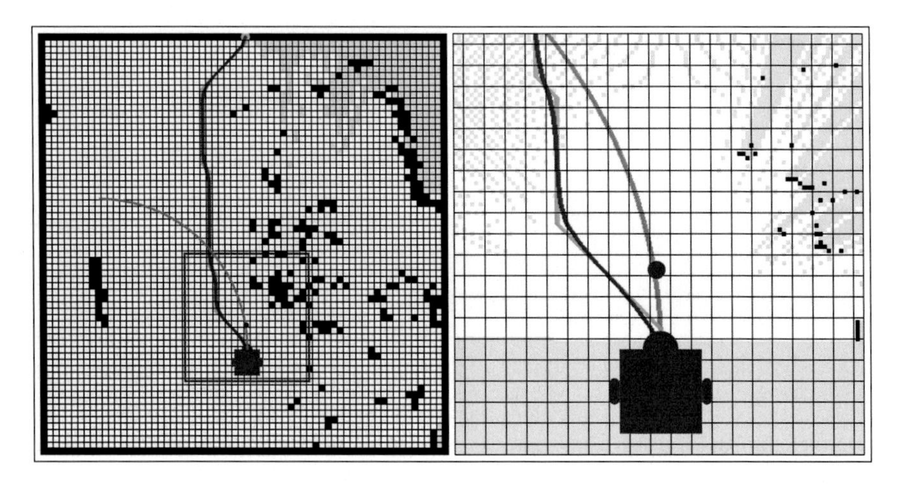

Fig. 12 The three stages of path trajectory and motion command generation. The figure displays the full local map on the *left* and zoomed-in part within the *red rectangle* on the *right* for clarity. The system first runs A* to generate a rigid path shown in *green* in both images. The system then deforms the rigid path (in *blue*) to optimally and smoothly avoid obstacles. Finally, the system uses Dynamic Window Approach (DWA) to generate the actual command, one that accounts for the current robot dynamics. That resulting command is shown as a curve in *orange*; here, the *orange* and *blue* trajectories differ significantly for the current time step, although they are aimed in the same direction, to avoid making a turn that would be so sharp that it might tip the robot over. In addition, the figure also displays the predicted location of the robot, noted by the *red disk*

modeling the robot shape to test whether any part of the robot would hit an obstacle. In the end, the system selects a motor command that is expected to best achieve the target heading and velocity, while staying clear of the obstacles. Figure 12 illustrates the three stages of the motion generation.

While other modules influence the robot motion in almost every time step, localization updates the intermediate goal in the local map infrequently. Most times, the goal location is placed in the canonical straight ahead position (observe Fig. 12). The goal is only moved during turns, or when the goal is inside the local map. For turns, the localization system sends a turn angle to the navigation system, which is the angle created by the involved incoming and outgoing edges in the global map. The navigation system then moves the goal to the appropriate location, triggering a path generation that forces the robot to turn in place immediately. It then allows the robot to rotate until its angle is aligned to the new goal direction, usually in less than 2 s. When the turn is completed, the local map is rotated so that the goal location moves to the canonical straight-forward position. The robot can then proceed normally.

6 Testing and Results

We rigorously test the system on Beobot 2.0 Siagian et al. (2011), pictured in Fig. 3, at the University of Southern California campus (USC) in Los Angeles. Beobot 2.0 has a sixteen-core 2.6 GHz distributed computing platform, which employs a subscribe/publish architecture to manage all functionalities, from sensor data acquisition, localization, navigation, to actuation. The source code is provided at Itti (2012). The system uses a subset of Beobot 2.0's sensors: a front-facing monocular camera and LRF (117 cm and 60 cm above the ground, respectively), an IMU, as well as wheel encoders. The camera outputs 320 × 240-pixel images, which are kept at the same dimension for road recognition, but scaled to 160 × 120 for vision localization.

We start with testing the localization system off-line by inputting videos of the environment recorded by a human-held camcorder and evaluating the gist-based segment classification and metric location estimation, in Sects. 6.1 and 6.2, respectively. We then test system on the robot, in real-world scenarios in Sect. 6.3.

6.1 *Off-Line Segment Classification Testing*

The off-line segment classification as well as subsequent metric localization is tested using data taken at three sites at the USC campus (Fig. 13). The first is the 38.4 × 54.86 m Ahmanson Center for Biological Research (ACB) building complex (first row of Fig. 13). Most of the surroundings are flat walls with little texture and solid lines that delineate different parts of the buildings. The second site is a 82.3 × 109.73 m adjoining Associate and Founders parks (AnF), which are dominated by vegetation (second row). The third site is an open 137.16 × 178.31 m Frederick. D. Fagg park (third row).

The model is extensively tested across multiple days and times of days, using an 8 mm hand held camcorder and carried by a person walking. We take data on up to six different times of the day, twice for each time to include most lighting conditions

Fig. 13 Examples of scenes from the Ahmanson Center for Biological Research (ACB) building complex, the adjoining Associate and Founders parks (AnF), and the Frederick. D. Fagg park, each displayed in its respective row

for several days. They cover the brightest (noon time) to the darkest (early evening) lighting, overcast versus clear, and changes in appearance due to temperature (hazy mid-afternoon). Note that the training and testing data for similar lighting conditions are always taken on different days to ensure the integrity of the results. Because of this, the system has to deal with missing or added objects, such as a park bench removed, or a service vehicles parked or a large storage box left off for the day.

The video clips are divided into segments, taking special care in creating a clean break between the two adjacent segments. That is, we stop short of the end of the current segment and wait a few moments before starting the next one. This ensures that the model is trained with data where ground-truth labeling (assigning a segment number to an image) is unambiguous. In all, there are 26,368 training and 13,966 testing frames for ACB, 66,291 training and 26,387 testing frames for AnF, and 82,747 training and 34,711 testing frames for FDF. The system uses the same classifier neural network structure, with nine output layer nodes for each segment. We then combine data from all sites and train a single classifier.

Table 1 shows the results, with an accumulated satisfactory performance of 13.05 % error or 86.95 % correctness. The column "False+" for segment x lists the number of incorrect guesses among all segment x guesses. Conversely, "False−" lists the number of incorrect guesses when the segment x images is inputted to the classifier. The errors are, in general, not uniformly distributed among segments, although the AnF segment errors appear to be more uniform. This is probably because the scenes in AnF have more overlap and are less structured, thus, prone to more accidental classification errors among the segments.

The results also show that the most difficult site is AnF with a total error of 15.79 %, about a 4 % performance drop compared to the other two. This is reasonable considering the visual challenges in AnF. On the other hand, increase in segment lengths does not appear to affect the results drastically. The results from FDF, which encompasses the largest area, and thus more testing frames, is better than the smaller ACB and FDF. This may be because the majority of the scenes within a segment do not change much, even in longer paths.

It is also important to note that proper lighting coverage in the training set is critical. When the system is tested with a set of data taken right after a training run, the error is between 9 and 11 %. However, if we exclude training images that have the same lighting condition as the testing data, the error usually triples (to about thirty to forty percent).

To gauge the system's scalability, we combine scenes from all three sites and train it to classify twenty seven segments. The only difference in the neural-network classifier is the output layer now consists of twenty-seven nodes. We use the same procedure as well as training and testing data (175,406 and 75,073 frames, respectively). Training takes much longer than the other experiments. The periodic structure of the confusion matrix (not shown here) reveals that, during training, the network first focuses on the inter-site classification before reducing the intra-site errors. The results for this classification is shown in the "combined" column of Table 1), with an overall error 13.55 %, which is not far worse than the average error of 13.05 %,

Table 1 Off-line segment estimation results

Segment	ACB		AnF		FDF		Accumulated		Combined	
	False+	False−	False+	False−	False+	False−	False+	False−	False+	False−
1	102/1608	90/1596	437/2617	957/3137	134/2775	710/3351	673/7000	1757/8084	1088/7083	2089/8084
2	256/1631	400/1775	538/2115	269/1846	460/3256	407/3203	1254/7002	1076/6824	1368/7044	1148/6824
3	116/1896	106/1886	355/3757	371/3773	634/3475	235/3346	1105/9128	712/9005	1723/10110	618/9005
4	43/1060	217/1234	485/2511	379/2405	96/3102	229/3235	624/6673	825/6874	350/6714	510/6874
5	287/1430	141/1284	399/3065	272/2938	47/2396	778/3127	733/6891	1191/7349	819/7574	594/7349
6	155/2142	100/2087	986/4627	583/4224	780/7068	341/6629	1921/13837	1024/12940	1872/12928	1884/12940
7	342/1706	317/1681	349/1909	392/1952	55/3311	455/3711	746/6926	1164/7344	915/6822	1437/7344
8	223/1279	43/1099	246/2814	287/2855	1387/6000	395/5008	1856/10093	725/8962	1113/9072	1003/8962
9	157/1213	267/1323	374/2982	659/3267	357/3058	400/3101	888/7253	1326/7691	924/7726	889/7691
Total	1681/13965		4169/26397		3950/34711		9800/75073		10172/75073	
Percent. (%)	12.04		15.79		11.38		13.05		13.55	

Table 2 Combined sites experimental results

Site	ACB	AnF	FDF	False−/Total	Pct. err (%)
ACB	12,882	563	520	1083/13,965	7.76
AnF	350	25,668	379	729/26,397	2.76
FDF	163	1433	33115	1596/34,711	4.60
False+	513	1996	899	3408	
Total	13,395	27,664	34014	75,073	
Pct. err (%)	3.83	7.22	2.64	4.54	

when training the sites separately. Notice also that the segment errors between the "accumulated" and "combined" rows change as well.

We also report the site-level confusion matrix (Table 2) for the combined network. It shows that the system can reliably pin a given test image to the correct site with only 4.54 % error (95.46 % classification). This is encouraging because the classifier can provide various levels of output. If the system is unsure of the actual segment location, it can at least rely on being at the right site.

6.2 Off-Line Metric Localization Testing

The system is tested at the same three sites as the above segment classification. Here the system localizes to a coordinate location the map. We also compare our system, which employs both local features (SIFT keypoints within salient regions and salient feature vector at the salient point) as well as global (gist) features, with two systems that use only salient regions and only gist features. The back-end Monte-Carlo localization modules for all three instances are identical. The SIFT-only system does not include the salient feature vector as a region signature, while the gist-only localization comparison may indicate how place recognition systems perform in metric localization task.

Because the data is recorded at approximately constant speed and with known start and end location, we can interpolate the ground-truth location assignments. Thus, the error is the measured distance (in meter) between the robot belief and the ground truth. That is, the longitudinal distance along the graph edges of the topological map.

Table 3 shows the results, with an overall error of 2.70 m. In general, the errors occurred because the identified salient regions are at the end of the segment and hardly change sizes even after a significant robot displacement. This is the case for the spike for FDF, as the system latches to a far away buildings (observe Fig. 13). Currently, the system uses the location of the matched database salient region as a hypothesis of where the robot currently is. Because the SIFT module can perform scale-invariant matching (with the scale ratio included as part of the result), the sys-

Table 3 Off-line localization estimation results

Segment	ACB		AnF		FDF		Total	
	Number frames	Error (m)	Number frames	Error (m)	Number frames	Error (m)	Number frames	Error (m)
1	1596	0.87	3137	2.36	3351	1.90	8084	1.88
2	1775	1.76	1846	3.05	3203	4.42	6823	3.36
3	1887	1.00	3773	2.93	3346	2.71	9006	2.44
4	1234	0.98	2405	2.28	3235	3.83	6874	2.78
5	1284	1.37	2938	2.39	3127	3.68	7349	2.76
6	2087	0.76	4224	2.50	6629	4.10	12,940	3.04
7	1681	0.68	1952	2.68	3711	3.28	7344	2.53
8	1099	0.83	2855	2.35	5008	3.38	8962	2.74
9	1323	0.59	3267	3.14	3101	3.29	7691	2.76
Total	13,966	0.98	26,397	2.63	34,711	3.46	75,073	2.70

tem limits the matching-scale threshold to between 2/3 and 3/2. This is not entirely effective as a scale ratio of 0.8 (the region found is smaller than the one matched in the database) can translate to a geographical difference of 5 m. When far away buildings are more salient, even if the robot moves toward them substantially, their size and appearance hardly change. Thus, although buildings are stable localization cues, they are not good for fine-grained location pin-pointing. The system would need regions that are closer (<3 m away).

Table 4 shows the comparison between SIFT-only (local features) system, gist-only (global features) system, and the presented bio-system, which uses both global and local features. The gist-only system cannot localize to the metric level because it can only pin-point location to the segment level. The SIFT-only system, on the other hand, performs almost as well as the presented system, although, there is a clear difference.

In the ACB site, the improvement is 42.53 %, from 1.72 m in SIFT-only to 0.98 m in the presented system, (one-sided t-test $t(27930) = -27.3134$, $p < 0.01$), while the AnF site is 18.65 %, from 3.23 to 2.63 m (one-sided t-test $t(52792) = -15.5403$, $p < 0.01$), and the FDF site is 23.74 % from 4.53 to 3.46 m (one-sided t-test $t(69420) = -32.3395$, $p < 0.01$). On several occasions, the SIFT-only system

Table 4 Model comparison experimental results

System	ACB	AnF	FDF	Total
	err. (m)	err. (m)	err. (m)	err. (m)
Gist	7.61	16.56	25.44	19.00
SIFT	1.71	3.22	4.53	3.55
Bio-system	0.98	2.63	3.46	2.71

completely misplaced the robot. In the presented system, whenever the salient region (SIFT and salient feature vector) matching is incorrect, the gist observation model is available to correct the mistakes. In contrast, the SIFT-only system can only make decision using one recognition module. Additionally, in kidnapped robot situations, where the robot is moved without informing the system (there 4 instances per run for ACB and AnF, and 5 for FDF), the presented system re-localizes faster because it receives many more observations (both global and local) than the SIFT-only system.

6.3 *Full Localization and Navigation Testing*

We construct three routes of 511.25, 615.48, and 401.16 m in length throughout the USC campus, traversing a wide variety of scenery. Beobot 2.0 logged over 3 h 17 min or 10.11 km of recorded experiment conducted within a 3 month winter period (December–February). The average robot speed during autonomous driving is 0.66 m/s (about half the average walking speed of people), with a maximum of 0.89 m/s. The difference is caused by a drop in battery charge as the experiment wears on. A video summary can be found in Siagian et al. (2013), while the system snapshot is displayed in Fig. 14.

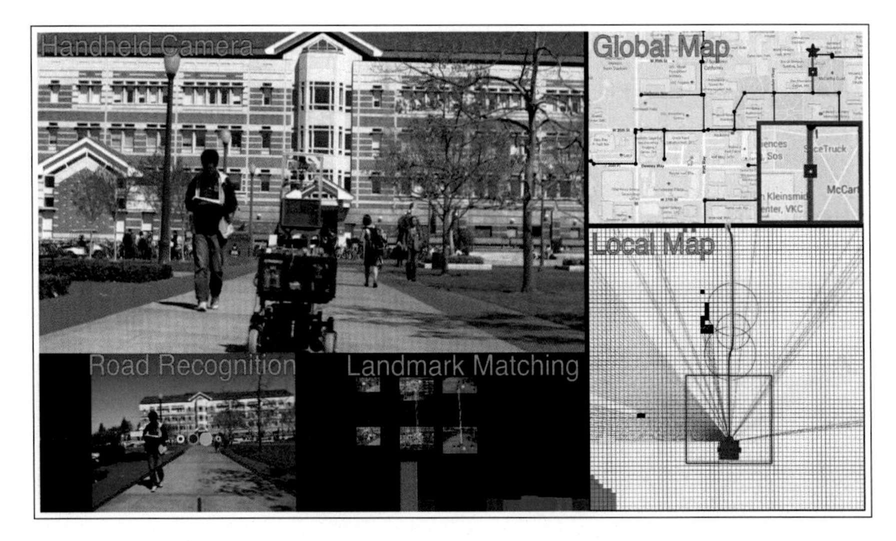

Fig. 14 A snapshot of the system test-run. *Top-left* (main) image displays the robot moving towards the goal. Below is the road recognition system which identifies the road heading (indicated by the *green disk*) and the robot's lateral position (*blue stub* on the *bottom* of the image). To the *right* are three salient landmark matches as well as the segment classification histogram below. The *top-right* of the figure shows the global topological map of the campus, with the current robot location indicated by a *red square*, and the goal by a *blue star*. It also draws the current segment in *pink*, matching the segment classification output underneath landmark matching. The *right bottom* of the figure displays the local grid occupancy map, which shows a robot path to avoid an approaching pedestrian

We selected routes that are fairly representative of the campus. They are also picked to vary the localization and navigation challenges and conditions, which are listed in Table 5. Each route consists of four segments, with intersections between them to test the system of which road to take. Scenes from each route are displayed on separate rows in Fig. 15.

Since the previous testing has shown that the system is robust in the presence of various lighting conditions, here the focus is on the effect of crowding on the system. Section 6.3.1 focuses on the localization system testing, while Sect. 6.3.2 is on navigation. In the former, the localization system is isolated by comparing its performance when the robot is manually driven versus fully autonomous. In addition, the robot is manually driven on a straight-line at different offsets from the center of the road to gauge how navigation error affects localization results. The navigation Sect. 6.3.2 evaluates how well the navigation system maintains the robot's lateral position throughout traversal, and how it is affected by obstacle avoidance.

6.3.1 Localization Testing

The localization accuracy is measured by comparing its output with manually obtained ground truth. The ground truth is recorded by sending signals to the system, indicating the true current robot location. The signals are sent at specific parts of the path and are spaced appropriately (as small as 3 m) to accurately record the robot's location throughout the experiment. The ground truth in between the signal points are interpolated because we assume the robot is running at approximately constant speed for the period. Note that these signals are only processed after testing, and not used during run-time. For these experiments, the system is manually initialized with a starting location. In Siagian and Itti (2009) it has been shown to be able to re-localize from kidnapped robot as well as random initialization. Here we decided not to do so because the system currently does not converge to the true location in a reasonable amount of time with 100 % certainty.

The localization box plots can be viewed in Fig. 16 for all three sites as well as the total, for both crowding condition and driving mode. Figure 17 shows an example run from the PROF route under low crowd condition for more detailed observation. To reduce the effect of dependence in the samples, the errors are averaged every 30 samples or within 1 s (localization output is 30 Hz) before creating the box plots. In addition, because the result distributions are not Gaussian, non-parametric statistical significance tests are used. Because of this, we do not test for interaction between effects, e.g. whether crowd level causes greater error for autonomous navigation than for manual driving.

The results show that the system is able to satisfactorily localize within 0.97 m (median) of the true location while driving autonomously and reaching goals of more than 400 m away. The main source of error, which is prevalent throughout all conditions, can be attributed to the scale granularity of the salient region matching. The disparity comes from a sustained spike in error that occurs when the robot is driving on a long segment and continually utilizes far away landmark at the end of the road.

Table 5 Experiment sites information

Site	Route length (m)	Pedestrian density (Person/m)	Weather/lighting condition frames	Speed (m/s)	Total training frames	Total testing frames	Total time (s)	Obstable avoidance time (s)
Library	511.25	0.19	4–6 p.m., partial sun	0.61	16,621	13,825	838.35	29.99
		0.97	12–2 p.m., sunny	0.44	16,688	34,979	1165.88	73.13
Professional	615.48	0.06	4–6 p.m., cloudy	0.73	21,488	20,000	840.79	0.00
		0.36	12–2 p.m., sunny	0.89	9702	20,773	692.37	46.10
Athletic	401.16	0.04	4–6 p.m., clear sky	0.77	19,022	11,017	522.21	13.67
		0.43	11 a.m.–6 p.m., sunny	0.68	9102	17,499	583.24	27.10

Fig. 15 Examples of scenes from the libraries (LIBR), professional schools (PROF), and athletic complex (ATHL) route, each displayed in its respective row. The figure also depicts various challenges that the robot encountered, such as pedestrians, service vehicles, and shadows. These images are adjusted to remove the *bluish tint* in the original images for better display

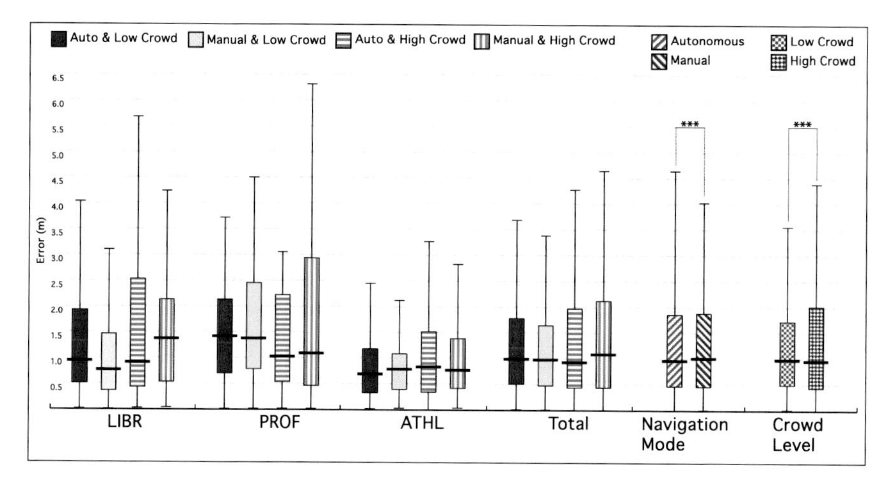

Fig. 16 Box plot of metric localization error from different routes, for both manual and autonomous navigation, and in both low and high crowd conditions. In addition, the figure also reports the total performance accuracies for each condition. The figure shows that the overall autonomous mode error median is 0.97 m, while medians of the various conditions remain low, from 0.75 to 1.5. Each box plot notes the median (the *thicker middle line*), the first and third quartile, and the end whiskers, which are either the minima/maxima or 1.5 * IQR (inter-quartile range). The *stars* above the *lines* joining various box plots note the statistical significance with *one star* indicates $p < 0.05$, *two stars* $p < 0.01$, and *three stars* $p < 0.005$

The system uses small salient regions of about one third the down scaled 160 by 120 input image. For landmarks that are near the robot, the small size regions are sufficient to pinpoint locations with high accuracy. However, far away landmarks can appear identical from a large range of viewing distances. An example is taken from the last segment of PROF route, where the system sustains errors above 3 m for more than 50 m (observe Fig. 17). As the robot is moving through a segment, the arch at

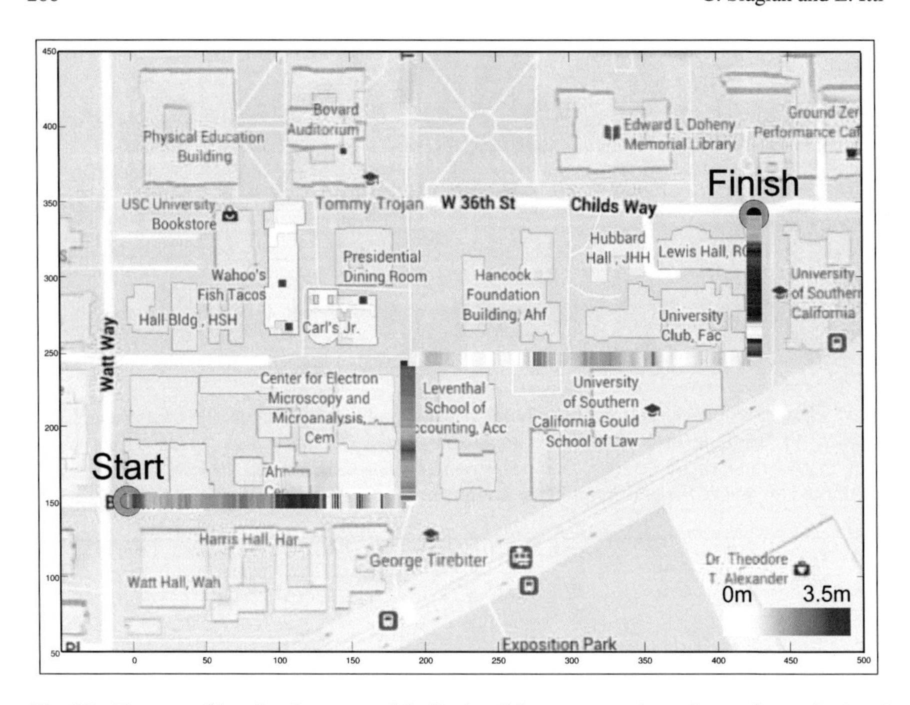

Fig. 17 Diagram of localization error while Beobot 2.0 autonomously navigates the professional schools (PROF) route during low crowd condition. The error is indicated by a gradation of *red color*, with the darker hue being higher error

the end of the corridor (shown in the last image of row two in Fig. 15) becomes very salient, rendering other parts less conspicuous by comparison.

The difference between low and high crowd conditions is statistically significant (Friedman test, $p < 0.005$) but with near equal medians (0.98 m low crowd vs. 0.96 m high crowd). Most times, the localization system is able to identify enough landmarks among the pedestrians just as well as when the crowd is much sparser. However, there are times when pedestrians do affect the system, particularly when they occlude the landmarks that are at ground level. These landmarks tend to be the ones that are closer to the robot. When this happens, the robot has to rely on landmarks that are higher above pedestrians. They are likely to be farther away, reducing the accuracy of the localization system, but would not completely misplaced the robot.

In addition, there is also a statistically significant localization performance difference between autonomous and manual-drive navigation (Friedman test, $p < 0.005$) with the autonomous drive median of 0.96 m being lower than manual driving median (1.00 m). This is quite surprising given that in autonomous drive, the robot is not always in the middle of the road due to an imperfect road recognition system. In manual drive, on the other hand, the robot moves steadily in the middle of the road, which is the same vantage point as during training.

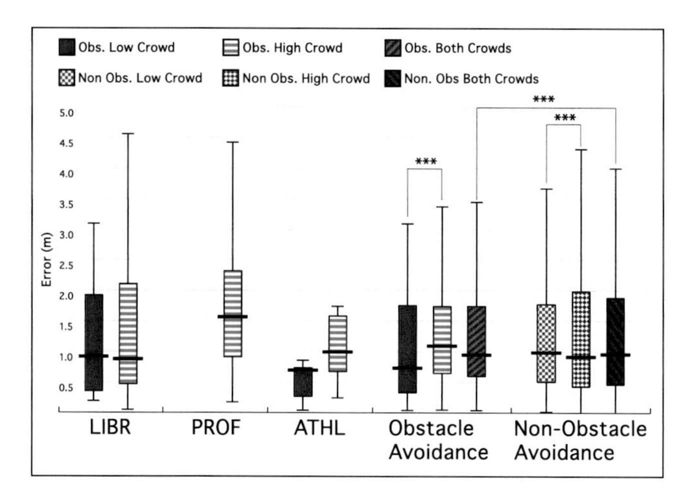

Fig. 18 Box plot of localization errors during obstacle avoidance on the different routes, and in both low and high crowd conditions. Note that, for the PROF route during low crowd, the robot did not execute a single obstacle avoidance maneuver

Another aspect to observe is during obstacle avoidance. In total, in our dataset there are 25 trajectory-changing obstacles that affect the robot's heading. This definition differentiates a select few from other nearby pedestrians seen in the camera but not close enough to warrant avoidance. Refer to Table 5 for the accumulated times that Beobot 2.0 spent avoiding obstacles, listed on the last column.

Figure 18 shows that there are statistically significant performance differences between low and high crowd conditions during obstacle avoidance maneuvers (Friedman Test, $p < 0.005$), between low and high crowd conditions during non-obstacle avoidance traversals (Friedman Test, $p < 0.005$), as well as between obstacle and non-obstacle avoidance traversals for both crowd levels combined (Friedman Test, $p < 0.005$). It should be pointed out that the median difference for the last comparison is only 1 cm, 0.95 m during obstacle avoidance versus 0.96 m during non-obstacle avoidance. This stability, which is also shown by the respective distribution shapes, demonstrates that the localization system can endure these significant but short perturbations. At times, an avoidance maneuver actually allows the robot to go back to the center of the road, improving the localization accuracy. In fact, on average, the localization error decreases from 1.39 to 1.09 m before and after avoidance.

For the most part, the amount of localization error carried by the system is manageable, especially in the middle of segments, where the task of the robot is mainly to proceed forward. However, the error becomes critical when the robot is trying to turn to the next segment at an intersection. Fortunately, considering that the robot is traveling on campus roads of 6–8 m in width, it never misses the road completely. Even when the robot needs to turn to narrower roads, it manages to do so because these roads tend to be in between buildings, allowing the LRF funnels the robot to the segment. An example is when the robot is entering the corridor at the last segment of PROF route. However, as a whole, turning is a weak point in the system that needs to be improved.

6.3.2 Navigation Testing

The accuracy of the navigation system is measured by its ability to maintain its position in the middle of all types of road. They range from clean roads with clear borders, roads with complex markings, to roads without clear boundaries. Navigation error is defined as the physical lateral distance of the robot center from the center of the road. The robot's lateral position is manually annotated when crossing certain points of the path. The trajectory in between is then reconstructed using odometry.

Figure 19 reports the result, which is divided by sites and crowding conditions. Overall, the system is able to stay within 1.37 m median of the center, which is acceptable given the challenges faced. However, clearly, there are a number of aspects that needs improving.

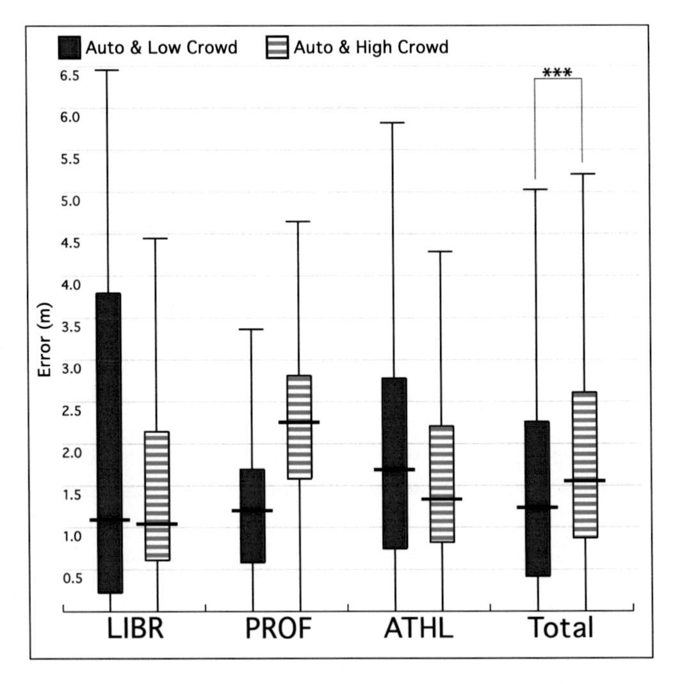

Fig. 19 Results of the navigation system in each site and under low and high crowding conditions. The error is measured as lateral deviation of the robot from the road center. The far *right bars* display the distributions of low versus high crowd performance, which is statistically significant (Friedman test, $p < 0.005$)

The data shows that there is a statistically significant difference between the low and high crowd conditions (Friedman test, $p < 0.005$), with the median of 1.23 m for low crowd versus 1.55 m for high. As mentioned in the previous section, excluding numerous slight changes of direction, there are only a low number (25 in total) of obvious obstacle avoidance situations. One reason is because the 7.2 m front extent of the local map allows the robot to register obstacles early enough to move away gracefully. In addition, the robot only encounters a small number of static obstacles because it is moving in the middle of the road. Also, because of the robot's appearance and relatively slower moving speed, pedestrians are aware of it approaching them. In some cases, the robot does have to avoid collision.

The starting point of an avoidance maneuver's window is when the robot first rapidly changes its direction, and ends when it stabilizes its lateral position back to the center. For some situations, however, the system error remains away from the center long after the maneuver is over. For example, during the LIBR test run in low crowd condition, the robot has to avoid two people, street signs, and a long bike rack in succession. Because the compound maneuver keeps the robot on the left flank of the road (3 m from the center) for so long, the system resets its current lateral position, thinking that the robot is at a new center of the road. Had the robot moved back to the center, as opposed to staying out on the flank for more than 70 m, the overall median would have been 0.87 m, lower than the actual 1.09 m error.

The road reset is added because the current road recognition system does not have the capability to recognize the true center robustly. Resetting the center whenever a new stable lateral position is established allows the system to react to changing boundary positions. If the system is able to estimate the road center, it would not only help for obstacle avoidance recovery, but also for turning at an intersection. Currently, the navigation system assumes that the original lateral position at the start of a segment is the center of the road.

Figure 20 shows navigation error during obstacle avoidance. There is statistical significance on all cases: navigation error during obstacle avoidance for low versus high crowd conditions (Friedman Test, $p < 0.005$), during non-obstacle avoidance traversals for low versus high crowd conditions (Friedman Test, $p < 0.005$), as well as between obstacle and non-obstacle avoidance traversals for both crowd conditions combined (Friedman Test, $p < 0.005$). Here the difference for the latter comparison is more pronounced, 1.58 m versus 1.36 m medians, respectively. In addition, the disparity is also supported by the average error increase between start and end of obstacle maneuver of 32 cm.

Even without obstacle avoidance, in the case of the PROF route during the low crowd condition (indicated by missing bar in Fig. 20), there is still error in the system. A major reason for this is because, in some areas, visual cues for road direction are sparse. An example is the water fountain area in the PROF route, where the road changes seamlessly to an open space before continuing to another road (observe the second image of the second row in Fig. 15). In the absence of these cues, and thus of a vanishing point, the robot simply maintains the last road IMU heading estimate while avoiding obstacles whenever needed. Despite this difficult stretch, however, the robot was still able to arrive at the target location, about 200 m further.

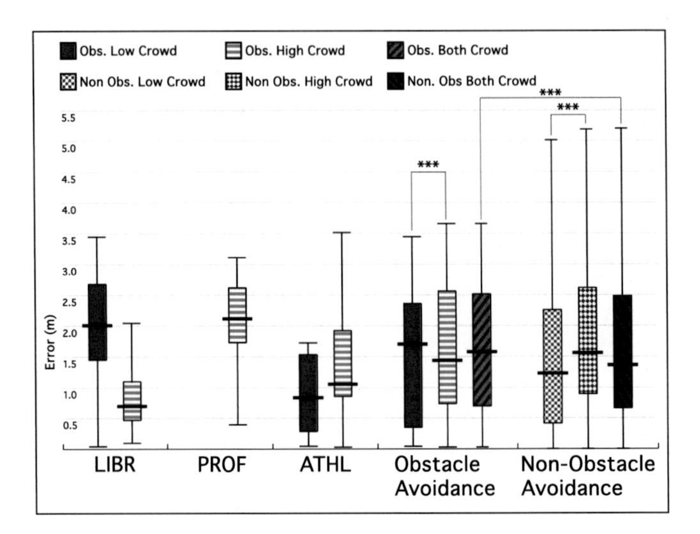

Fig. 20 Navigation errors during obstacle avoidance from different routes, and in both low and high crowd conditions. Note that the result for the PROF route during low crowd is missing because the robot did not execute a single obstacle avoidance maneuver during the time

7 Overall Performance Evaluation

We present a real-time biologically inspired visually-guided autonomous localization and navigation system. The system is ideal for service robot tasks in unconstrained pedestrian environments, with broad potential societal impact. We demonstrate the system's efficacy using a wheeled robot, and by driving it through challenging and long-range routes of at least 400 m at the University of Southern California (USC) campus. In a total of 10.11 km of documented testing, the robot shows its robustness in navigating among crowds, turning at intersections, and stopping at assigned destinations.

We also show that the gist features succeed in classifying a large set of images from 27 segments, despite its computational simplicity. The combined-sites experiment shows that the number of differentiable scene groupings can be quite high. Even if a few large objects are added or removed from the location, the overall gist of the scene is largely unaffected because these objects are small compared to the whole scene. The complementary gist and saliency features are also beneficial for mobile robot vision localization. As suggested by the study of human visual cortex, they are implemented in parallel using shared raw feature channels (color, intensity, orientation). These features are utilized by the biologically-inspired multi-level metric topological map localization. That is, the efficiently-computed gist features first provide segment classification information. Through the saliency model, the system then matches salient regions in the scene to refine the location to metric coordinate. The road recognition module keeps the robot on or near the middle of the road. By doing so the robot's viewing perspective is kept at a single general direction.

Consequently, the vision localization system does not have to be trained from other perspectives. Such requirements would put the recognition system past its limits as the landmark database would become prohibitively large if we were to include views of each landmark from all directions.

There are a number of localization issues to improve upon, such as not having to initialize the location of the system. This requires adjustments in the landmark recognition search parameters during uncertain times. Another extension is adding Simultaneous Localization and Mapping (SLAM) to create a metric topological map. There are a number of SLAM algorithms that aim to create such hierarchical map. A system by Marder-Eppstein et al. (2011) connects local overlapping LRF-based occupancy maps to a single global topological map. Similarly Pradeep et al. (2010) constructs a topological map by associating stereo-based local submaps. Note that these systems have only been tested indoors, and adaptation to outdoor environments may not be trivial. We are also particularly interested in creating a map with meaningful nodes that represent important geographical points, similar to Ranganathan and Dellaert (2011), which identifies unique LRF signatures produced by junctions in indoor environments. The key is to robustly do the same outdoors, where the presence of surrounding walls cannot be assumed. One way is by using online gist feature grouping technique Valgren and Lilienthal (2008), where large changes in feature space correspond to moving into a new segment.

The road recognition module needs to recognize the true road center or, at least, avoid traveling on the side of the road. This can be done by recognizing the road boundaries using changes in visual appearance and/or proximity profile. Another way is by using the visual landmarks, which carries image coordinates that can be used to adjust the target lateral position. This technique is also important for turning more properly in the intersection, where often times the robot does not start a segment in the middle of the road after a turn. There are also situations where there is no road to follow, for example, in open areas such as a square or a large hall. In this case the system simply maintains its current forward heading set by the last seen road, while trying to avoid obstacles. A better solution is for the robot to have a richer understanding of the space to navigate in more complex area.

The robot may also fail to detect obstacles in its path, particularly, thin objects (e.g., poles, fences) and objects with holes (bicycle approached at certain angles). There are also occasions where the planar LRF is placed too high above the ground to detect shorter or protruding object parts such as the seat of a bench. However, we find that there is no correct LRF height because, if it is placed low, the robot may attempt to drive underneath an obstacle and hit it, but if it is placed too high, the robot may miss short obstacles (a gimbal-mounted scanning laser may be necessary to address this problem). Furthermore, the system is currently unable to properly treat negative obstacles such as a ditch or hole, making it unsafe to navigate on roads with hazardous boundaries such as sidewalks, ramps, and cliffs. We are, however, encouraged that the robot is able to localize and navigate through crowded environments. A point of improvement would be to move more naturally among people

Trautman and Krause (2010), for example using a GPU-based pedestrian recognition Benenson et al. (2012), or characterizing how people move as a group Henry et al. (2010).

8 Discussions and Conclusions

Mobile robot localization and navigation has been a great fit for interdisciplinary collaboration between robotics and neuroscience. There are a number of biological inspirations that help advance the field, from low level visual stimuli processing to high level architecture. However, an emphasis in mobile robot research that we also found in biological systems is a need to balance latency and accuracy. We found this in the usage of gist and saliency in performing coarse-to-fine localization. Moving forward, the two sub-problems to focus on are creating a more complete recognition process and representation of space.

For the former, a robust recognition module must make use of a variety of rich low-level features from many domains to recognize a target object under adverse conditions. For example, the current road recognition only utilizes road contours, which do not describe of the road completely. The module should also look for consistent color and pattern themes within the road region, such as using histogram to differentiate the road from its flanking areas Chang et al. (2012). In addition, contextual priors from other modules can help pare down possibilities and improve the odds of obtaining a correct result. We can utilize a human (or pedestrian) recognition module to locate where people are and to identify appropriate road patches for travel.

The second direction of research is improving the spatial representation, especially for more complex paths or spaces Beeson et al. (2010), Kuipers et al. (2004), such as an open area. Open areas such as a plaza present a challenge in terms of constructing a topological map because the robot is free to move in any direction. In this case, a grid map would be more appropriate. However, if there are points of interest within the area like a statue or a bench, an overlapping grid and topological representation may actually be needed. Open areas also pose a problem for road recognition because the needed road markings may not be present. In this case navigation has to be able to identify a target object (e.g., the statue in the plaza) or path to exit, and drive straight towards it. This is an issue for sizable areas, where the targets are not visible from all parts of the area. In this case more cues must be incorporated to the map. What would be ideal is for the robot to be able to decipher the global function of the space, and create a more appropriate map where all the pertinent target locations are included.

Acknowledgments This work was supported by the National Science Foundation (grant numbers CCF-1317433 and CNS-1545089), the Army Research Office (W911NF-12-1-0433), and the Office of Naval Research (N00014-13-1-0563). The authors affirm that the views expressed herein are solely their own, and do not represent the views of the United States government or any agency thereof.

References

Achanta R, Hemami S, Estrada F, Ssstrunk S (2009) Frequency-tuned salient region detection. In: IEEE international conference on computer vision and pattern recognition (CVPR)

Ackerman C, Itti L (2005) Robot steering with spectral image information. IEEE Trans Robot 21:247–251

American Honda Motor Co Inc. (2009). Asimo—the world's most advanced humanoid robot. http://asimo.honda.com/. Accessed 15 July 2009

Bay H, Tuytelaars T, Gool LV (2006) Surf: speeded up robust features. In: Proceedings of European conference on computer vision (ECCV), pp 404–417

Beeson P, Modayil J, Kuipers B (2010) Factoring the mapping problem: mobile robot map-building in the hybrid spatial semantic hierarchy. Int J Robot Res 29:428–459

Benenson R, Mathias M, Timofte R, Gool LV (2012) Pedestrian detection at 100 frames per second. In: Proceedings of IEEE computer society conference on computer vision and pattern recognition (CVPR). IEEE Computer Society, Providence, RI, USA, pp 290–2910

Biederman I (1982) Do background depth gradients facilitate object identification? Perception 10:573–578

Bruce N, Tsotsos J (2006) Saliency based on information maximization. In: Weiss Y, Scholkopf JPB (eds) Advances in neural information processing systems, vol 18. MIT Press, Cambridge, MA, USA, pp 155–162

Blanco JL, Gonzalez J, Fernndez-Madrigal JA (2006) Consistent observation grouping for generating metric- topological maps that improves robot localization*. In: ICRA. Barcelona, Spain

Chang CK, Siagian C, Itti L (2012) Mobile robot monocular vision navigation based on road region and boundary estimation. In: Proceedings of IEEE/RSJ international conference on intelligent robots and systems (IROS), pp 1043–1050

Chubb C, Sperling G (1988) Drift-balanced random stimuli: a general basis for studying non-Fourier motion perception. J Opt Soc Am 5:1986–2007

Dalal N, Triggs B (2005) Histograms of oriented gradients for human detection. In: CVPR

Epstein R, Stanley D, Harris A, Kanwisher N (2000) The parahippocampal place area: perception, encoding, or memory retrieval? Neuron 23:115–125

Frintrop S, Jensfelt P, Christensen H (2006) Attention landmark selection for visual slam. In: IROS. Beijing

Fox D, Burgard W, Thrun S (1997) The dynamic window approach to collision avoidance. IEEE Robot Autom Mag 4:23–33

Fox D, Burgard W, Dellaert F, Thrun S (1999) Monte carlo localization: efficient position estimation for mobile robots. In: Proceedings of sixteenth national conference on artificial intelligence (AAAI'99)

Gao D, Mahadevan V, Vasconcelos N (2008) On the plausibility of the discriminant center-surround hypothesis for visual saliency. J Vis 8:2301–2311

Goncalves L, Bernardo ED, Benson D, Svedman M, Ostrowski J et al (2005) A visual front-end for simultaneous localization and mapping. In: ICRA, pp 44–49

Henry P, Vollmer C, Ferris B, Fox D (2010) Learning to navigate through crowded environments. In: Proceedings of IEEE international conference on robotics and automation (ICRA), pp 981–986

Hyvrinen A (1999) Fast and robust fixed-point algorithms for independent component analysis. IEEE Trans Neural Netw 10:626–634
Itti L (2012) iLab Neuromorphic Vision C++ Toolkit (iNVT). http://ilab.usc.edu/toolkit/. Accessed 15 Dec 2012
Itti L, Koch C, Niebur E (1998) A model of saliency-based visual attention for rapid scene analysis. IEEE Trans Pattern Anal Mach Intell 20:1254–1259
Itti L, Dhavale N, Pighin F (2003) Realistic avatar eye and head animation using a neurobiological model of visual attention. In: Bosacchi B, Fogel DB, Bezdek JC (eds) Proceedings of SPIE 48th annual international symposium on optical science and technology, vol 5200. SPIE Press, Bellingham, WA, pp 64–78
Itti L, Koch C (2001) Computational modelling of visual attention. Nat Rev Neurosci 2:194–203
Katsura H, Miura J, Hild M, Shirai Y (2003) A view-based outdoor navigation using object recognition robust to changes of weather and seasons. In: IROS. Las Vegas, NV
Kuipers B (2008) An intellectual history of the spatial semantic hierarchy. In: Jefferies M, Yeap AWK (eds) Robot and cognitive approaches to spatial mapping, vol 99. Springer, pp 21–71
Kuipers B, Modayil J, Beeson P, Macmahon M, Savelli F (2004) Local metrical and global topological maps in the hybrid spatial semantic hierarchy. In: Proceedings of IEEE international conference on robotics and automation (ICRA), pp 4845–4851
Liu T, Yuan Z, Sun J, Wang J, Zheng N et al (2011) Learning to detect a salient object. IEEE Trans Pattern Anal Mach Intell 33:353–367
Li F, VanRullen R, Koch C, Perona P (2002) Rapid natural scene categorization in the near absence of attention. In: Proceedings of National Academy of Science, pp 8378–8383
Lowe D (2004) Distinctive image features from scale-invariant keypoints. Int J Comput Vis 60:91–110
Macé MJ, Thorpe SJ, Fabre-Thorpe M (2005) Rapid categorization of achromatic natural scenes: how robust at very low contrasts? Eur J Neurosci 21:2007–2018
Marat S, Phuoc TH, Granjon L, Guyader N, Pellerin D et al (2009) Modelling spatio-temporal saliency to predict gaze direction for short videos. Int J Comput Vis 82:231–243
Marder-Eppstein E, Berger E, Foote T, Gerkey B, Konolige K (2010) The office marathon: robust navigation in an indoor office environment. In: Proceedings of IEEE international conference on robotics and automation (ICRA), pp 300–307
Marder-Eppstein E, Berger E, Foote T, Gerkey B, Konolige K (2011) Kurt Konolige, Eitan Marder-Eppstein, Bhaskara Marthi. In: Proceedings of IEEE international conference on robotics and automation (ICRA), pp 3041–3047
Matsumoto Y, Inaba M, Inoue H (2000) View-based approach to robot navigation. In: IEEE-IROS, pp 1702–1708
McNamara TP (1991) Memory's view of space. In: Bower GH (ed) The psychology of learning and motivation: advances in research and theory, vol 27. Academic Press, pp 147–186
Milford M, Wyeth G (2008) Mapping a suburb with a single camera using a biologically inspired slam system. IEEE Trans Robot 24:1038–1053
Milford M, Wyeth G (2010) Persistent navigation and mapping using a biologically inspired slam system. Int J Robot Res (IJRR) 29:1131–1153
Montemerlo M, Becker J, Bhat S, Dahlkamp H, Dolgov D et al (2008) Junior: the stanford entry in the urban challenge. J Field Robot 25:569–597
Montemerlo M, Thrun S, Koller D, Wegbreit B (2002) Fastslam: a factored solution to the simultaneous localization and mapping problem. In: AAAI
Moravec H, Elfes A (1985) High resolution maps from wide angle sonar. In: Proceedings of IEEE international conference on robotics and automation (ICRA), vol 2, pp 116–121
Muralidharan K, Vasconcelos N (2006) A biologically plausible network for the computation of orientation dominance. In: Weiss Y, Scholkopf JPB (eds) Advances in neural information processing systems (NIPS), vol 18. MIT Press, Cambridge, MA, USA, pp 155–162
Muralidharan K, Vasconcelos N (2010) On the connections between sift and biological vision. Front Syst Neurosci

Murrieta-Cid R, Parra C, Devy M (2002) Visual navigation in natural environments: from range and color data to a landmark-based model. Auton Robots 13:143–168
Oliva A, Schyns P (1997) Coarse blobs or fine edges? evidence that information diagnosticity changes the perception of complex visual stimuli. Cogn Psychol 34:72–107
Oliva A, Schyns P (2000) Colored diagnostic blobs mediate scene recognition. Cogn Psychol 41:176–210
Oliva A, Torralba A (2001) Modeling the shape of the scene: a holistic representation of the spatial envelope. Int J Comput Vis 42:145–175
Potter MC (1975) Meaning in visual search. Science 187:965–966
Pradeep V, Medioni G, Weiland J (2010) Robot vision for the visually impaired. In: Proceedings of IEEE computer society conference on computer vision and pattern recognition (CVPR). IEEE Computer Society, pp 15–22
Quinlan S, Khatib O (1993) Elastic bands: connecting path planning and control. In: Proceedings of IEEE international conference on robotics and automation (ICRA), pp 802–807
Rahtu E, Kannala J, Salo M, Heikkil J (2010) Segmenting salient objects from images and videos. In: ECCV, pp 366–379
Ramisa A, Tapus A, de Mantaras RL, Toledo R (2008) Mobile robot localization using panoramic vision and combination of local feature region detectors. In: ICRA. Pasadena, CA, pp 538–543
Ranganathan A, Dellaert F (2011) Online probabilistic topological mapping. In: Int J Robot Res (IJRR) 30:755–771
Renniger L, Malik J (2004) When is scene identification just texture recognition? Vis Res 44:2301–2311
Rensink RA (2000) The dynamic representation of scenes. Vis Cogn 7:17–42
Rosin PL (2009) A simple method for detecting salient regions. Pattern Recogn 42:2363–2371
Rutishauser U, Walther D, Koch C, Perona P (2004) Is bottom-up attention useful for object recognition? In: CVPR, vol 2, pp 37–44
Sanocki T, Epstein W (1997) Priming spatial layout of scenes. Psychol Sci 8:374–378
Se S, Lowe DG, Little JJ (2005) Vision-based global localization and mapping for mobile robots. IEEE Trans Robot 21:364–375
Serre T, Wolf L, Bileschi S, Riesenhuber M, Poggio T (2007) Robust object recognition with cortex-like mechanisms. IEEE Trans Pattern Anal Mach Intell 29:411–426
Siagian C, Chang CK, Voorhies R, Itti L (2011) Beobot 2.0: cluster architecture for mobile robotics. J Field Robot 28:278–302
Siagian C, Chang CK, Itti L (2013) Beobot 2.0. http://ilab.usc.edu/beobot2. Accessed 15 Dec 2012
Siagian C, Chang C, Itti L (2013) Mobile robot navigation system in outdoor pedestrian environment using vision-based road recognition. In: Proceedings of IEEE international conference on robotics and automation (ICRA). Both first authors contributed equally
Siagian C, Itti L (2007) Biologically-inspired robotics vision Monte-Carlo localization in the outdoor environment. In: Proceedings of IEEE/RSJ international conference on intelligent robots and systems (IROS)
Siagian C, Itti L (2008) Storing and recalling information for vision localization. In: IEEE international conference on robotics and automation (ICRA). Pasadena, California, pp 1848–1855
Siagian C, Itti L (2007) Rapid biologically-inspired scene classification using features shared with visual attention. IEEE Trans Pattern Anal Mach Intell 29:300–312
Siagian C, Itti L (2009) Biologically inspired mobile robot vision localization. IEEE Trans Robot 25:861–873
Thorpe S, Fize D, Marlot C (1995) Speed of processing in the human visual system. Nature 381:520–522
Thrun S (2011) Google's driverless car. Talk was viewed at http://www.ted.com/talks/sebastian_thrun_google_s_driverless_car.html. Accessed 1 Sept 2012
Thrun S (1998) Learning metric-topological maps for indoor mobile robot navigation. Artif Intell 99:21–71

Thrun S, Bennewitz M, Burgard W, Cremers A, Dellaert F et al (1999) MINERVA: a second generation mobile tour-guide robot. In: Proceedings of IEEE international conference on robotics and automation (ICRA)

Torralba A (2003) Modeling global scene factors in attention. J Opt Soc Am 20:1407–1418

Torralba A, Murphy KP, Freeman WT, Rubin MA (2003) Context-based vision system for place and object recognition. In: Proceedings of international conference on computer vision (ICCV). Nice, France, pp 1023–1029

Trautman P, Krause A (2010) Unfreezing the robot: navigation in dense, interacting crowds. In: Proceedings of IEEE international conference on intelligent robots and systems (IROS), pp 797–803

Treisman A, Gelade G (1980) A feature-integration theory of attention. Cogn Psychol 12:97–137

Turner RS (1994) In the eye's mind: vision and the Helmholtz-Hering controversy. Princeton University Press

Tversky B (2003) Navigating by mind and by body. In: Spatial cognition, pp 1–10

Tversky B, Hemenway K (1983) Categories of the environmental scenes. Cogn Psychol 15:121–149

Ulrich I, Nourbakhsh I (2000) Appearance-based place recognition for topological localization. In: Proceedings of IEEE international conference on robotics and automation (ICRA), pp 1023–1029

Ungerleider LG, Mishkin M (1982) Two cortical visual systems. In: Ingle DJ, Goodale MA, Mansfield RJW (eds) Analysis of visual behavior. MIT Press, Cambridge, MA, pp 549–586

Valgren C, Lilienthal AJ (2008) Incremental spectral clustering and seasons: Appearance-based localization in outdoor environments. In: Proceedings of IEEE international conference on robotics and automation (ICRA). Pasadena, CA, pp 1856–1861

Walther D, Koch C (2006) Modeling attention to salient proto-objects. Neural Netw 19:1395–1407

Wang J, Zha H, Cipolla R (2006) Coarse-to-fine vision-based localization by indexing scale-invariant features. IEEE Trans Syst Man Cybern 36:413–422

Willow Garage (2009) PR-2—Wiki. http://pr.willowgarage.com/wiki/PR-2. Accessed 15 July 2009

Wolfe J (1994) Guided search 2.0: a revised model of visual search. Psychon Bull Rev 1:202–238

Zhang L, Tong MH, Marks TK, Shan H, Cottrell GW (2008) Sun: a Bayesian framework for saliency using natural statistics. J Vis 8:231–243

Zhang W, Kosecka J (2005) Localization based on building recognition. In: IEEE workshop on applications for visually impaired, pp 21–28

Attention and Cognition: Principles to Guide Modeling

John K. Tsotsos

Abstract Interest in the modeling of visual attention and cognition is strong with the number of models growing quickly. It thus becomes important to try to consolidate what all this activity has demonstrated in terms of what principles may be abstracted from the collective experience that can guide future research. This is not a straightforward task; many have tried and have little to show for it. Here, a different view is presented, one that attempts to combine multiple perspectives on the problem. The novelty is that in contrast with the vast majority of past work, there is an explicit assertion that no single principle can capture the complexities of human attentional and cognitive behavior. There are several principles, each defined in a particular context, with interactions among them. Many previous authors have stated principles that in fact are more correctly considered as modeling philosophies or requirements and these will be so distinguished. The development of a model of human visual cognition is dependent on the choice of which experimental observations act as constraints during its development (Tsotsos 2014). Those constraints provide a means to select solutions among those potential ones that satisfy the principles. Here, we begin by proposing a set of elements that may be considered as the components of attention, without any claims or completeness or optimality, and these will act as the first level set of constraints on any modeling activity. A look at specific models of attention and cognitive architectures will reveal a variety of principles, philosophies and requirements that have been shown to be important. This will be followed by the introduction of the Survival Requirement as a replacement for any over-arching principle of optimality because there still is a need for a selection criterion for choosing among competing solutions. The presentation will conclude by considering how progress on open problems in neuroscience may be facilitated by considering this list of principles, philosophies and requirements.

J.K. Tsotsos (✉)
Department of Electrical Engineering and Computer Science, York University, 4700 Keele St., Toronto, ON M3J1P3, Canada
e-mail: tsotsos@cse.yorku.ca

Q. Zhao (ed.), *Computational and Cognitive Neuroscience of Vision*,
Cognitive Science and Technology, DOI 10.1007/978-981-10-0213-7_12

1 Introduction

Any quest to specify the biological and computational principles that underlie human attentional and cognitive behaviors is not only an enormous and poorly defined adventure, but also must navigate a path that is littered with failed past attempts and blind alleys. The need to understand—and the potential impact of an understanding—keeps us optimistic, and thus here, we present another attempt.

Since this is a paper about principles for modeling, perhaps the best place to begin is to define what is meant by a *principle*. Principles describe the fundamental nature of something or universal properties and relationships between things. Principles explain the *why* and *how* of various phenomena and can be tested via the scientific method. A principle is a truth, and like a law, it describes a relationship through which a system's contents yield results under given conditions. A principle is expressed as a concise verbal statement, as a mathematical formulation, as a graphical depiction of a set of processes and communications, and sometimes using the broadest of methods of expression at our disposal, the language of computation.[1] It must always apply under the same conditions, and implies a causal relationship among its elements. Principles must be confirmed and broadly agreed upon through the process of inductive reasoning, a kind of reasoning that constructs or evaluates general propositions that are derived from specific examples. A statement of a modeling philosophy—such as a desire to develop models that have biological realism—is not a principle because it makes no falsifiable predictions about the domain of interest, namely, biology. Although not always straightforward, modeling philosophies will be distinguished from modeling principles. There is yet another type of statement, that is neither a principle nor a philosophy, that is relevant here. A principle states a truth that must underlie modeling and must be included within the model foundation. A philosophy states a belief whose application should yield positive benefits to a modeling enterprise impacting how design choices are made. A requirement states a specification that must be achieved by the modeling activity. Several examples of each will be presented.

This presentation will attempt to summarize a broad array of basic principles as they apply to attention and cognition within vision, specifically human vision. Since the title uses the word principle in the plural, the reader may suspect that more than one principle will be presented. If we took our cue from modern physics, perhaps the search for a unifying single principle should be our goal. Some have indeed taken up this challenge. For example, Friston's free energy principle claims this

[1]Computer science, broadly defined, is the theory and practice of representing, processing, and using information and encompasses a body of knowledge concerning algorithms, communication, languages, software, and information systems. In a nice paper, Peter Denning (Denning 2007) claimed that it offers a powerful foundation for modeling complex phenomena such as cognition. The language of computation is the best language we have to date, he claims, for describing how information is encoded, stored, manipulated, and used by natural as well as synthetic systems. As such, the language of computation subsumes mathematics and formal logic which are nevertheless critical tools for expressing appropriate elements of theories.

characteristic (Friston 2010) in that it purports to unify action, perception and learning. Free energy is optimized in this view and can be formulated as energy minus entropy or alternatively, as surprise plus a divergence term. An alternative formulation is to set up a Multi-Objective Optimization Problem (MOOP) whose goal is to find a vector of decision variables that satisfy constraints and optimize a vector function whose elements represent the objective functions. These functions form a mathematical description of performance criteria that are usually in conflict with each other. Hence, the term 'optimize' means finding such a solution which would give values of the objective functions that are acceptable to the decision maker. For example, in aircraft design, there are many objectives, such as the desire to maximize range, passenger volume, payload mass and cruise speed while minimizing fuel consumption and costs, among others. They key to such formulations is the availability of the complete set of performance criteria, objective functions, acceptability values, and objective functions. Although for man-made artifacts this may be possible, our understanding of human brain function at this point in time does not lend itself easily to this requirement.

In Tsotsos (1990, 2011) an analysis of visual processing was presented with respect to its computational complexity with the conclusion that the relationship between processing time and size of input has an inherent exponential nature. The specifics of this relationship point to the fact that no optimal solution exists save for the evolution of a much larger brain. The reason that no single solution exists has to do with two main elements. First, it is formally provable that there are many sub-problems that human perception—and thus intelligence—must deal with that are computationally intractable. This means that no optimal solution exists in any realization, and specifically, not for the human brain. The second element concerns the multivariate nature of the problem. In vision there are obvious variables—pixels and features—yet in the realization of visual processing in the brain there are many more dealing with details such as neural packaging, heat management, and nutrient and waste transport, to more computational issues such as memory, connectivity, transmission speed, and so on. No single mechanism could possibly optimize all of this effectively. In any case, as is well known in all engineering tasks, simultaneously optimizing many interrelated variables can only be dealt with by defining sufficient criteria for a solution. Optimizing one variable often leads to suboptimal values for another, and as a result a globally satisficing solution is what is sought in practice. This is what was referred to in the previous paragraph with respect to multi-objective optimization problems. Even if we restrict consideration to visual processes, it is very difficult to be precise about the nature of the objective function for this optimization, although it is clear that minimizing processing time, minimizing neural cost, and maximizing correctness of perception are primary ingredients.

It is similarly unlikely that there exists a single unifying principle considering the task of modeling attention and cognition. Formal optimality concerns seem to just not be on the right pathway because of their mathematical requirements such as control over time to convergence or appropriate structure of solution space, let alone the aforementioned computational complexity issues. This is clearly not a proof of this paragraph's opening sentence. However, since the optimization route has been examined previously, perhaps a different approach is needed and that is what is presented here.

The steps we will take in developing principles will first involve describing what the elements of attentive processing might be. This tentative set is the target of modelers, to discover the overall computational architecture that can explain all of the elements in the context of how they might assist an agent in the performance of specific tasks. Next, some of the principles that have been employed to date to realize attentive architectures will be discussed. Principles that apply to the modeling of cognition should have some overlap with those for attention and these are explored next. This is followed by the introduction of the Survival Requirement as a replacement for any over-arching principle of optimality because there still is a need for a selection criterion for choosing among competing solution. The presentation will conclude by considering how progress on open problems in neuroscience may be facilitated by the principles, philosophies and requirements described in the paper.

2 Biological Elements of Attention

Before beginning the process of defining modeling principles, it is useful to know exactly what is being modeled. With respect to visual attention, there is unfortunately little agreement. The variety of theories and models of visual attention and the number of relevant experimental papers that have been presented over the course of the past few decades are numbing to say the least. It would not be unreasonable to think that any attempt at complete and fair acknowledgment of all this research is doomed to failure, if for no other reason than space limits.

Classic views of visual attention present a 2-stage process in that there is a pre-attentive stage followed by an attentive stage (e.g. Treisman and Gelade 1980). At the other extreme, there is the view of Tsotsos (2011) where attention is a set of many mechanisms within the categories of suppression, selection and restriction. Survey papers or volumes, of which there are many good ones (Pashler 1998; Itti et al. 2005; Carrasco 2011; Nobre and Kastner 2013, for example) do not really help much in detailing what exactly attention may be in a way that it is amenable to theory and model. But, to make this paper somewhat concrete, an attempt must be made to specify exactly what an attention model should encompass. Currently, there is no model that covers all of the observed characteristics of attention. The following list—acknowledged up front as most probably incomplete if not also inadequate—presents the most acknowledged characteristics of attentional processing (for the list of seminal citations for each see Tsotsos 2011):

Alerting—The ability to process, identify and move attention to priority signals.
Attentional Footprint—Optical metaphors describe the 'footprint' of attentional fixation in image space and the main ones are *Spotlight, Zoom Lens*, *Gradient,* and *Suppressive Surround.*

Binding—The process by which visual features are correctly combined to provide a unified representation of an object.
Covert Attention—Attention to a stimulus in the visual field without eye movements.
Disengage Attention—The generation of the signals that release attention from one focus and prepare for a shift in attention.
Endogenous Influences—Endogenous influence is an internally generated signal used for directing attention. This includes domain knowledge or task instructions.
Engage Attention—The actions needed to fixate a stimulus whether covertly or overtly.
Executive Control—The system that coordinates the elements into a coherent unit that responds correctly to task and environmental demands.
Exogenous Influences—Exogenous influence is due to an external stimulus and contributes to control of gaze direction in a reflexive manner. Most common is perhaps the influence of abrupt onsets.
Inhibition of Return—A bias against returning attention to a previously attended location or object.
Neural Modulation—Attention changes baseline firing rates as well as firing patterns of neurons for attended stimuli.
Overt Attention—Also known as *Orienting*—the action of orienting the body, head and eyes to foveate a stimulus in the 3D world. Overt fixation trajectories may be influenced by covert fixations.
Post-Attention—The process that creates the representation of an attended item that persists after attention is moved away from it.
Pre-attentive Features—The extraction of visual features from stimulus patterns perhaps biased by task demands.
Priming—Priming is the general process by which task instructions or world knowledge prepares the visual system for input. *Cueing* is an instance of priming; perception is speeded with a correct cue, whether by location, feature or complete stimulus. Purposefully ignoring has relevance here also and is termed *Negative Priming*. If one ignores a stimulus, processing of that ignored stimulus shortly afterwards is impaired.
Recognition—The process of interpreting an attended stimulus, facilitated by attention.
Salience/Conspicuity—The overall contrast of the stimulus at a particular location with respect to its surround.
Search—The process that selects the candidate stimuli for detection or other tasks out of the many possible locations and features in cluttered scenes.
Selection—The process of choosing one element of the stimulus over the remainder. Selection can be over locations, over features, for objects, over time, and for behavioral responses, or even combinations of these.
Shift Attention—The actions involved in moving an attentional fixation from its current to its new point of fixation.

Time Course—The effects of attention take time to appear and this is observed in the firing rate patterns of neurons and in behavioral experiments, showing delays as well as cyclic patterns.
Update Fixation History—The process by which the system keeps track of what has been seen and processed, and how that representation is maintained and updated, that then participates in decisions of what to fixate and when.

Each of these elements is supported by a rich set of experimental evidence and it is the sum of all this evidence that ideally acts as constraints for the ultimate theory or model of visual attention, in the sense referred to by Tsotsos (2014). Tsotsos, following Zucker (1981), wrote that computational models of perception have two essential components, representational languages for describing information and mechanisms that manipulate those representations. The many diverse kinds of constraints on theories discovered by experimental work, whether psychophysical, neurophysiologic, or theoretical, impact the representations and their manipulation mechanisms. All of these different kinds of constraints are needed, or the likelihood of discovering the correct explanation is seriously diminished. In other words, careful understanding of one's problem, with potential solutions circumscribed by the appropriate set of constraints—from psychophysics, neurophysiology, computation, and theory—is a critical precursor to success.

The state-of-the-art is quite far from achieving such an all-encompassing theory of attention however, not in small part due to the fact that large integrative models of attention are few. One hypothesis for an architecture that hopes to integrate all of these elements of attention within an attention executive and task-driven control structure is the STAR model of Tsotsos and Kruijne (2014). This is based on the Selective Tuning model of attention (Tsotsos 2011; Rothenstein and Tsotsos 2014) and extended to include an attention executive, working memory, task guidance components, and more. Parts of these new elements derive inspiration from Ullman's Visual Routines (Ullman 1984) for some aspects of task-based control. Visual routines are updated and renamed Cognitive Programs (CP's), and take a form and function more consistent with our modern understanding of biological vision and attentive processes. Basically, CP's represent a sequence of transformations from one representation to another in service of particular exogenous tasks. They include not only transformations but also decision stages that are required to choose transformations or actions, draw conclusions and monitor task execution progress. A specification of a CP will involve specific variables, representation of retinotopic arrays of perceptual information, attentional and task required control signals and their settings, and more. Another nice example of an integrative model is the work of Beuth and Hamker (2015) where they focus on explaining single cell recordings in multiple brain areas and present computational circuits of attention involved in spatial- and feature-based biased competition, modulation of the contrast response function, modulation of the neuronal tuning curve, and modulation of surround suppression. Such integrative efforts will play an increasingly larger role in the future towards elucidating our understanding of attentional processing.

3 Principles for Modeling Attention

Now that we have at least a first order description of what aspects of attention need to be modeled, we can begin to ask the question about principles of attention, that is, do the characteristics enumerated in the previous section arise as a result of some small set of basic principles? The massive literature on attention reveals its most enduring characterization, that of capacity limit or bottleneck. In other words, there is too much to process and the system cannot handle it, thus limitations appear. Thus, the first principle that acts as a foundation for modeling attention is that of reducing information load. Due to the nature of the published literature, a review will not appear here. In any case, there exist many published reviews (for example, see Tsotsos and Rothenstein (2011), Frintrop et al. (2010), Kimura et al. (2013), Borji and Itti (2013), Bruce et al. (2015), Bylinskii et al. (2015), which in total review over 140 distinct models). Both the Borji and Itti and the Bylinskii et al. papers attempt to define the dimensions of a model taxonomy, and although some progress is made, the attempts are not fully satisfactory. Nevertheless, the single agreed-upon aspect of attentive processing is that attention is needed to deal with the amount of information that the brain receives, and as a result we propose a definition of attention which can stand as a modeling principle as well, following Tsotsos (2011):

> *Attention has the goal of reducing computational load. Attention is a dynamic tuning of processing in reaction to the task, goals and input of the moment, in some ways automatic and in others volitional.*

In other words, as the needs of an agent or the characteristics of its environment change, the system reacts by adjusting its resources so that they may best be applied towards those needs.

The particular mechanisms of dynamic tuning fall within three classes as described in Tsotsos (2011), and updated here in Fig. 1. Most can be directly connected to the elements of attention listed in Sect. 2. Each of these mechanisms is naturally accompanied by its required parameters. For example, a region of interest could be specified by a location and extent in a retinotopic representation of the visual stimulus; an exogenous cue can be represented by a set of contiguous retinotopic locations and features; a location-based inhibition of return can be specified by a retinotopic location, spatial extent, and rate of decay; and so on. Each of these mechanisms can be defined in a similar manner. What these specifications provide is the set of parameters that an attention executive system must control in order to tune the visual system from moment to moment.

The elements within the classes *Increase Signal-to-Noise Ratio* (SNR) and *Reduce Search Space* directly reduce the dimensionality of the problem to be solved. However, the goal of the *Reduce Search Space* mechanisms is not to reduce the dimensionality of input data; rather, it is to reduce the dimensionality of the data with respect to the task of the moment and with respect to its resulting combinatorial impact. This use of dimensionality differs from Bellman's Curse of

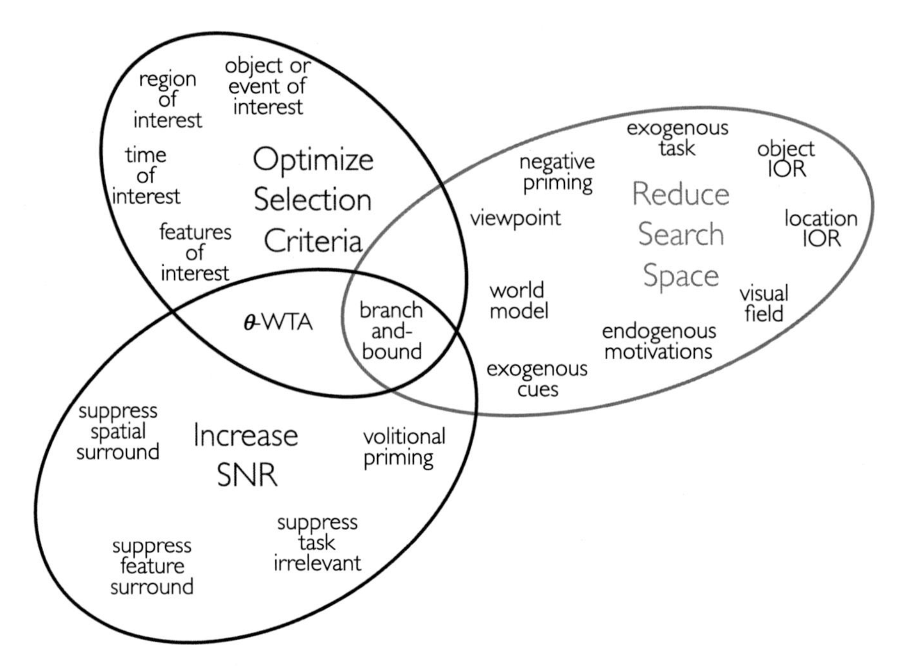

Fig. 1 A taxonomy of the different tuning mechanisms included in the Selective Tuning model of visual attention. For details see Tsotsos (2011)

Dimensionality (Bellman 1957). Bellman referred to various phenomena that arise when analyzing and organizing data in high-dimensional spaces that do not occur in low-dimensional settings such as the three-dimensional physical space of everyday experience. The common theme of these problems is that when the dimensionality increases, the volume of the space increases so fast that the available data become sparse. This sparsity is problematic for any method that requires statistical significance. In order to obtain a statistically sound and reliable result, the amount of data needed to support the result often grows exponentially with the dimensionality. Also organizing and searching data often relies on detecting areas where objects form groups with similar properties; in high dimensional data however all objects appear to be sparse and dissimilar in many ways which prevents common data organization strategies from being efficient. The 'curse of dimensionality' is not a problem of high-dimensional data, but a joint problem of the data and the algorithm being applied. It arises when the algorithm does not scale well to high-dimensional data, typically due to needing an amount of time or memory that is exponential in the number of dimensions of the data. However, it is not so obvious that this applies to visual information processing in the brain. The differences seem to be at least three. The input representation is fixed by the size and processing nature of the retina and not by the dimensionality of the visual world or our knowledge of it. There is no requirement (nor evidence) that processing is completely veridical or optimal—it must only suffice for the task at hand. Finally, although the general form of the problem indeed has an inherent exponential nature, it is not necessarily the problem the brain is solving. To be sure, there is an exponential nature to vision, but it is due to the combinatorics of selecting visual entities that drive human action (Tsotsos 2011). However, it would be a hopeless strategy for evolution to attempt to develop a sufficiently large brain for the general problem. Instead the strategy appears to have been to evolve a brain that makes approximations to the general problem whose impact is not detrimental to the survival of the species, thus changing the nature of the problem solved (see Tsotsos 1990, 2011 for further discussion).

The mechanisms in the class *Increase Signal-to-Noise Ratio* reduce the search space in a different manner. Depending on the task, different aspects of the data are considered 'signal' and 'noise' even for identical input sets. Recall the famous Yarbus task where subjects are given different questions to ask of the same photograph and exhibit very different scanpaths for each question (Yarbus 1967). Classical dimensionality reduction methods do not make this critical distinction.

The class *Optimize Selection Criteria* plays a different role. The decision process underlying selection is addressed here. It may involve the selection of one option among many; it may involve the selection of a few options to consider in parallel from among many; it may involve a more probabilistic decision rather than binary; it may involve more suppression of poor choices than selection of good choices; it may involve a temporal ordering of which options to explore; and potentially other versions. Each possibility may be appropriate for some kinds of tasks yet not for others and thus the determination of best selection criterion is necessarily task-dependent. Whereas the *Reduce Search Space* class includes the many domains where selection may occur (space, time, world model, viewpoint), the *Optimize Selection Criteria* class provides ways for how that selection is accomplished.

The main classes of attentional tuning, namely, Increase SNR, Optimize Selection Criteria and Reduce Search Space, will stand as principles because there is substantial experimental evidence to support them, and also for most of the sub-mechanisms in the figure. As just one example, see the review by Borji and Itti (2013) that focuses on the many approaches to selection of region of interest and their connection to human eye movement patterns (Fig. 1).

4 Principles for Modeling Cognition

This section likely addresses the most poorly defined aspect of this exercise. Much has been written about what constitutes cognition from multiple disciplines. It is not within the scope of this paper to cover this full breadth of discussion; rather, a few highlights will be touched where they can be sensibly integrated with the remainder of the principles discussed here.

In Artificial Intelligence, the most central relevant and well-studied concept is that of Rational Action. Rational Action, carried out by a rational agent, maximizes goal achievement given the agent's current knowledge, the agent's ability to acquire new knowledge, and the current computational and time resources available to the agent (Russell and Norvig 2010). The fact that constraints on resources exist points to the fact that it is likely that no optimal solution is available for all instances of a problem. Certainly in vision, it has been shown by a number of authors that sub-problems within vision have exponential computational complexity (for a short survey see Tsotsos 2011). This means that one needs to consider one of the many available heuristics (Garey and Johnson 1979, detail a variety of strategies and heuristics for dealing with NP-Complete problems and these are as applicable here as for theoretical computer science problems). One of these heuristics is to focus on approximate solutions, that is, solutions that are not optimal in some way but that are *good enough* for the current needs. *Satisficing* is a strategy that entails searching through the available alternatives until an acceptability threshold is met. This differs from optimal decision-making, an approach that attempts to find the best alternative available. The term *satisficing*, (a combination of *satisfy* and *suffice*), was introduced by Herb Simon in 1956. Satisficing can take more than one form. If one knows a particular problem requires such an approximation but has the luxury of time, then one can spend as much time as one likes to find this acceptable solution among all the possible ones. One the other hand, if time is limited, perhaps strictly limited by the need to act before something else occurs, then a different sort of search would occur, one that would find a *just in time* solution, the best one within the time limit. If time is extremely tight, then an almost reflexive response is needed, perhaps the first one that comes to mind. Clearly, external tasks and situations as well as internal motivations play an important role in determining the right sort of approximation to employ. Different from this strategy is the one where subsets of the full problem are defined where optimal procedures apply without infeasible characteristics. Here, the first step is to determine when such a problem is presented. Then, the most appropriate solution can be deployed. A rational agent, then, attempts to achieve its current goal, given its current constraints, by applying these selection methods to choose among its many possible solution paths. This is exactly like the Optimize Selection Criteria principle for visual attention described earlier.

Moving to Cognitive Science a variety of ideas can be found. O'Reilly (1998) presents five principles for biologically based computational models of cortical cognition:

i. biological realism
ii. distributed representations
iii. error-driven learning
iv. Hebbian learning
v. bidirectional activation propagation and inhibitory competition.

Although he calls these principles, at least one, namely *i*, seems to be more correctly considered as a modeling philosophy, and does not fit the working definition given in the first section in that it makes no predictions about the nature of domain, human vision and cognition.

Just and Varma (2007) have proposed 5 principles as important:

i. Thinking is the product of the concurrent activity of multiple brain areas that collaborate in a large-scale cortical network.
ii. Each cortical area can perform multiple cognitive functions, and conversely, many cognitive functions can be performed by more than one area.
iii. Each cortical area has a limited capacity of computational resources, constraining its activity.
iv. The topology of a large-scale cortical network changes dynamically during cognition, adapting itself to the resource limitations of different cortical areas and to the functional demands of the task at hand.
v. The communications infrastructure that supports collaborative processing is also subject to resource constraints, construed here as bandwidth limitations.

These are closer to what we seek. For our purposes, a small re-phrasing might be appropriate. The first two really advocate that computation in the brain cannot be partitioned into independent modules. The fourth is very much like the attention principle presented in the previous section. And the third and fifth refer to resource issues also as mentioned earlier. The fact there is this overlap is pleasing because we see how the same principles can apply to both the modeling of attention and of cognition.

Varma (2011), in a very nice review of cognitive architectures, emphasizes additional modeling concepts. When comparing models, one should prefer models that account for large numbers of empirical regularities using small numbers of computational mechanisms, i.e., the principles of empirical coverage and parsimony. He adds:

i. Successful architectures possess subjective and inter-subjective meaning, making cognition comprehensible to individual cognitive scientists and organizing groups of like-minded cognitive scientists into genuine communities.
ii. Successful architectures provide idioms that structure the design and interpretation of computational models.
iii. Successful architectures are strange: They make provocative, often disturbing, and ultimately compelling claims about human information processing that demand evaluation.

Varma says that cognitive architectures cluster into paradigms based on shared computational mechanisms such as production systems, connectionist methods and the exemplar paradigm. Again, although completely valid and important points, these too seem more like goals of modeling or criteria by which models are evaluated rather than principles or truths about the natural world of cognitive behavior.

The variety of cognitive architectures is almost as numbing as the number of attention models (see Samsonovich 2010, 2012). Surveying the large number of systems developed to date, it is interesting to note what each includes by way of perceptual and attentive capabilities (they all have some level of cognitive capacity). Quite a number have no perception at all, while many accept only simulated input. Several also use no vision but rely on sonar or laser. Those that do have visual perception components often include object recognition or support for path planning and navigation and other task-specific components. With respect to attention, some systems include eye movements and selection of regions-of-interest. As a group, these systems show concerted effort to integrate perception with reasoning and task execution, mostly guided by biological realities. Samsonovich presents three main challenges for cognitive architectures. He suggests that they should be human-compatible and useful as society members, self-sustainable, and human-extending. These are, like Varma's points, excellent long-term goals for research but they are also not principles underlying cognition.

Recently, and more directly relevant to the attention and cognition goals of this presentation are the massive neuronal network simulations that have become possible not in small part due to increased computing power and large engineering efforts. Zylberberg et al. (2010) develop a large-scale neural system that embodies attention in the form of a router whose job is to set up the precise mapping between sensory stimuli and motor representations, capable of flexibly interconnecting processors and rapidly changing its configuration from one task to another. Eliasmith et al. (2012) describe another large-scale neural model, impressive for its ability to generalize performance across several tasks. The entire vision component is modeled using a Restricted Boltzmann Machine as an auto-encoder, but attention is not used. The major brain areas included in the model are modeled using abstract functional characterizations and are structured in a feed-forward processing pipeline for the most part. Both of these, and in fact most major proposals, view the visual system as a passively observing, data-driven classifier of some sort, exactly the kind of computational system that Marr had envisioned (Marr 1982) but not of the kind indicated by modern neurobiology (see for example Kravitz et al. 2013). The kind of attentive, cognitive architecture that we seek to model is not the one observed in an anaesthetized, passive observer (i.e., Hubel & Wiesel's cat); our goal system is a dynamic, responsive, and living process where the observer plays key roles in deciding what is looked at, when it is looked at, how is it looked at, why it is looked at and how what is looked at will be used. This may summarized by the principle of active perception, for which there is a wealth of experimental and computational evidence (e.g., Tsotsos 1992; Findlay and Gilchrist 2001).

5 The Survival Requirement

Even though a large number of candidate principles, philosophies and requirements have been described so far, we are still missing some method that permits these to be combined into a single system. To be sure, the details of system integration or implementation are beyond the scope of this paper. However, as noted earlier, the development of a model of human visual cognition is dependent on the choice of which experimental observations act as constraints during its development (Tsotsos

2014). Those constraints provide a means to select solutions among those potential ones that satisfy the principles. We can sketch a strategy that might play this selection role.

Something like the multi-objective optimization problem framework mentioned earlier seems to be needed. Suppose that is it possible to enumerate the many different variables that play roles in attention and cognition tasks. This is not an easy assumption, however, the goal here is to present a potential research avenue as opposed to a complete solution. At least for the visual attention problem, such an enumeration is derivable from the control signals required of an attention executive (see Tsotsos and Kruijne 2014). Some examples of the variables that are part of some attention elements was given in Sect. 2; taken to its logical completion, that set of variables is what is required here. Further, assume that it is possible to quantify the well-being of an organism or agent; this is not unlike the assumption made by genetic/evolutionary algorithms where solutions are selected through a *fitness-based* process, where fitter solutions, as measured by a fitness function, are typically more likely to be selected. Here, let us assume that this function measures how likely it is that an organism or agent will survive after its next action at time t, and this is represented by $\mathbb{S}(t)$. An action is anything that results as a response from the agent given its current situation (and may be null). The kinds of actions referred to here are those generated by mechanisms such as visual routines (e.g., Ullman 1984; Rao 1998; Yi and Ballard 2009) or more broadly cognitive programs (Tsotsos and Kruijne 2014). For simplicity in this discussion, assume that survival $\mathbb{S}(t)$ can be given as a real number. Let the relevant variables—the ones that play a role in attention and cognition, such as described in the previous sections—be represented by a_i where $1 \leq i \leq k$. k is likely a large number. $\mathbb{S}$, then, is a function Z of these variables:

$$\mathbb{S}(t) = \mathrm{Z}(\alpha_1, \alpha_2, \ldots, \alpha_k, t') \quad (.)$$

where $t - t' = \mathrm{T}$, the time required by the system to determine its response to the environmental and task condition present at time t.

If an agent is considering a particular action, where that action impacts its internal variables, a mapping exists from the settings of those variables to the survival value of that action. It is assumed that such a mapping is learned over the lifetime of the agent. The function also explicitly gives the time at which the response is provided. Let τ be the maximum time permissible for a response in order for the survival measure to be valid. τ is clearly task and situation dependent and its value must be dynamically set by whatever executive system is controlling the agent. This leads to the first constraint on the computation:

$$\mathrm{T} \leq \tau$$

where T represents the time elapsed before the agent responds to its situation. In other words, this is the time constraint that includes the time to sense, the time to compute the interpretation of any input signals, the time to decide on and compute the response and the time to execute the response.

In general, given a particular situation and task, an agent would seek to maximize its survival, that is, to ensure that any response it generates would make its value of $\mathbb{S}$ is strictly non-decreasing, i.e., monotonically increasing:

$$\frac{\partial \mathbb{S}}{\partial t} \geq 0 .$$

This is the second constraint associated with survival. The combinatorics of a search over this k-dimensional space to determine appropriate settings of all variables can be horrendous. Although not proved here, one should not expect such a problem to be a tractable one. It has a similar overall form to the sensor planning problem proved NP-hard by Ye and Tsotsos (1996). As a result, in the search for a good solution, an organism or computational agent can use heuristics to assist. For example, one may consider more expensive changes last, consider small changes first, one may not consider changes to operating characteristics unless no other option can be found, one can assume that some subset of variables will remain unchanged thus effectively reducing the dimensionality of the search space, and so on.

The last element required here is guidance for how to include the mechanisms that set each of the variables a_i. Certainly, and primarily because of the time constraint, we cannot invoke a general optimization process for this task. What is needed is something closer to Simon's Satisficing strategy. A model produces a *good-enough* solution if it permits the agent in which it is embodied to succeed in its task. A good-enough solution is not the same regardless of the situation in which the agent finds itself. On the contrary, the acceptability threshold depends strongly on the specifics of the environment, task, and resource constraints of the moment. The agent should not needlessly waste resources over-optimizing a solution that makes no difference to the overall outcome but should make certain enough resources are used to enable it to accomplish its goals of the moment. In other words, the acceptability threshold is not a simple, static single threshold but a complex, dynamic entity that depends on the environment in which the agent must behave and accomplish the tasks of interest. There is no reason for a solution that goes beyond being good enough to fulfill the goal of the moment; it would waste resources. What we need is a way to specify which variables matter and which do not, and for those that matter, which settings are sufficient or not. All of this must be computed in a dynamic manner contingent on the context of the moment. Assume there exists such a function *M()*, whose value can be T (true—the setting of the variable is sufficient), F (false—the setting of the variable is not sufficient), or U (irrelevant—the variable is irrelevant). Here, an irrelevant value signifies that the setting of a particular variable has no impact on the agent's survival. This can be stated as follows:

$$\forall j, C(t) = \Gamma: (M(a_j, t) \neq F), \quad 1 \leq j \leq k$$

Γ is the specification of the context C at time t, and represents the set of the task instructions, world knowledge applicable to the task, and environmental circumstances. The function *M()* is not so unusual a function; classic planning programs within AI can provide such information.

Simple examples abound of where we, in our everyday lives, apply this principle. Suppose you are at a dinner meeting, enjoying a conversation with your companion. Out of the corner of your eye you see your wine glass and without any further consideration or interruption of your conversation or train of thought, you reach over and grasp it. There are a multitude of possible grasp positions as well as pathways along which the glass could travel to your lips. But it doesn't matter which is chosen as long as the goal of taking a sip of wine is fulfilled without spilling. There is no optimization involved here, but there is the need to find a good enough solution.

This Survival Requirement plus its set of constraints may suffice to point to a direction for future development, a development which remains a challenge.

6 Discussion

The previous sections have presented some principles that underlie the modeling of visual attention and cognition. It goes without saying that it is unlikely these would be the definitive words on the topic. It is highly unlikely that these principles form the complete set, and it is highly likely that many are simply not useful or just wrong. Perhaps a more reasonable question to pose is whether or not these are useful to help move efforts forward.

One way of determining if the above principles are at all useful could be to consider how they might help achieve solutions for open problems in the field. Adolphs (2015) presents one view of the open problems in neuroscience. He has a very interesting list, full of challenges as well as hope since he puts solutions on a timeline. Here, we look at this list from a different viewpoint. It likely is true that when he says that data and predictions are not enough to denote success, he means that modeling and simulation are also needed for building a brain as he notes in his second meta-question. Thus his unsolved problems address the full spectrum of problems and include those that modeling presents.

Adolphs places a focus on algorithms; but here perhaps a more computational viewpoint is necessary than appears in his presentation. In the list of unsolved problems, representation appears only in the context of abstract thought. This is too narrow a role. Rather, representation is the key everywhere. Without well-defined representations, algorithms cannot be designed let alone tested. The issue is not relevant for abstract thought alone; it is relevant for all levels of processing. What do the signals that emerge from the optic bundle represent? What about the feedforward signals from area V1, or the feedback signals from V4 or the communication from MT to V4? Once representations are proposed, algorithms can be designed to transform one representation to the next in a processing stream and can be tested to see how well they perform those tasks. And in order to select a representation, a hypothesis must be made as to what each visual area, or neural assembly, or neuron, is supposed to compute, what 'problem' it is solving. Thus, it would not be too strong a statement to stress that developing solutions must involve these elements. But this is not new: Marr said exactly the same with his three levels of analysis: computational, representation and algorithm, and implementation (Marr 1982). A first philosophy for modeling is to use Marr's three levels as a guide. That is, to decompose the components of a solution into Marr's levels and to address them in that order. The strategy has been used effectively ever since not only in vision modeling but throughout computational neuroscience.

We might add a fourth item to Marr's three, namely, the requirement that a solution must also satisfy a complexity level analysis (Tsotsos 1990). This was originally proposed as a supplement to Marr's first level of analysis, the computational level. Marr stated his first level as providing answers to the questions: What is the goal of the computation? Why is it appropriate? and, What is the logic of the strategy by which it can be carried out? Unfortunately, Marr, perhaps not realizing that it is not difficult to pose computational solutions that are physically unrealizable, missed an important issue. As argued in Tsotsos (1990), there are a large number of perfectly well-defined computational problems whose general solution is intractable—unrealizable on available physical resources or requiring time longer than the age of the universe. Even worse, there are well-defined problems that are undecidable meaning there provably exists no algorithm to determine the result. For this reason, the fourth level, the complexity level, is intended to ensure the logic of the strategy for solving the problem is actually possible as well as realistic with its resource requirements. Varma (2011) points out the resource constraint issue, as do most authors working on visual attention models. It is a pervasive problem and thus must be included in the list of principles that guide modeling.

Algorithm performance can be characterized in yet another way. How does a particular algorithm behave in the face of time pressure, or incomplete information, or any non-ideal environmental factor? Good algorithms are those that degrade gracefully. This does not refer only to a tolerance to faults or damage as it does in software or hardware design. Rather, it refers in addition, to the ability of the system to behave sensibly in non-ideal circumstances. The requirement of graceful degradation needs to also be in a modeler's repertoire.

Let us return to Adolphs' list of open problems and the question of how the modeling principles describe here can help move progress forward. He lists a number of problems that he claims should be solvable within 50 years; let's focus on these here:

i. How do circuits of neurons compute?
ii. What is the complete connectome of the mouse brain (70,000,000 neurons)?
iii. How can we image a live mouse brain at cellular and millisecond resolution?
iv. What causes psychiatric and neurological illness?
v. How do learning and memory work?

vi. Why do we sleep and dream?
vii. How do we make decisions?
viii. How does the brain represent abstract ideas?

It is doubtful that modeling can help much for solutions to problems ii, iii, iv, or vi. However, the goals of cognitive modeling clearly include developing and testing answers to questions i, v, vii, and viii. In order to argue that the principles presented here have utility for these questions, it is useful to summarize what has been discussed so far in terms of modeling philosophies, requirements and principles:

Modeling Philosophies

- Satisfy Marr's 3 levels of analysis:

  i. computational level
  ii. representational and algorithmic level
  iii. implementation level

- Maximize empirical coverage
- Occam's Razor (among competing hypotheses that predict equally well, the one with the fewest assumptions should be selected)

Modeling Principles

- Attention acts to dynamically reduce computational load
- Attention dynamically tunes vision and cognition in reaction to task, goals and input

  i. increase signal-to-noise ratio
  ii. optimize selection criteria
  iii. reduce search space

- Rational action
- Perception is active (there is no passive observer in the brain)
- Representations are distributed as are processes that use them
- Visual attention and visual cognition are properties of a network of processes
- Learning is error-driven and Hebbian
- Competition and bidirectional information transmission are ubiquitous

Modeling Requirements

- Satisfy complexity level analysis:

  i. the problem to be solved is provably decidable and tractable
  ii. the proposed algorithms are tractable
  iii. the required resources match the available resources

- Design for graceful degradation
- Achieve Satisficing behavior
- Adhere to the Survival requirement

It should be obvious that these modeling directives can indeed contribute to future research; the major question really is whether or not they are sufficient on their own. The answer will come only with continued large-scale integrative modeling efforts. But it is important these efforts not be under-constrained. That is, that must try to connect many experimental results. A model that provides an explanation for one result is not interesting; it leaves too many variables of the full system open to incorrect interpretations. The principles, philosophies and requirements may seem a bit irrelevant for such a narrow modeling effort; they only show their power if the model is sufficiently large and complex. Models should be explicit with respect to which representations are important and why, explicit with respect to the mechanisms for transforming one representation into another and what that transformation means, and how decisions are made in the context of an agent's current task. It will be interesting to see what effect these principles, philosophies and requirements might have on the next generation of models.

Acknowledgments JKT gratefully acknowledges support from the following sources without which this research could not have been conducted: the Natural Sciences and Engineering Research Council of Canada; the Canada Research Chairs Program; the Air Force Office of Scientific Research, Air Force Material Command, USAF under Award No. FA9550-14-1-0393.

References

Adolphs R (2015) The unsolved problems of neuroscience. Trends Cogn Sci 19(4):173–175
Bellman RE (1957) Dynamic programming. Princeton University Press
Beuth F, Hamker F (2015). A mechanistic cortical microcircuit of attention for amplification, normalization and suppression, *Vision Research*, in press
Borji A, Itti L (2013) State-of-the-art in visual attention modeling. IEEE Trans Pattern Anal Mach Intell 35(185–207):110–126
Bruce ND, Wloka C, Frosst N, Rahman S, Tsotsos, JK (2015). On computational modeling of visual saliency: Examining what's right, and what's left. *Vision research* (in press)
Bylinskii Z, DeGennaro EM, Rajalingham R, Ruda H, Zhang J, Tsotsos JK (2015). Towards the quantitative evaluation of visual attention models. *Vision Research*, in press
Carrasco M (2011) Visual attention: The past 25 years. Vis Res 51(13):1484–1525
Denning PJ (2007) Computing is a natural science. Commun ACM 50(7):13–18
Eliasmith C, Stewart TC, Choo X, Bekolay T, DeWolf T, Tang Y, Rasmussen D (2012) A large-scale model of the functioning brain. Science 338(6111):1202–1205
Findlay JM, Gilchrist ID (2001) Visual attention: the active vision perspective. Vis Atten, 83–103. Springer New York
Frintrop S, Rome E, Christensen HI (2010) Computational visual attention systems and their cognitive foundations: a survey. ACM Trans Appl Percept (TAP) 7(1):6
Friston K (2010) The free-energy principle: a unified brain theory? Nat Rev Neurosci 11 (2):127–138
Garey M, Johnson D (1979) Computers and intractability: a guide to the theory of NP-completeness. Freeman, San Francisco
Itti L, Rees G, Tsotsos JK (eds) (2005) Neurobiology of Attention, Elsevier Press
Just MA, Varma S (2007) The organization of thinking: what functional brain imaging reveals about the neuroarchitecture of complex cognition. Cogn Affect Behav Neurosci 7(3):153–191

Kimura A, Yonetani R, Hirayama T (2013) Computational models of human visual attention and their implementations: A survey. IEICE Trans Inf Syst 96(3):562–578
Kravitz DJ, Saleem KS, Baker CI, Ungerleider LG, Mishkin M (2013) The ventral visual pathway: an expanded neural framework for the processing of object quality. Trends Cogn Sci 17 (1):26–49
Marr D (1982) Vision: A computational investigation into the human representation and processing of visual information. Henry Holt and Co., Inc., New York, NY, USA
Nobre K, Kastner S (eds) (2013) The Oxford handbook of attention. Oxford University Press
O'Reilly RC (1998) Six principles for biologically based computational models of cortical cognition. Trends Cogn Sci 2(11):455–462
Pashler H (ed) (1998) Attention. Psychology Press, East Sussex, UK
Rao S (1998) Visual routines and attention, PhD Dissertation, MIT, February
Rothenstein AL, Tsotsos JK (2014) Attentional modulation and selection—an integrated approach, Public Libr Sci (PLoS) ONE 9(6): e99681
Russell S, Norvig P (2010) Artificial intelligence: a modern approach, 3rd edn. Pearson Education Inc., Upper Saddle River, NJ
Samsonovich AV (2010) Toward a unified catalog of implemented cognitive architectures. BICA 221:195–244
Samsonovich AV (2012) On a roadmap for the BICA Challenge. Biol Inspired Cogn Archit 1:100–107
Treisman A, Gelade G (1980) A feature integration theory of attention. Cogn Psychol 12:97–136
Tsotsos JK (1990) Analyzing vision at the complexity level. Behav Brain Sci 13–3:423–445
Tsotsos JK (1992) On the relative complexity of passive vs active visual search. Int J Comput Vis 7(2):127–141
Tsotsos JK (2011) A computational perspective on visual attention. MIT Press, Cambridge MA
Tsotsos JK, Rothenstein A (2011) Computational models of visual attention. Scholarpedia 6 (1):6201
Tsotsos JK, Kruijne W (2014) Cognitive programs: software for attention's executive. Front Psychol, 5
Tsotsos JK (2014) It's all about the constraints. Curr Biol 24(18):R854–R858
Ullman S (1984) Visual routines. Cognition 18(1–3):97–159
Varma S (2011) Criteria for the design and evaluation of cognitive architectures. Cogn Sci 35 (7):1329–1351
Yarbus AL (1967) Eye movements and vision. Plenum Press, New York
Ye Y, Tsotsos JK (1996) 3D sensor planning: its formulation and complexity. In: Kautz H, Selman B, (eds) Proceedings 4th International Symposium on Artifi cial Intelligence and Mathematics. January 3–5, Fort Lauderdale, FL
Yi W, Ballard D (2009) Recognizing behavior in hand-eye coordination patterns. Int J Hum Robot 6(3):337–359
Zucker, S. W. (1981). Computer Vision and Human Perception: An Essay on the Discovery of Constraints. Proc. 7th Int. Conf. on Artificial Intelligence. August 24–28, Vancouver, BC, 1102–1116
Zylberberg A, Slezak DF, Roelfsema PR, Dehaene S, Sigman M (2010) The brain's router: a cortical network model of serial processing in the primate brain. PLoS Comput Biol 6(4): e1000765

Computational Neuroscience of Vision: Visual Disorders

Clement Tan

Abstract The purpose of this chapter is to serve as a primer for researchers in vision not actively engaged in clinical eye care. It aims to give a broad overview of challenges in altered visual function from the perspective of clinicians. Rather than focussing on specific disease entities, it aims to describe the challenges in terms of functional changes.

1 Introduction to Normal Anatomy and Physiology (Forrester et al. 2008; Kaufman et al. 2003)

The process of vision begins with light entering the eye. Light is focused by the cornea and lens onto the retina. The retina converts the light energy into an electro-chemical signal. This signal is transmitted down the optic nerve to the brain. The primary visual cortex in the occipital lobe of the brain is the initial termination of the visual pathway. From there, the signal is passed to other parts of the brain for interpretation of the signal.

1.1 The Cornea (Fig. 1)

This transparent structure is made up of 5 layers.

The epithelium is the most superficial layer and is made up of columnar, polyhedral and squamous cells. These cells are capable of regenerating and the cells that divide are found in the periphery of the cornea in an area known as the limbus. Bowmans layer is the basal lamina on which the columnar cells of the epithelium

C. Tan (✉)
Department of Opthalmology, National University of Singapore
and National University Hospital, 1E Kent Ridge Road,
NUHS Tower Block, Level 7, Singapore 119228, Singapore
e-mail: Clement_WT_TAN@nuhs.edu.sg

Q. Zhao (ed.), *Computational and Cognitive Neuroscience of Vision*,
Cognitive Science and Technology, DOI 10.1007/978-981-10-0213-7_13

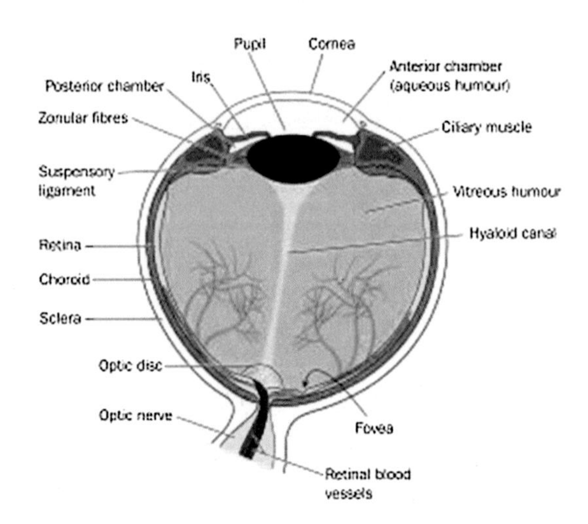

Fig. 1 Anatomy of the eye

rest. The stroma is the major part of the cornea, accounting for 90 % of its thickness. It is made up of collagen (principally type I). Descemet's membrane represents the basement membrane of the endothelium and is important surgically because it can become detached from the stroma. The endothelium is a single layer of cells that does not regenerate when damaged. It is responsible for actively pumping water out of the stroma into the anterior chamber and does so via a carbonic anhydrase mediated pathway.

The major requirement for the cornea is remaining transparent. The fibrils of the cornea stroma are of fairly uniform in diameter and arranged in a quasi-random configuration with each collagen fibril about 200 nm from the next. Any disruption in the usual arrangement of the cornea results in the loss of transparency.

1.2 The Lens

The lens is a biconvex structure that is suspended behind the iris by zonular ligaments to the ciliary (muscle) body. The contraction and relaxation of the ciliary muscle brings about changes in the lens curvature that allows light from objects of different distances to be brought into focus on the retina.

1.3 The Retina (Fig. 2)

Light striking the retina triggers a photochemical cascade that ultimately translates a light signal (photons) into an electrico-chemical signal (a process known as phototransduction). The cells that detect light and convert the photon energy into the

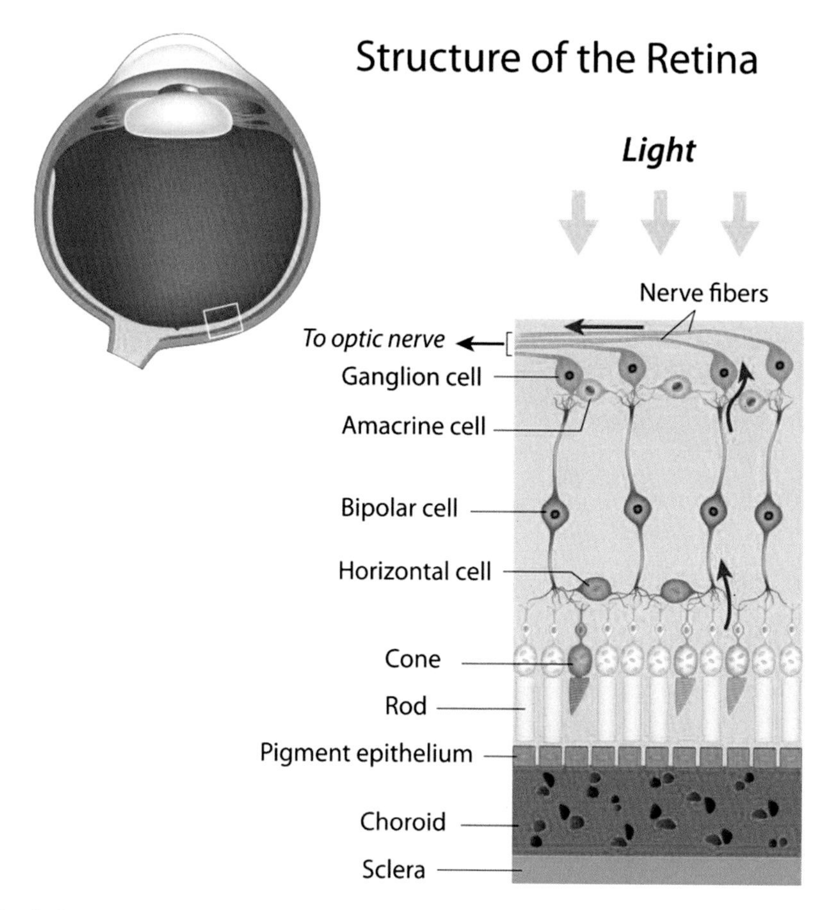

Fig. 2 Structure of the retina

electrico-chemical signal are the photoreceptors. There are 2 populations of these—named rods and cones for their distinctive appearance on microscopy. Cones are responsible for visual acuity (resolving ability) and colour vision. The are located mainly in the macula. Rods are able to detect small quanta of light and the rod system is principally responsible for vision in low light conditions and also for motion detection. Described simply, light falling on the retina triggers hyperpolarization of the photoreceptor cell and this leads to depolarization of the bipolar cell and subsequently of the retinal ganglion cell (RGC). Each RGC has a receptive field—or—an area of retina (number of photoreceptor cells) that it covers. There are approximately 120 million rod and 6 million cone photoreceptors in the retina. These converge onto about 1.2 million retinal ganglion cell neurons. A single photoreceptor to bipolar cell to retinal ganglion cell relationship exists for cones near the fovea but elsewhere, the ganglion cells have a large receptive field.

1.4 *The Visual Pathway Between the Eye and the Primary Visual Cortex the Optic Nerve (Fig. 3)*

The visual input from the retina is mapped to neurons in the the visual pathway. This mapping (retinotopic arrangement) is preserved all the way to the primary visual cortex. So it is possible to know from visual field testing, which part of the visual pathway is likely to have been involved.

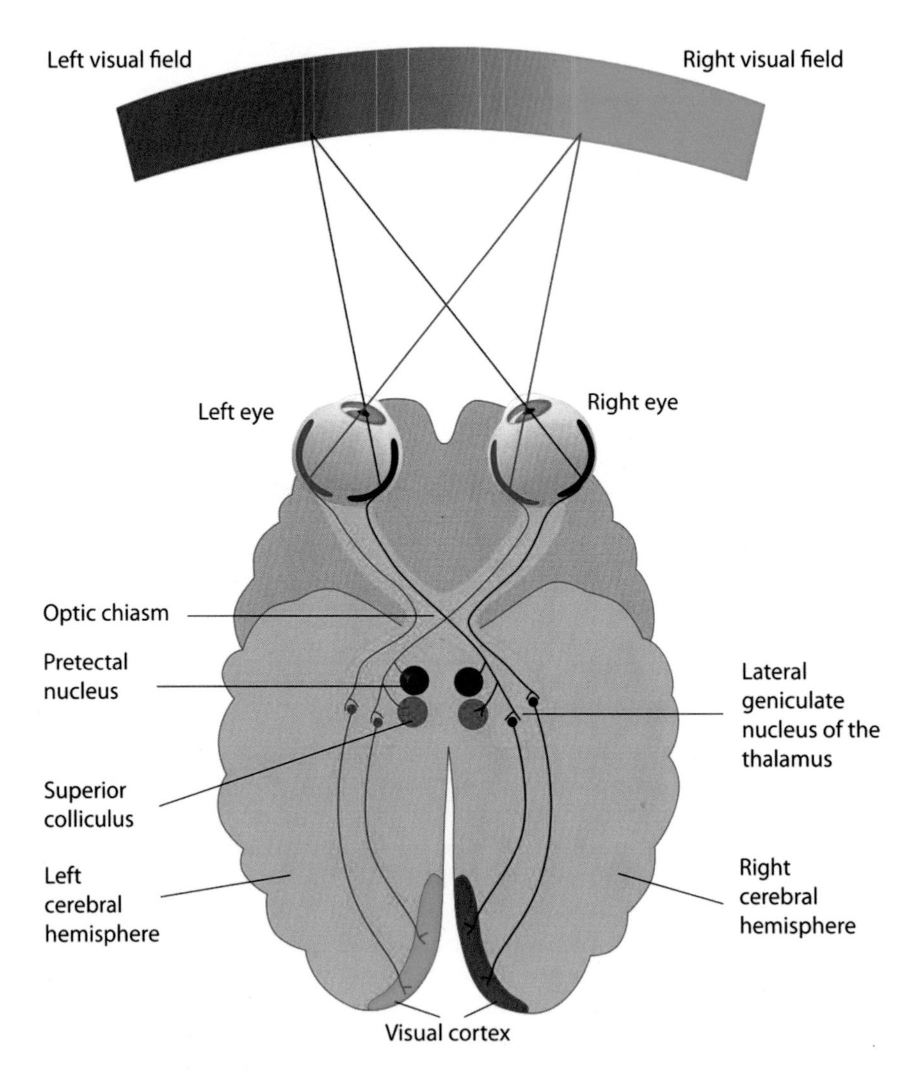

Fig. 3 Visual pathway. *Note* the right visual cortex receives input from the left visual field and vice versa

About half of the fibers of the optic nerve (which originate from the nasal retina) cross over to the contralateral side at the optic chiasm to form the optic tract with nerve fibers from the temporal retina of the fellow eye. This decussation results in the visual input from the right half of a visual scene being transmitted in the left side of the brain and vice versa.

The optic tract terminates in the lateral geniculate nucleus (LGN) in the midbrain. Here, axons of the RGCs synapse with neurons that make up the optic radiation. The optic tract neurons that transmit information originally from the inferior retina pass into the temporal lobe of the brain where they loop forwards around the anterior horn of the lateral ventricle before passing posteriorly to join the neurons serving the superior retina which have passed directly posteriorly to the parietal lobe. These merged fibers now pass to the occipital lobe where they terminate in synapses with the neurons in the primary visual cortex.

While it is tempting to think of retinal images as represented in the visual cortex simply on the basis of a point-to-point topographical representation as the retinotopic arrangement would suggest, in reality, there is further subdivision and complexity to the representation of visual input. Three pathways are known to exist for the transmission of visual information. They are named for cell populations in the LGN that are anatomically and functionally distinct. The parvocellular pathway deals with spatial resolution and colour. It links mainly the foveal cones with cells in layers 3-6 of the LGN. The magnocellular pathway deals with motion detection and luminance. It links the rods with layers 1 and 2 of the LGN. The koniocellular cells are found in the interlaminar area of the LGN. Their role in visual perception is not fully elucidated.

1.5 Cortical Visual Areas

Neurons in the visual cortex respond best to a subset of stimuli in their receptive field. This is known as neuronal tuning and helps explain the functions of the various visual areas (Fig. 4).

The primary visual cortex (Broadman area 17, area V1, striate cortex) is located on the medial aspect of the occipital lobe, within the calcarine sulcus. Histologically, V1 is, like other parts of the cerebral cortex, made up of 6 layers. The neurons of the lateral geniculate nucleus (which make up the optic radiation) terminate in layer IV. Area V1 has projects to the association visual cortices through two streams. The dorsal stream is associated with motion, object location and control of the eyes (e.g. saccadic eye movement) and arms. The ventral stream is concerned with form recognition and object representation; and is involved with long-term memory.

Anterior to V1 is V2 (secondary visual cortex, prestriate cortex). It is the first of the visual association areas receiving feedforward information from V1 and sending feedback information to it. It is involved in the processing of orientation, spatial frequency and colour (ventral stream).

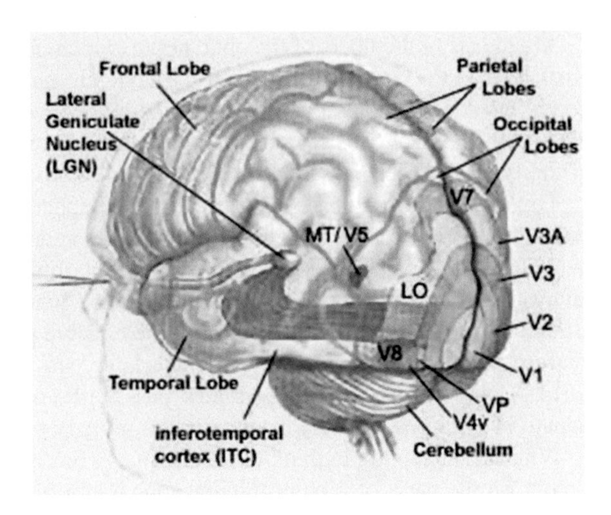

Fig. 4 Cortical visual area

Area V3 lies in front of V2. V3 is sometimes described as a "complex" because its exact extent is still a matter of controversy. It is located within Broadman area 19. V3 receives input from V2 and V1 and projects to the parietal lobe. It is also involved in the dorsal stream.

Area V4 lies anterior to V2. Like V2 it is part of the ventral stream, receiving inputs from V1 and V2. It is tuned for orientation, spatial frequency, colour and for objects of intermediate complexity.

Area V5 is also know as visual area MT (middle temporal). It has inputs from V1, V2, V3, the koniocellular region of the lateral geniculate nucleus and the inferior pulvinar. It has a role in motion perception and complex stimuli.

Area V6 is located and receives input from V1. It is tuned for orientation of visual contours, and the direction of motion. It self motion and wide field stimulation.

1.6 The Extraocular Muscles

There are 6 muscles surrounding each eye that are responsible for moving the eye. The major directions of movement for each eye are: adduction (moving towards the nose), abduction (moving towards the ear), elevation (moving the eye up), depression (moving the eye down), intorsion (moving the superior pole of the eye towards the nose), extortion (moving the superior pole of the eye towards the ear). The extra ocular muscles and the movements they each bring about may be summarised in the following table (Fig. 5).

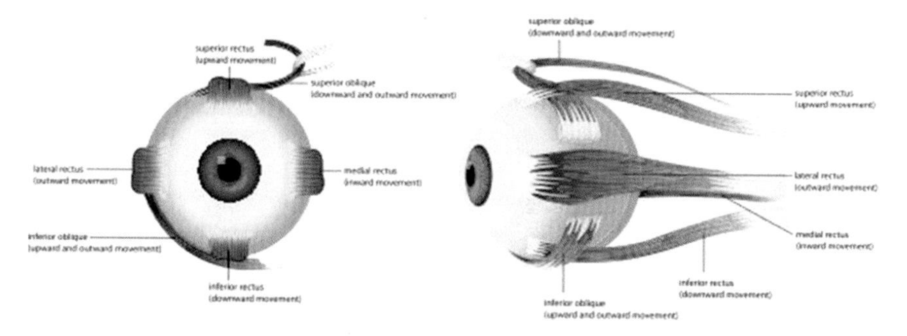

Fig. 5 Extraocular muscles

	Primary action	Secondary action	Tertiary action
Medical rectus	Adduction	–	–
Lateral rectus	Adduction	–	–
Superior rectus	Elevation	Intorsion	Adduction
Inferior rectus	Depression	Extorsion	Adduction
Superior oblique	Intorsion	Depression	Adduction
Inferior oblique	Extorsion	Elevation	Adduction

The medial rectus, superior rectus, inferior rectus and inferior oblique muscles receive innervation via the third cranial nerve. The superior oblique is supplied by the fourth cranial nerve and the lateral rectus by the sixth cranial nerve.

The eyes normally move together as a pair. Loss of alignment between the 2 eyes causes double vision (diplopia) unless the brain suppresses the image from one eye (usually a result of childhood squint).

2 Assessment of Visual Function and Disorders of Measurable Visual Function (Bye et al. 2013; Forrester et al. 2008)

The aspects of vision that are commonly measured by in the clinical context are:

1. Visual acuity
2. Colour vision
3. Visual fields
4. Contrast sensitivity
5. Binocular vision

These are usually measured by psychophysical tests. That is, tests that require a response from the subject. They are also largely measured in controlled conditions.

2.1 *Visual Acuity*

This is the measure of the individual's ability to discriminate 2 objects separated in space. In clinical practice, this is an empirical value based on the assumption that the cone photoreceptor can discriminate 2 objects subtended by an angle of 1 min of arc (measured from the nodal point of the eye).

2.1.1 Testing of Visual Acuity

Classically, optotypes (test charts) used to test visual acuity are constructed based on this principle with targets on the optotype of decreasing size. The most common example of this is the Snellen chart (Fig. 6), where each target (letter or number) subtends a visual angle of 5' and the bars on the letters an angle of 1' of arc. Standard normal visual acuity is described as 6/6 or 20/20 (100 % or 1.0) and this refers to the ability to resolve a target that subtends an angle of 5' at either 6 m or 20 feet. The LOGMAR (LOGarithm of the Minimal Angle of Resolution) is designed in a more precise manner with definitive sizing and spacing of letters (Fig. 7). It provides a more quantitative measure of visual acuity and is the standard for clinical trials.

Spatial acuity better than 100 % (up to 4" of arc) can be measured if the test involves discriminating the offset between 2 very fine lines. This is known as Vernier acuity or hyperacuity. It is well recognised as being present but is not usually measured in clinical practice.

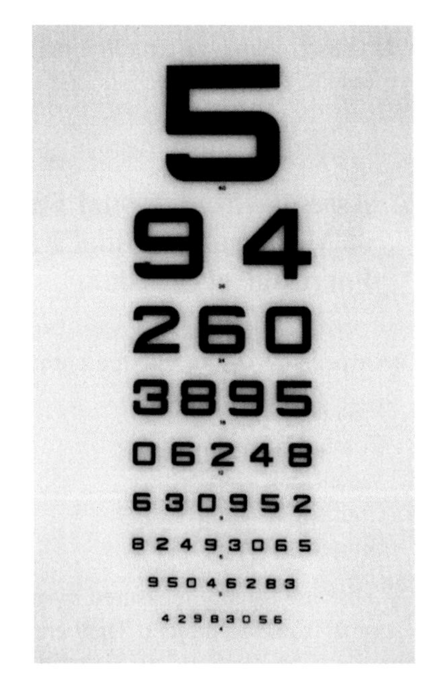

Fig. 6 Snellen chart

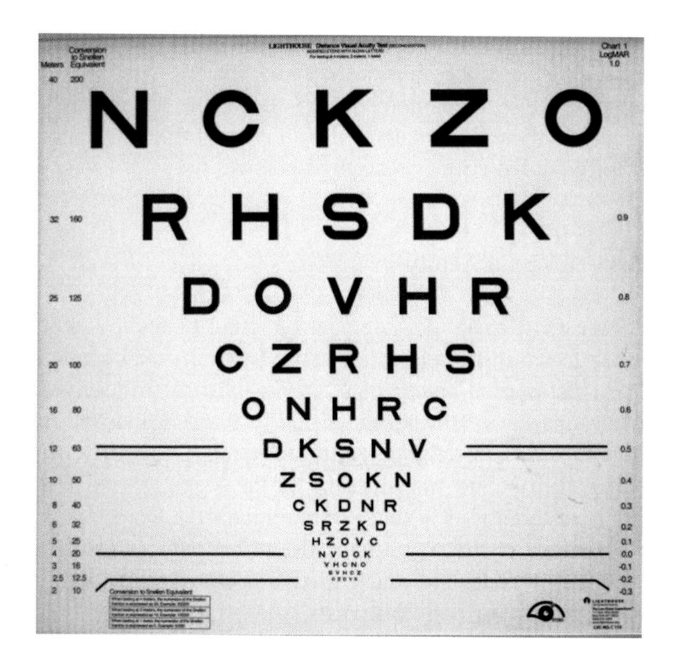

Fig. 7 LOGMAR chart

2.1.2 Physiological Factors that Affect Visual Acuity (Bye, Modi and Stanford 2013)

Development and ageing. It is estimated that visual acuity is not fully developed till about 6 months of age. In the aging eye, increased intraocular scatter reduces visual acuity by reducing contrast.

Retinal eccentricity. Visual acuity is best at the centre of the fovea. Away from the fovea, cones are further apart with increased retinal summation areas, so visual acuity falls.

Luminance. Visual acuity is constant over a side range of photopic luminances. High luminances cause an unexplained reduction in acuity.

Contrast. Visual acuity is better with higher contrast.

Pupil size. Visual acuity is constant for pupil size between 2.5 and 6 mm. Point-spread function widens at pupil size less than 2.5 mm and above 6 mm.

Exposure duration. Visual acuity reduces with decreasing exposure.

Eye (or target) movement. Significant movement of a retinal image causes a reduction in visual acuity.

Meridonal variation. It is observed that when testing function at different retinal meridians, horizontal and vertical meridians are favoured generally.

Interaction effects. If targets a spaced too close together, visual acuity drops. This is know as crowding.

Binocular viewing. Two eyes are better than one as this increase the chances of the highest level of visual processing of the same image.

Refractive error. This causes defocus. It is debatable whether this is a physiological phenomenon (hence normal) or a pathological one (an abnormality or a form of illness). Refractive error is discussed below.

2.1.3 Disorders of Visual Acuity

Visual acuity is the most basic measurement of visual function and also the most common construct tested in the eye care setting. Loss of visual acuity (inability to resolve objects at the normal acuity of 1' of arc) results from any disorder that interrupts the transmission of the visual image to the retina, from disorders that disrupt the phototransduction mechanism in the retina (principally in the macula) and disorders that disrupt the transmission of the signal to the brain. It is no surprise, then, that restoration of visual acuity remains the major focus of eye care professionals. Treatment is directed at the cause of the reduced visual acuity in the first instance. However, there are many instances when visual acuity cannot be restored by drugs, surgery or optical devices.

The most common condition that affects visual acuity is refractive error. This, described simply is the inability of the optical system of an otherwise eye to bring an image to focus on the retina. This defocus blur can be corrected with optical devices such as spectacles and contact lenses that essentially shift the focal point so that images fall clearly on the retina (Fig. 8).

In normal vision, the image is focussed on the retina. In hyperopia, the eyeball is too short and image is focussed behind the retina and convex lenses are required to shift the image forward to the retina. A similar situation exists in presbyopia—but in this case, the problem is not one of eyeball length but of inability of the crystalline lens to change its shape to bring the image to focus. In myopia, the eyeball is too long and images are focussed in front of the retina. Concave lenses required to focus the image onto the retina.

2.2 Colour Vision

The perception of colour begins in the retina and is made possible by the presence of cone photoreceptors that are sensitive to light of different wavelengths. There are 3 major populations of cones. These are described as red, green and blue cones or as long, medium and short wavelength cones, named for the wavelength of light that they are maximally sensitive to. Cone signals are transmitted through the parvocellular pathway to the brain and colour is interpreted in the parieto-occipital cortex.

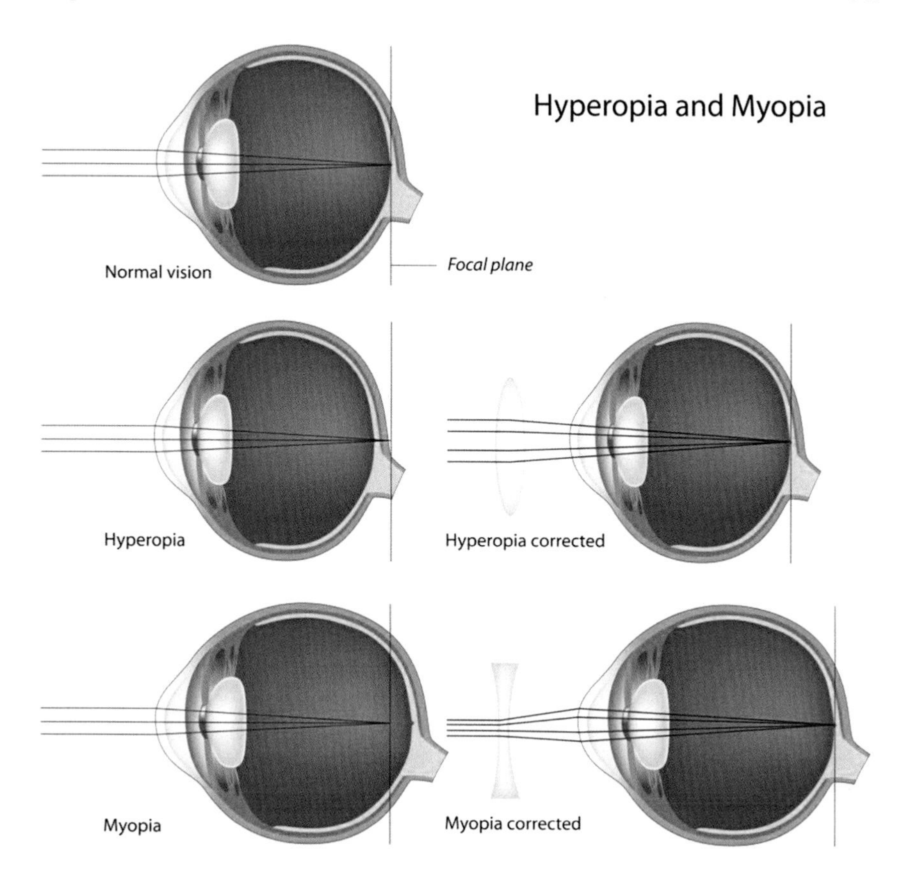

Fig. 8 Refractive error

2.2.1 Testing of Colour Vision (Fig. 9)

Colour vision is tested via pseudoisochromatic plates, colour matching (e.g. Nagel anomaloscope) and hue discrimination/colour arrangement (Farnsworth 100 Hue) tests.

2.2.2 Disorders of Colour Vision

These may arise from genetic defects of the cone photoreceptor pigment, disorders of the macula, optic nerve and brain. Inability to see specific parts of the colour vision spectrum does not consistently suggest the site of the abnormality. No therapy exists at present to restore lost colour vision.

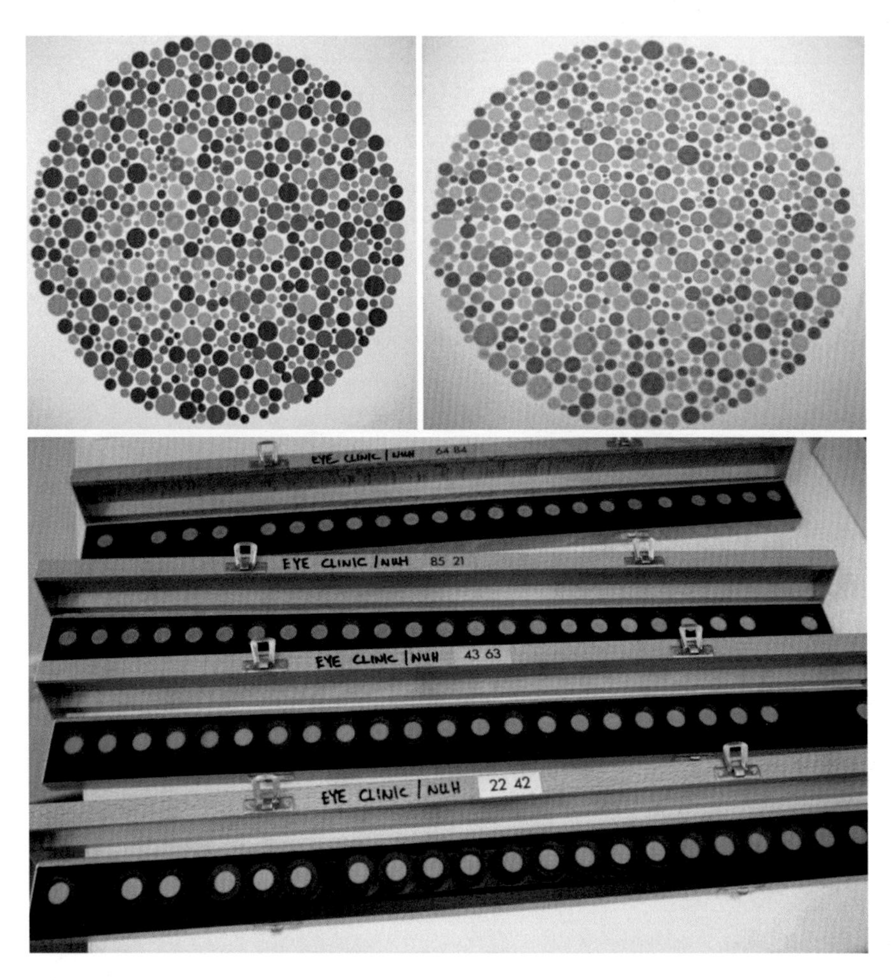

Fig. 9 Pseudoisochromatic (Ishihara) plates and Farnsworth 100 Hue test

2.3 Visual Fields

The visual fields are the spatial breadth of vision. The normal field of vision for one eye extends from fixation 60° nasally, 100° temporally, 60° superiorly and 75° inferiorly. Binocularly, the visual fields extend 120° horizontally and 135° vertically. The visual fields observe a retinotopic arrangement all the way from the retina to the occipital lobe. The highest resolution of the visual field is from the central visual field served mainly by the macula. This is a result of the smaller receptive fields of the retinal ganglion cells in the macular region, where close to the fovea, there is a one-to-one relationship between photoreceptor and ganglion cell. The result of this is that the central 10° of vision is represented by 55 % of the primary visual cortex. The central 30° of vision is represented by 80 % of the primary visual cortex. Hence testing the central 30° of the visual field may be sufficient in many instances to detect defects arising from cortical disease.

2.3.1 Testing of Visual Fields

Visual fields may be measured in a number of ways. The most quantitative of these is static automated perimetry. This measures ability of the subject to detect a small stimulus (target) presented briefly (0.2 s) at random locations across the visual field. The luminance of the stimulus is varied to determine the minimum luminance required for it to be seen.

2.3.2 Disorders of the Visual Fields

Visual field defects may arise from any disorder of the visual pathway. However, if the problem is one of the optical media, then once the opacity is removed, the visual fields should return to normal. Damage to the retina or optic nerve produces visual defects in one eye only. Damage to the visual pathway beyond the optic nerve (from the optic chasm to the occipital cortex) usually produces visual field loss in both eyes. Because of the anatomy of the visual pathway as detailed above, the pattern of visual field loss is often helpful in suggesting the site of the damage.

At the optic chiasm, where the ganglion cell axons serving the nasal retina cross over to the contralateral side. The nasal retina "sees" the temporal visual field. Hence lesions of the optic chiasm cause a bitemporal visual field defects because it is principally the crossing fibres which are affected. The visual pathway after the optic chiasm consists on each side of nerve fibres conveying information about the visual world on the contralateral visual field. Lesions of the optic tract, optic radiation and occipital lobe result in homonymous visual field defects—loss of vision on the contralateral half of the visual world. So a patient who has damage to the right optic tract, optic radiation or occipital lobe might have difficulty seeing objects on his left. Patients frequently adapt to this by increasing their scanning to the non-seeing hemifield, but it is difficult to quantitate the degree to which they do. Some patients may also adopt specific head postures to overcome the visual field defects.

Patients with visual field loss may sometimes find it difficult to express their visual symptoms if the loss is minor and confined to one eye or if it is bitemporal or homonymous.

2.4 Contrast Sensitivity

Contrast sensitivity can be thought of as the ability to distinguish thin white lines against a uniform background illumination. It can be measured using a sinusoidal grating (a spatial pattern of alternating dark and light bands where the average luminance remains the same but the contrast between the shaded areas (troughs) and light areas (peaks) changes. The degree of contrast (C) is described by the following formula:

$$C = (I_{max} - I_{min}) / (I_{max} + I_{min})$$

where I_{max} is the luminance at the peaks (light area) and I_{min} that at the troughs (shaded area).

2.4.1 Testing of Contrast Sensitivity

Contrast sensitivity can be tested by responses to gratings of different contrast shown (Fig. 10).

2.4.2 Factors that Affect Contrast Sensitivity

Contrast sensitivity is affected by luminance, bar width, length, grating motion, phase shifts, grating orientation (most sensitive in vertical and horizontal directions) and wavelength. It is interesting to note that contrast sensitivity appears to induce more electrical signal responses in the M cells (linked to rod function) than in P cells (cone function).

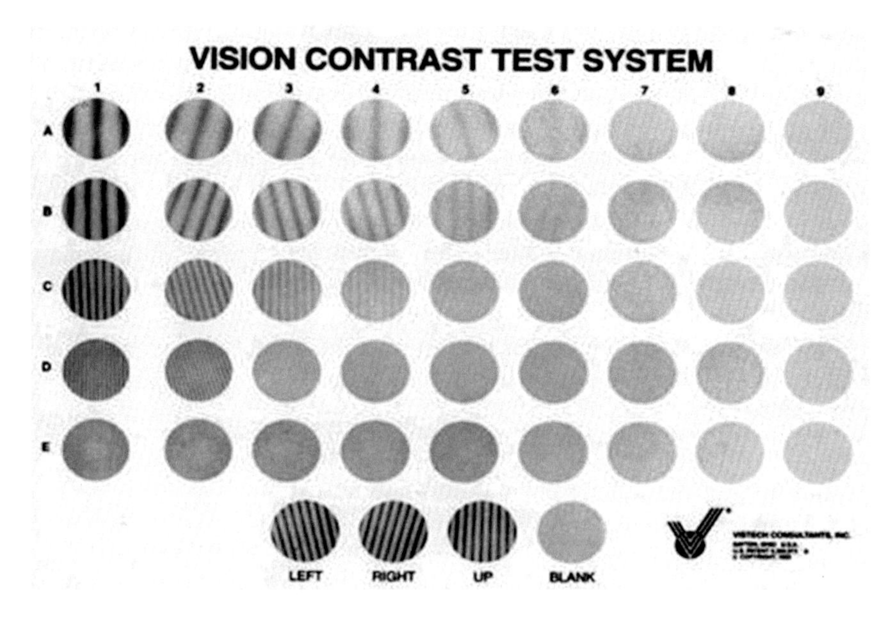

Fig. 10 Contrast sensitivity test

2.4.3 Disorders of Contrast Sensitivity

Poor contrast sensitivity may be the result of many visual disorders such as cornea disease, cataract, retinal disease and optic nerve disease.

2.5 *Binocular Vision*

The eyes function as a pair and this allow for the specific function of stereo vision. In addition, using 2 eyes increases the chances of the highest level of visual processing of the same image and this is thought to be why binocular visual acuity can be better than monocular.

2.5.1 Test of Binocular Vision

Binocular vision is tested principally by testing for suppression (to see if both eyes are looking at the same time) and by testing for stereopsis. The field of binocular single vision (the range of eye movements where single vision is maintained without breakdown to diplopia) is separately measured.

2.5.2 Disorders of Binocular Vision

Any condition that alters the normal movement of the eyes or prevents fusion (the ability to put together the images from each eye) may bring about loss of binocular visual function. Loss of alignment of the 2 eyes relative to each other may bring about diplopia (double vision); reduced vision in one eye may bring about loss of stereopsis; brain injury, stroke of serious illness may cause convergence insufficiency (inability to pull both eyes towards the nose—which is required for near vision), leading again to loss of binocular vision function.

3 Disorders of Cortical Visual Function—Loss of Function (Bye et al. 2013; Liu et al. 2010; Miller et al. 2004)

The areas of the brain responsible for different visual functions is described above. Damage to these areas can produce a variety of clinical scenarios when function is lost. Because of their relative specificity, these losses can help clinicians to localize the site of the lesion causing loss of function. Conversely, lesions of various areas of the brain help us better understand the function of those parts of the brain.

3.1 Acquired Alexia

This is the loss of reading ability in persons who were previously literate. This "word blindness" may exist with loss of writing ability (alexia with agraphia), may be seen in lesions of the left angular gyrus (parietal lobe). In this location, patients may also have loss of mathematical ability (acalculia), right-left disorientation and finger agnosia. Alexia with agraphia may also be seen with Broca's (nonfluent) aphasia, a result of lesions in the dominant frontal lobe.

Alexia may also exist on it's own (alexia without agraphia) in lesions that are almost always in the left hemisphere, particularly the medial and inferior occipitotemporal region. One explanation of this is the disconnection of visual information from linguistic processing centers. Patients exhibiting this condition are able to write, but are then not able to read what they have just written.

3.2 Prosopagnosia

This is the loss of ability to recognise familiar faces. It is frequently associated with homonymous visual field defects, achromatopsia (loss of colour vision) or hemiachromatopsia (loss of colour vision in one hemifield) and topographagnosis (getting lost in familiar surroundings). These deficits together indicate damage to the medial occipital lobe and surrounding structures. Patients with prosopagnosia rely on voices or nonfacial visual cues to identify people.

3.3 Cerebral Achromatopsia

Lesions of the lingual and fusiform gyri (occipital lobe) cause deficits of colour perception. Bilateral lesions are generally required to produce achromatopsia. However, unilateral lesions can produce achromatopsia in the contralateral visual field. Colour sorting (e.g. Farnsworth-Munsell 100 Hue) or matching tests are best for testing for impaired colour perception. Colour naming is not adequate as residual colour perception may allow approximate categorisation of colours (guessing) even though the subject is unable to discriminate hue and saturation.

3.4 Akinetopsia

Loss of motion perception has been reported in lesions of the occipitotemporal area. Motion perception is difficult to assess as there are few clinical tests. Reported symptoms include difficulty perceiving differences in speed and direction of motion,

3.5 *Blindsight and Residual Vision*

Individuals with loss of vision due to lesions of the optic radiations or visual cortex may demonstrate some residual visual function in their blind fields (blindsight) or awareness of visual stimuli (residual vision). It is difficult to test these remaining functions with the same level of precision as regular vision testing. The pathway for blindsight seems to involve the subcortical pathway from retina to superior colliculus. Localisation, perception of motion, form, colour and contrast are all reported.

4 Disorders of Cortical Visual Function—Positive Visual Phenomena (Liu et al. 2010; Miller et al. 2004)

Most of the foregoing discussion has focussed on the deficits of visual function. Whether reduced visual acuity, loss of colour vision, visual field defects, loss of contrast sensitivity, loss of binocular vision or "negative" cortical visual phenomena—all refer to reduced function. Positive visual phoneme involve false images—perceiving what isn't there. It may related to measurable loss of visual function, such as visual field defects. The altered perception may be due to damage to areas of the brain adjacent to the visual pathway or to release phenomena arising from loss of vision.

4.1 *Visual Distortion*

Images may be perceived as too small (*micropsia*), too large (*macropsia*) or distorted (*metamorphopsia*). All three may occur as a result of retinal disease, particularly metamorphopsia. Less often, they have been reported in lesions of the occipital lobe (all), parental lobe (metamorphopsia), in seizures, and in psychiatric patients (micropsia).

4.2 *Visual Hallucinations*

This is the phenomenon of perceiving (seeing) without any actual external visual stimuli. They may be simple (elementary), consisting of geometric shapes, patterns, lines, flashes, lights; or complex, consisting of recognisable objects and figures such as animals and people. Hallucinations may occur in altered mental states, such as those related to the use or withdrawal of drugs or intoxicants; in cognitive dysfunction, such as dementia, Parkinson's disease. Some specific hallucinations are described below:

a. Visual seizures.
   Epilepsy involving the occipital lobe produce mostly elementary hallucinations (squares, circles, zig-zag lines). Seizures arising in the temporal lobe (occipitotemporal, anterior temporal) are more likely to produce complex hallucinations (animals, objects, words) though simple ones may also be possible.
b. Release hallucinations
   A patient who has significant visual loss (usually binocular) may experience simple or complex visual hallucinations. They are thought to be "released" in the visual cortex because the absence of sensory input (vision). One form of this release hallucination is known as the Charles Bonnet syndrome and this presents with vivid, complex visual halluciations in the setting of visual loss.
c. Visual phenomena associated with migraine
   Patients who suffer from migraine experience a range of visual hallucinations. Some are simple—spots, lines, sparkling lights, blind spots. These may precede headache or may present on their own. Some more complex visual phenomena are also associated with migraine. These include micropsia, macropsia and Alice in Wonderland syndrome (distorted body image). Symptoms of visual perseveration (see below) may also be present in migraine. The visual phenomena associated with migraine may also be present in more serious disease such as seizures, arteriovenous malformations and tumours of the brain, so care must be taken to exclude these before attributing the symptoms to migraine.

4.3 Visual perseveration

This is the persistence, recurrence or duplication of a visual image and can take several forms.

If a recently viewed image does not fade away after it has been seen but remains in the visual field, this is known as palinopsia. *Palinopsia* has been reported in lesions of the parieto-occipital lobe, medial occipital lobe and occipitotemporal lobe. It is also know to be drug-induced and has been reported in metabolic and psychiatric conditions.

If instead, multiple copies of a seen object are simultaneously perceived, this is known as *cerebral diplopia or polyopia*. It is distinguished from the more common strabismic diplopia by being present with monocular viewing. Unlike polyopia from refractive causes, it does not disappear when a pinhole is applied. Cerebral polyopia is reported in lesions of the occipital lobe.

References

Bye L, Modi N, Stanford M (2013) Basic sciences for ophthalmology. Oxford University Press, Oxford

Forrester JV, Dick AD, McMenamin PG, Roberts F (2008) The eye: Basic sciences in practice. Saunders, Edinburgh

Kaufman PL, Alm A, Adler FH (2003) Adler's physiology of the eye: Clinical application. Mosby, St. Louis

Liu GT, Volpe NJ, Galetta S (2010) Neuro-ophthalmology: Diagnosis and management. Saunders Elsevier, Philadelphia

Miller NR, Newman NJ, Biousse V, Kerrison JB (2004) Walsh and Hoyt's clinical Neuro-ophthalmology, 6th edn. Lippincott Williams and Wilkins, Philadelphia

Printed in the United States
By Bookmasters